Water Security in the Mediterranean Region

NATO Science for Peace and Security Series

This Series presents the results of scientific meetings supported under the NATO Programme: Science for Peace and Security (SPS).

The NATO SPS Programme supports meetings in the following Key Priority areas: (1) Defence Against Terrorism; (2) Countering other Threats to Security and (3) NATO, Partner and Mediterranean Dialogue Country Priorities. The types of meeting supported are generally "Advanced Study Institutes" and "Advanced Research Workshops". The NATO SPS Series collects together the results of these meetings. The meetings are co-organized by scientists from NATO countries and scientists from NATO's "Partner" or "Mediterranean Dialogue" countries. The observations and recommendations made at the meetings, as well as the contents of the volumes in the Series, reflect those of participants and contributors only; they should not necessarily be regarded as reflecting NATO views or policy.

Advanced Study Institutes (ASI) are high-level tutorial courses to convey the latest developments in a subject to an advanced-level audience

Advanced Research Workshops (ARW) are expert meetings where an intense but informal exchange of views at the frontiers of a subject aims at identifying directions for future action

Following a transformation of the programme in 2006 the Series has been re-named and re-organised. Recent volumes on topics not related to security, which result from meetings supported under the programme earlier, may be found in the NATO Science Series.

The Series is published by IOS Press, Amsterdam, and Springer, Dordrecht, in conjunction with the NATO Emerging Security Challenges Division.

Sub-Series

A.	Chemistry and Biology	Springer
B.	Physics and Biophysics	Springer
C.	Environmental Security	Springer
D.	Information and Communication Security	IOS Press
E.	Human and Societal Dynamics	IOS Press

http://www.nato.int/science
http://www.springer.com
http://www.iospress.nl

Series C: Environmental Security

Water Security in the Mediterranean Region

An International Evaluation of Management, Control, and Governance Approaches

edited by

Andrea Scozzari
Institute of Geosciences and Earth Resources
National Research Council (CNR)
Pisa, Italy

and

Bouabid El Mansouri
University Ibn Tofail
Kenitra, Morocco

Published in Cooperation with NATO Emerging Security Challenges Division

Proceedings of the NATO Advanced Research Workshop on
Environmental Security: Water Security, Management and Control
Marrakech, Morocco
31 May – 2 June 2010

Library of Congress Control Number: 2011935527

ISBN 978-94-007-1625-4 (PB)
ISBN 978-94-007-1622-3 (HB)
ISBN 978-94-007-1623-0 (e-book)
DOI 10.1007/978-94-007-1623-0

Published by Springer,
P.O. Box 17, 3300 AA Dordrecht, The Netherlands.

www.springer.com

Printed on acid-free paper

All Rights Reserved
© Springer Science+Business Media B.V. 2011
No part of this work may be reproduced, stored in a retrieval system, or transmitted in any form or by any means, electronic, mechanical, photocopying, microfilming, recording or otherwise, without written permission from the Publisher, with the exception of any material supplied specifically for the purpose of being entered and executed on a computer system, for exclusive use by the purchaser of the work.

Preface

This book contains 24 Chapters, that represent most of the outcomes of a NATO – funded Advanced Research Workshop (ARW) on "Environmental Security: Water Security, Management and Control" (http://www.es-ws.eu). This three-day ARW has explored different aspects of the environmental security assessment process, focusing on the assessment, monitoring and management of resources, with an overview of the related scientific knowledge.

The role of water in our communities, from local to regional and up to global levels, poses a series of key questions about climate change, about the anthropogenic impact on the environment, and about all the interconnected actions and events that affect the availability and quality of the resource. All these questions share a common demand for more scientific knowledge and information.

In this particular context the disciplinary boundaries are fading, and there's a growing need to create broader connections and wider collaborative interdisciplinary groups, aimed at building an integrated knowledge-base at the service of stakeholders and of the whole society, in order to face up the new dynamics of human-hydrologic systems.

Following this concept, contributions of multiple disciplinary backgrounds, such as Law studies, Hydrology, Monitoring and Information Technologies, Geophysics, Geochemistry, Environmental Sciences, Hydraulic Engineering, Systems Engineering, Economic and Social studies, converge and interact in this workshop.

The issue of managing and securing water supplies represents a general challenge in the Middle East North African region (MENA), and is often seen from the perspective of water supply managers and technologists. The fusion of so many diverse backgrounds, as in our meeting, offers an opportunity to produce a step ahead in networking the various expertises, by stimulating discussion and creating new collaborative opportunities.

The availability and access to the water resources has always been a factor of development and gave birth to the former civilizations, such as those that were built on the main rivers of the ancient history: Pharaonic Civilisation in Egypt and Babel Civilisation, thanks to the resources of the Nile, Euphrates and Tiger. The availability of water resources in these zones led to the stabilization and safety of the human aggregates. The availability of water resources and the environmental security are thus a factor of stabilization and blooming of the human societies. It is then understandable how, when these resources are not available, the inverse effect may happen: i.e. the instability and even the conflict, both at intra-national and at international level.

The consciousness of the humanity woke up quite recently, to throw an alarm bell concerning the necessity of the implementation of programs, whose objective is the elaboration of a rational management of water resources. Several meetings at the global scale were held to discuss environmental problems and particularly those relative to water security: Rio (1992), Kyoto (1998), Johannesburg (2002). Humanity is now aware of the rarity of water because of overexploitation and pollution, and also the extreme hydrological phenomena that the planet is experiencing are related by most of the scientific community to man-induced climate changes.

In recent years there has been a growing need for the development of tools for environmental security assessment and for the management of environmental security issues. In this context, emerging risks about water resources and their relationship with more general environmental security concerns are particularly relevant. Thus, the assessment of critical environmental factors is a necessary step for the prevention of conflict.

Decision-makers who understand the scientific aspects of emerging water risks are in a better position to design policy and legal frameworks that promote sound environmental management and protection, which in turn may serve to minimise environmental stresses and security challenges. Moreover, a better integration of scientific and policy expertise can strengthen institutions to intervene in cases where environmental damage has already occurred and reduce the likelihood of human harm and economic loss. Meaningful collaboration among scientific and policy experts is thus crucial for the formulation of policy frameworks that effectively assess, avert, and eliminate environmental stress, reducing security threats at the same time.

In this particular context, one of the aims of this ARW has been to address the need for policymakers to access, understand, and apply policy-relevant data and scientific information, as they anticipate and respond to threats to domestic and regional water supplies.

Effective policy frameworks also depend on governance approaches that create a basis for interested parties to become engaged in the process of making and implementing policies and actions relating to the environment. At a domestic level, public participation can help shape policies to address resource claims and avoid degradation and scarcity.

All the participating parties can serve an important role in monitoring resource availability and providing early warning of potential scarcity. They can also promote compliance with applicable norms designed to minimise environmental degradation, particularly where they are given the right to seek redress in case the rules have been violated. Participatory compliance and dispute resolution systems provide a socially-acceptable way to minimise environmental harm and to manage conflict even where some dispute is inevitable.

The ARW wanted to strengthen the ability of participants to engage in interdisciplinary policymaking at the same time that it modelled participatory policy dialogue, even by using particular tools, such as the simulation of a hypothetical water security problem. The simulation engaged participants to work interactively on a problem-solving exercise, by assembling cross-disciplinary groups, in order to enhance networking and collegiality among participants. The interactive work shows how each specific expertise and role can help give shape to a course of action that assures long-term water security.

Chapters of this book are organised to bring the reader through the various policy-related, science-related and technology-related aspects in the context of water security. The first Chapters (1, 2 and 3) deal with environmental security issues and policy aspects, focusing on the MENA region. The following Chapters (4, 5, 6 and 7) go through an overview of remote sensing applications and electromagnetic techniques for water monitoring, including an electro-active method for the transportation of contaminants. An original approach to the study of solid particles transport is shown in Chap. 8, while Chaps. 9 and 10 switch to some challenging aspects of geochemical applications, such as the artificial recharge of aquifers and the geologic storage of CO_2 with its related water issues.

Threatens to the quality of water are discussed in Chaps. 11–15, starting from an example of an integrated geological-hydrogeological-geochemical approach for the protection of high quality resources (11), followed by an overview of trace contaminants in surface water (12). The following Chapters introduce general problems about water quality by using selected case studies. Chapter 13 discusses the worldwide risk of arsenic contamination in water, taking as an example a specific case study in Italy, while Chap. 14 deals with the many aspects of the study and remediation of an abandoned industrial site, by studying a zinc plant located in Siberia (Russia). Finally, Chap. 15 stresses the significance of boron contamination presenting a case study in Greece.

Chapters 16–20 present particular frameworks of water resources management and optimisation in different geographic areas, such as: the Jordan Valley and the lowering of the Dead Sea level (16), the southern oases of Egypt and the long-term sustainability of groundwater exploitation (17), the water resources management and protection in Algeria (18), the scenarios of groundwater availability and quality in North Aquitania (France) by numerical models (19) and the approach to the national water management in Morocco compared to other experiences across the Mediterranean (20). Chapter 21 discusses the environmental management and policy recommendations in agriculture, taking as an example the specific context of the Bulgarian economic, institutional and natural environment.

Chapters 22–24 deal with technological aspects and specific experiences in water treatment and desalination, starting from the experience of autonomous desalination systems within the ADIRA project in Morocco (22), followed by an overview of the current desalination technologies focused on the Egyptian situation (23), closing with the description of a water treatment system based on electrodialysis and its perspective application (24).

Editors of this volume worked in order to transfer this multidisciplinary vision to the context of this book, and this has been possible thanks to the efforts of all our valuable Authors and Contributors.

Andrea Scozzari
Bouabid El Mansouri

Editorial Note and Acknowledgements

Co-directors of this meeting have firstly to acknowledge NATO for sponsoring and funding this event and for supporting the publishing of this volume. Also, we have to be thankful to the Science for Peace and Security Program of NATO for their subsequent help and great availability.

Besides, Co-directors want to express their gratitude to the other institutions supporting this event:

- The University Ibn Tofail, Kenitra, Morocco
- The CNR Institute of Geosciences and Earth Resources (CNR-IGG), Pisa, Italy
- The Ministry of High Education, Executive Training & Scientific Research, Morocco
- The Moroccan Committee of the International Association of Hydrogeologists
- The BRGM - Water Division, Orléans, France
- The ONEP-IEA (Office National de l'Eau Potable), Morocco

A particular acknowledgement goes to Tarik Bahaj (Moroccan Ministry of High Education, Executive Training & Scientific Research), with whom the early idea of this ARW was firstly developed.

Co-directors have also to acknowledge all the colleagues that contributed to the local logistics and organisation, in particular Lahoucine Hanich and his students, whose availability and supportive attitude made possible the smooth development of this event.

Also, Co-directors have to be grateful to Eric Dannenmaier for organising and moderating the water security simulation, with an informal and effective style that has been greatly appreciated. Colleagues from the Foundation for Environmental Security and Sustainability completed the right mixture for success, taking their experience both in this particular session and in all the other interactive phases of this workshop. Editors of this book hope that some of the collegiality and interaction that was gained in those days in Marrakech can be reflected by the style and contents of this volume.

Co-directors (which are also editors of this book) must also acknowledge Emanuela Ferro and Cristina Rosamilia for their strong support and incredible availability in providing editorial assistance.

Again, the editors are indebted to all the reviewers for the precious help and suggestions provided, that guaranteed the quality of this book. In addition to the editors, the reviewing phase has been supported by:

Magdy AbouRayan, Mansoura University, Egypt
Mohamed Azaroual, BRGM - Water Division, France
Rosa Cidu, University of Cagliari, Italy
Elissavet Dotsika, NCSR "Demokritos", Greece
Yousif Kharaka, USGS, USA
Luigi Marini, University of Genoa, Italy
Barbara Nisi, CNR-IGG, Italy
Fausto Pedrazzini, CNR-IFC, Italy
Stefano Vignudelli, CNR-IBF, Italy

Finally, we wish to give our heartful thanks to all the speakers, participants and Committee members, that have contributed to give life and success to this event, both under the organising and under the scientific point of view. Thanks to everyone, and sorry if we forgot someone!

Andrea Scozzari
Bouabid El Mansouri

Contents

1 **A View of the Main Environmental Security Issues and Policies** 1
  Fausto Pedrazzini

2 **Water Security: Identifying Governance Issues and Engaging Stakeholders** 11
  Eric Dannenmaier

3 **European and International Funding Programmes for Environmental Research with African Countries** 21
  Thomas Ammerl

4 **Space Technology for Global Water Resources Observations** 31
  Jérôme Benveniste

5 **Remote Sensing for Environmental Monitoring: Forest Fire Monitoring in Real Time** 47
  Abel Calle, Julia Sanz, and José-Luis Casanova

6 **Electromagnetic Methods and Sensors for Water Monitoring** 65
  Francesco Soldovieri, Vincenzo Lapenna, and Massimo Bavusi

7 **Subsurface Permeability for Groundwater Study Using Electrokinetic Phenomenon** 87
  Alexander K. Manstein and Mikhail I. Epov

8 **Solid Particles Transport in Porous Media: Experimentation and Modelling** 97
  Hua-Qing Wang, Nasre-Dine Ahfir, Abdellah Alem,
  Anthony Beaudoin, Ahmed Benamar, Abdel Ghadir El Kawafi,
  Samira Oukfif, Samiara El Haddad, and Hui Wang

9 **Challenges of Artificial Recharge of Aquifers: Reactive Transport Through Soils, Fate of Pollutants and Possibility of the Water Quality Improvement** 111
  Mohamed Azaroual, Marie Pettenati, Joel Casanova,
  Katia Besnard, and Nicolas Rampnoux

10	**Geologic Storage of CO_2 to Mitigate Global Warming and Related Water Resources Issues** Yousif K. Kharaka and Dina M. Drennan	129
11	**Environmental Security Issues Related to Impacts of Anthropogenic Activities on Groundwater: Examples from the Real World** Brunella Raco, Andrea Cerrina Feroni, Simone Da Prato, Marco Doveri, Alessandro Ellero, Matteo Lelli, Giulio Masetti, Barbara Nisi, and Luigi Marini	153
12	**Overview on the Occurrence and Seasonal Variability of Trace Elements in Different Aqueous Fractions in River and Stream Waters** Rosa Cidu	163
13	**A Worldwide Emergency: Arsenic Risk in Water. Case Study of an Abandoned Mine in Italy** Luca Fanfani and Carla Ardau	177
14	**Acid Mine Drainage Migration of Belovo Zinc Plant (South Siberia, Russia): A Multidisciplinary Study** Svetlana Bortnikova, Yuri Manstein, Olga Saeva, Natalia Yurkevich, Olga Gaskova, Elizaveta Bessonova, Roman Romanov, Nadezhda Ermolaeva, Valerii Chernuhin, and Aleksandr Reutsky	191
15	**Distribution and Origin of Boron in Fresh and Thermal Waters in Different Areas of Greece** Elissavet Dotsika, Dimitrios Poutoukis, Wolfram Kloppmann, Brunella Raco, and David Psomiadis	209
16	**Environment and Water Resources in the Jordan Valley and Its Impact on the Dead Sea Situation** Wasim Ali	229
17	**Hydrochemistry and Quality of Groundwater Resources in Egypt: Case Study of the Egyptian Southern Oases** Anwar A. Aly, Abdelsalam A. Abbas, and Lahcen Benaabidate	239
18	**Means of Mobilization and Protection of Water Resources in Algeria** Mohammed Kadri, Ahmed Benamar, and Brahim Bendahmane	255
19	**Regional Model of Groundwater Management in North Aquitania Aquifer System: Water Resources Optimization and Implementation of Prospective Scenarios Taking into Account Climate Change** Dominique Thiéry, Nadia Amraoui, Eric Gomez, Nicolas Pédron, and Jean Jacques Seguin	275

20	**Challenges and Strategies for Managing Water Resources in Morocco Comparative Experiences Around the Mediterranean Sea** .... Mokhtar Bzioui	291
21	**Environmental Management in Bulgarian Agriculture – Modes, Efficiency, Perspectives** .... Hrabrin Bachev	301
22	**Autonomous Desalination and Cooperation. The Experience in Morocco Within the ADIRA Project** .... Vicente J. Subiela and Baltasar Peñate	319
23	**Desalination Technologies as a Response to Water Strategy Problems (Case Study in Egypt)** .... Magdy AbouRayan	339
24	**Advanced Water Treatment System: Technological and Economic Evaluations** .... Artak Barseghyan	353

Chapter 1
A View of the Main Environmental Security Issues and Policies

Fausto Pedrazzini

Abstract The natural environment is normally in an equilibrate status but it can easily be altered by different phenomena, some of them induced directly or indirectly by men. The effects of those phenomena could lead to negative processes, like permanent land degradation, depletion of water resources and desertification. As a consequence the environment becomes unstable and populations are exposed to increased competition for essential food and water resources. Such a competition could involve neighbour communities and countries, thus creating conditions for tensions and conflicts. The consequences of the degradation of the environment have an immediate negative effect on the ecosystem as a whole; but this evidence has to be expanded by taking into account the effects of a degraded environment on the dynamic of populations, on their activities and on their development. An equilibrate environment is by definition safe since it provides the essential resources and services to all its living components. In case of natural or man-induced disasters the environment is no longer safe since their direct consequences (floodings; forest fires; accidents in power plants and in industrial sites) put populations in serious danger. In addition to these events, other factors like bad management or overexploitation of an ecosystem and the climate change, could create very adverse environmental conditions (water scarcity; land degradation; desertification etc.), which ultimately contribute to destabilise entire populations; their lifestyle and their development. Such a process is frequently the origin of tensions and conflicts among populations and between neighbouring communities and countries, particularly when these are sharing the same river basins and/or competing for the same basic resources. This is the typical scenario to which the

F. Pedrazzini (✉)
CNR Institute of Clinical Physiology, Pisa, Italy
e-mail: fausto.pedrazzini@ifc.cnr.it

A. Scozzari and B. El Mansouri (eds.), *Water Security in the Mediterranean Region*,
NATO Science for Peace and Security Series C: Environmental Security,
DOI 10.1007/978-94-007-1623-0_1, © Springer Science+Business Media B.V. 2011

concept of Environment Security should be referred. It is a relatively new and non traditional field for analysis and assessments, but unfortunately it is already matter of serious concern in many part of the world.

Keywords Environmental security • Water security

1.1 Introduction

Nowadays, the term "security" is interpreted in many different ways. Its meaning ranges from a strict definition of safety from armed conflict, to a more general interpretation of "human security", which includes social, political, environmental and other aspects.

The relation between the natural environment and the security of populations has been the topic of much research and the content of many scientific publications; but only recently it is becoming an important subject of international environmental analysis. Referring to Environment Security means to approach an evolving concept for which an established definition does not exist yet [11].

Environment is characterised by natural phenomena and by threats of various origin, which could alter its original features and lead to emergency situations or even long-term variations which change the natural status of large regions.

The environment conditions and their variations could affect security of geopolitical areas and their populations, beyond what could be expected.

The pollution of the basic environmental resources (air, water and soil) has an immediate negative effect on the health, the safety and the well being of humans, animals and all biological lives in an ecosystem, but even a non-polluted environment could be affected by a variety of transformations which alter its original features.

This happens when there is a reduction of agricultural practices and of the related land management; when there is an excessive concentration of buildings or, on the contrary, when entire territories are abandoned due to urbanisation or migrations. Under these circumstances, land becomes less and less productive and more easily eroded since it is less able to absorb meteorological events. A negative process is therefore taking place and in some cases this could be irreversible. When this occurs, the consequent adverse conditions are a serious threat for populations and for their peaceful development [8]. Unfortunately such a scenario is already a reality in many regions of the world and there are strong probabilities that it could even worsen due to the effects of climate change.

For these reasons, adequate preventive measures should be planned and implemented at the appropriate national and international levels, while the local and the regional authorities should be concerned to put into practice specific initiatives and proper management which are essential for the conservation of an equilibrate environment [2].

1.2 Discussion

The basic concept underlining this discussion is an *Equilibrate Environment* versus a *Degraded Environment*. While in an *Equilibrate Environment* resources and services are ready and available for all organisms, including humans and their settlements, a *Degraded Environment* implies low availability of resources; insufficient services, possible tensions/conflicts and eventually displacements of populations.

What defines an *Equilibrate Environment?* An environment is equilibrate when all its components (including humans) interact in an optimal way. In addition, these conditions contribute to create a *Secure Environment* since all its components go along smoothly. On the contrary, when these conditions are not met, the environment becomes unstable and its *Security.* is under threat.

What does actually mean *Environmental Security?* Essentially we refer to the security of the Environment, when alterations of the considered environment affect its equilibrium and give origin to tensions and conflicts within and outside the same area. However there is not an agreed definition by the scientific community on this concept, which is a continuous evolving. When the security of the environment is taken into consideration, other relevant situations and phenomena should be taken into account, like for example nuclear accidents (Tchernobyl; Three Mile Island); ecological disasters (Seveso, Bhopal); natural extreme events (earthquakes, tsunami, forest fires, flooding); intentional and/ or criminal disruption of natural resources. All these are events which have a direct and dramatic effect on the *Security* of both the environment and the populations living on it.

1.2.1 Factors Which Directly Affect Environmental Security

The following paragraphs discuss the basic factors which have directly consequence on the environment and its security.

1.2.1.1 Water Availability/Scarcity

The availability of fresh water and its quality is essential for agriculture, running industries, providing energy, and ensuring health and sanitation. However water resources are very unevenly distributed and according to evaluations of the UN Commission for Sustainable Development, more than 40% of countries lie within water stress zones. A substantial portion of fresh water in basins is shared by two or more nations. Water scarcity can therefore lead to tensions and conflicts [5]. During the past half century there have been more than 450 water-related disputes of hostile sorts, and on 37 occasions rival countries have fired shots, blown-up dams, or undertaken some other form of violent action [7].

However, the perspective of "water wars" seems unlikely if the issue is approached as a direct cause-effect relationship; on the contrary the environment security aspects should be related to the social, political, economical and cultural contexts [9].

Another aspect of water scarcity is human security which seems likely to cause security threats both nationally and internationally. Insufficient food production due to water shortages leads environmental problems like deforestation and desertification and of consequence poverty and malnutrition, factors which may induce migration flows either domestically or into neighbouring countries, giving origin to environmental and/or poverty refugees often ending up in equally poor countries [5].

Not only can water shortages produce the seeds of conflict, but conflict can also cause water shortages. People fleeing armed conflict bring heavier water demand. For example in 2006 the number of Eritrean refugees seeking asylum in Sudan increased by 30%. This sudden increase has caused further pressure on limited water resources in Sudan. In addition conflicts have a direct effect on water resources such as the pollution of water. For example during the Rwandan genocide, corps in wells, rivers and streams polluted water resources, then carried risks of transmitting infectious diseases. The Danube river was also polluted during the conflicts in the former Yugoslavia, both in Bosnia and even more in the war over Kosovo [5]. This is a worldwide concern and under these circumstances, some initiatives should be taken into consideration carefully, like stabilising water demand; integrating the management of watersheds; implementing active cooperation and solidarity between populations for the proper management of water resources.

Some other important issues should be also assessed, starting by the consideration that water policies are still focused on supply, rather than on an efficient exploitation and distribution [12]. As a matter of fact, transport losses and household leaks have a very negative effect of the efficiency of water use. In addition the overall picture concerning water supply could become critical due to the uncertainties of climate change. Finally it should not be forgotten that water resources could be the target of a terrorist attack either by intentionally polluting them or by altering the water management/supply.

1.2.1.2 Land Availability/Degradation

The quality of land and soil directly depends on the equilibrate availability of water. Water scarcity is the first step of the land degradation process; a phenomenon which could have dramatic consequences including desertification and migrations. These issues were the basis of a book written by Rubio et al. [10] in which water scarcity, land degradation and desertification in the Mediterranean region were discussed. Particularly it was evaluated to what extent water and land management affects populations in a way that reduces their security and also may lead to conflicts within the community and may even cross political boundaries, thus affecting regional political stability.

Land degradation implies the reduction of soil fertility and may give origin to the process of desertification. Some 24 billion tons of fertile soil disappear annually.

One third of Earth's land surface (4 billion hectares) is threatened by desertification, and over 250 million people are affected by it [6].

There is an increased attention on desertification, defined as the soil degradation in a arid or sub-humid environment, which implies the reduction or loss of the biological productivity of agricultural and irrigated land, as well as pasture and forests. Another characteristic of desertification is its persistence [1]. Consequences of desertification (famine poverty, migration) have a direct impact on the stability of populations and as stated by a study of the United Nations University in 2006, "desertification is one of the most important challenges for the global environment because it is a phenomenon which might reduce and inhibit the sustainable development which had started in different parts of the world. Such a process might destabilise the society by increasing the poverty and by creating the conditions for migration induced by altered environmental conditions, besides inducing increased pressure on those areas which are not degraded yet" [6].

Since environmental security is ultimately matter of availability of resources, a direct relation with food availability for populations can also be established. The recent rising of food prices (also caused by environmental mismanagement) provoked riots and violent protests in many countries. By taking into account that 13% of the world's population is undernourished due to extreme poverty, while up to two billion people lack food security intermittently (source: FAO), it can be concluded that the consequences of a degraded environment directly affect the basic requirements of populations and their stability.

An example of interactions between land use and conflicts is typically the Darfur conflict. At this purpose the UN Secretary general made some statements on June 2007. He said that the Darfur crisis began as an ecological crisis, arising at least in part from climate change... It is no accident that the violence in Darfur erupted during the drought... When Darfur's land was rich, farmers welcomed herders and shared their water... For the first time in memory, there was no longer enough food and water for all. Fighting broke out.

In addition to the risks of land degradation and desertification, natural or man-induced disasters are devastating for the environment, particularly in fragile regions of the world. Earthquakes, flooding, heat waves, forest fires, nuclear and industrial accidents have long lasting consequences on entire regions and their populations, including security concern. Prevention against these phenomena is essential, as well as national and international regulations, solidarity and emergency programmes.

On this issue, international organisations could play an important role and become platforms for trans-national collaborations, aimed at evaluating to what extent the not proper management of the environment could affect the life conditions and the security of populations living in a specific region. A few years ago, several international organisations (UNDP, UNEP, OSCE and NATO) launched a cooperative partnership called ENVSEC aimed at addressing environmental risks to security and at fostering stability through environmental cooperation. ENVSEC operates in a number of regions (Central Asia, Southern Caucasus, the Balkans, Eastern Europe) by conducting regional assessments which form the basis for regional initiatives like creating regional maps highlighting issues and areas where environmental problems influence security or are possible source for trans-boundary environmental cooperation [3].

A number of events were organised in the recent years with a specific focus on environmental issues and security:

- Desertification in the Mediterranean Region: a Security Issue. A NATO Advanced Research Workshop co-Directed by W.G. Kepner of the U.S. Environmental Protection Agency and by J.L. Rubio of the Centro de Investigaciones sobre Desertificacion, Valencia, Spain; held in Valencia on December 2003
- Desertification: a Security Threat? – Analysis of Risks and Challenges. A conference organised on the occasion of the World Day to Combat Desertification, by the German federal Foreign Office in Berlin on June 2007
- Water Scarcity, Land Degradation and Desertification in the Mediterranean Region – Environmental and Security Aspects A NATO – OSCE workshop co-directed by U. N. Safriel, the Hebrew University of Jerusalem, Israel, and by J.L. Rubio of the Centro de Investigaciones sobre Desertificacion, Valencia, Spain; held in Valencia on December 2007
- Environment, Forced Migration & Social Vulnerability. An International Conference organised by the UNU Institute for Environment and Human Security (UNU-EHS) in Bonn, Germany, on October 2008.

All these events highlighted the consequences of the environmental degradation towards the status of populations, by taking into account the effects on living conditions, including the basic requirements in food and resources, but also more general issues like social and political, stability, migrations, safety and security of populations. It can be assumed that a general concern is raising internationally on these issues and taking a more higher place in the agenda of governments and politicians, NGOs and all those subjects interested in the well being of populations in an equilibrate environment.

1.2.1.3 Climate Change and Extreme Natural Events

Land and soil degradation have become a global problem, with very strong implications for land cover and land use. Thus combating this phenomenon has to be forward looking and has to address sustainable land use and integrated development, in a future with pronounced climate change perspectives. Already today we have many regions in the world with extreme climatic conditions. For instance we see prolonged droughts like in Northern Africa, extreme rainfall in areas like Indonesia and the Amazon Basin and hurricanes in the tropics [4].

Climate changing is an evolving picture, but it is nowadays widely accepted that the increasing concentration of the so-called greenhouse gases in the atmosphere is altering the Earth's radiation balance and causing the temperature to rise. This process in turn triggers a chain of events which leads to changes in the hydrological cycle components such as rainfall intensity and frequency, evapo-transpiration rate, river flows, soil moisture and groundwater recharge. The consequences include the lost of soil fertility and land productivity, but also salinisation, acidification, contamination, damage to life in soils and other negative impacts on human activities.

In its fourth assessment report the Intergovernmental Panel on Climate change (IPCC) stated, among other items, that globally, both atmospheric temperatures and sea-surface temperatures are on the rise, with projected increases in temperature varying considerably according to the emission scenario used. Precipitation patterns will also change: typically, already dry areas will become even drier and already wet areas will become even wetter. These changes, along with the increasing occurrence of extreme events, will have major implications on future land use. Another key factor in the future is glacial retreat, which is going to affect freshwater availability in areas such as the Andes and also Asia. What further aggravates this situation is that many countries affected by climate change already bear the heavy burden of soil degradation. For instance, soil degradation is very high in large areas of China, while it is accelerating fast in Central Asia and North Africa [4].

In the same year (2007) the German Advisory Council on Global Change (WBGU) issued a report addressing the topic "Climate Change as a security Risk" in which six threats to international stability and security could potentially be caused by climate change:

- An increase in the number of weak and fragile states as a result of climate change. For example, costs of adaptation to climate change could be too high, or perhaps adaptation strategies cannot be tackled for other reasons
- Increased risk for the global economy and its development
- Risks of growing international distributional conflicts between the main drivers of climate change and those most affected. It is mainly the rich societies who are causing climate change but the poor societies suffering the most from the impacts, thus increasing the growing equity gap
- Consequently industrialised countries might actually lose their legitimacy as global governance actors because of climate change impacts
- Climate change might trigger and intensify migration with the emergence of environmental refugees and migrants
- The climate-induced conflicts will potentially give a new dimension to what is considered a classic security policy.

So what can be done today? The first aspect is to foster a cooperative setting for a multi-polar world; a global political governance system to cope with future problems. The second aspect is to consider climate policy not only as environmental policy, but as security policy as well, by avoiding dangerous climate change. The third aspect is then helping and supporting developing countries – particularly those most affected by climate change – to cope with and adapt to climate-induced changes [4].

Climate change is ultimately an issue which goes across many other primary concerns of the modern society like energy production, storage & efficient use; economy and industrial production & development; agriculture and better use of water & available resources. All these are issues which need a global approach and the coherent commitment of governments and international organisations.

1.3 Conclusions

The integrity of the Environment is an essential factor for the well-being and for the sustainable development of populations, particularly in less-favored regions and in more fragile ecosystems. It is recognised that environmental issues do not respect political and/or cultural borders, but they follow the general rules of natural phenomena and events. The factors which may alter the equilibrium of the Environment are various and interact at different levels regardless national and international boundaries. The security of the Environment refers to a concept by which the overall functions and services of the environment are at the disposal of the populations living on it. The practical effects of the disruption of the virtuous cycle which starts with a stable Environment which provides resources for sustaining and maintaining the well being of populations, might not be fully perceived, particularly with respect to the term security, normally associated with criminal and military scenarios. But as I tried to elaborate in this paper, the possible tension and conflicts originating from a degraded environment are becoming additional threats to the stability of entire regions. For such a reason such an issue is taking an high rank in the agenda of governments, public institutions and international organisations.

References

1. Adeel Z, Safriel U et al (2005) Ecosystems and human well being desertification synthesis. Millennium Ecosystem Assessment, World Resource Institute, Washington, DC
2. Blue Plan (2005) In: Benoit G, Comeau A (eds.) A sustainable future for the Mediterranean: the Blue Plan-s environment and development outlook. Earthscan, London
3. Borthwick F (2009) International organisations cooperation around environment and security. In: Stec S, Baraj B (eds.) Energy and environmental challenges to security. Springer, Dordrecht
4. Buchman N (2007) Impact of climate change on land use and desertification: a new security risk? In: Desertification: a security threat? Specialist conference report, Federal Foreign Office, Berlin
5. Coskun BB (2007) Growing dangers: emerging and developing security threats. In: NATO Review, winter 2007. Available at http://www.nato.int/docu/review/2007/issue4/english/analysis5.html
6. De Kabermatten G (2009) Grounding security, securing the ground. In: Rubio JL, Safriel U et al (eds.) Water scarcity, land degradation and desertification in the Mediterranean region. Springer, Dordrecht
7. Myers N (2004) Environmental security: what's new and different? Background paper for The Hague conference on environment, security and sustainable environment
8. Pedrazzini F (2008) Water scarcity, land degradation and desertification as factors for social and political instability. In: Environment, forced migration & social vulnerability, international conference proceedings. www.efmsv2008.org
9. Renner M (2006) Introduction to the concepts of environmental security and environmental conflict. Institute for Environmental Security, The Hague. Available at http://www.envirosecurity.org/ges/inventory/IESPP_I-C_Introduction.pdf
10. Rubio JL (2009) Desertification and water scarcity as a security challenge in the Mediterranean. In: Rubio JL, Safriel U et al (eds.) Water scarcity, land degradation and desertification in the Mediterranean region. Springer, Dordrecht

11. Schrijver N (1989) International organisation for environmental security. Secur Dialogue 20(2):115–122
12. Thibault HL et al (2009) Facing water crisis and shortages in the Mediterranean. In: Rubio JL, Safriel U (eds.) Water scarcity, land degradation and desertification in the Mediterranean region. Springer, Dordrecht

Documents Analysed

NATO Science for Peace and Security Programme. Available at: www.nato.int/science

Millennium Ecosystem Assessment (2005) Ecosystems and human well being & desertification synthesis. World Resources Institute, Washington, DC

Desertification: a security threat? – Analysis of risks and challenges. In: A Conference on the occasion of the World Day to Combat Desertification, Berlin, 26 June 2007. Available at: www.gtz.de

OSCE, The Environment for Europe process. Available at: http://www.environmentforeurope.org/institutions/osce.html

The United Nations University-Institute for Environment and Human Security UNU-EHS. Homepage http://www.ehs.unu.edu/

ENVSEC, Environment and Security Initiative. Homepage http://www.envsec.org/

UNCCD. Homepage www.unccd.int

Institute for Environmental Security. Homepage http://www.envirosecurity.org

United Nations Environment Programme, UNEP. Homepage www.unep.org

Chapter 2
Water Security: Identifying Governance Issues and Engaging Stakeholders

Eric Dannenmaier

Abstract This paper examines the concept of environmental security and the relevance of governance frameworks (policies, laws, and institutions) for mitigating water scarcity and degradation concerns in the Middle East and North African (MENA) region (This region consists of Algeria, Bahrain, Djibouti, Egypt, Iran, Iraq, Israel, Jordan, Kuwait, Lebanon, Libya, Malta, Morocco, Oman, Qatar, Saudi Arabia, Syria, Tunisia, the United Arab Emirates, West Bank and Gaza, and Yemen). The paper highlights the importance of governance frameworks that support meaningful risk assessment, policy analysis, strategic planning, and policy implementation in coordination with experts in the scientific community and other relevant stakeholders.

The paper begins by providing a core definition of environmental security, which is affected by anthropogenic drivers categorized as resource demand, resource depletion, and resource degradation. It also offers a dynamic model for assessing security-relevant consequences that arise through the interaction of these drivers with natural events or conditions and population/technology concerns (collectively "environmental stressors") within a local, national and regional economic, social, and political context. Under this model, anthropogenic triggers act directly or in combination with natural conditions or events to threaten security through a range of vectors. These consequences may escalate or interact – driven by social, political or economic factors – often in a nonlinear fashion.

The author posits that governance frameworks may intervene within the broader economic, political and social context to minimize the cause, avoid the consequence, or prevent escalation. The magnitude and impact of any specific security threat depends on the ability of governance frameworks to avoid and/or respond.

E. Dannenmaier (✉)
Indiana University School of Law, 530 West New York Street, Inlow Hall Suite 337,
Indianapolis, IN 46202, USA
e-mail: edan@iupui.edu

Those states with adequate governance frameworks are more likely to recognize environmental stresses and minimize their impact.

Environmental security challenges will ultimately be met by states on the basis of their national priorities and traditions, but the author suggests that careful coordination by state officials with key civil society stakeholders and relevant members of the scientific community is needed to animate agencies that have not been traditional partners and to strengthen the legitimacy and efficacy of institutional responses.

Keywords Environmental security • Water security • Governance • Environmental stressors • Civil society • Stakeholders

2.1 Introduction

Historically, the security of states in the MENA region has been dependent on water resources. As the Arab Water Council has noted, "water has played a dominant role in determining human activities, settlements, socioeconomic interactions and growth in the Arab countries [defined to include 22 states in North Africa and the Middle East] more than in any other part of the world." [1]. MENA governance institutions have long sought to manage a scarce water supply distributed among a growing population, and water scarcity [5] is anticipated to increase dramatically in the region in the near future. By one recent estimate, "per capita water availability is set to fall by 50 percent by 2050." [16]. In addition to population growth, the impact of climate change on scarcity could be profound. The UN Development Report for 2008 notes, for example, that "In North Africa even modest temperature increases could dramatically change water availability. For example, a 1°C increase could reduce water runoff in Morocco's Ouergha watershed by 10% by 2020. If the same results hold for other watersheds, the result would be equivalent to losing the water contained by one large dam each year." [14]. While these predictions sound dire, they are likely optimistic. At present, climate negotiators have only tentatively embraced a commitment to keeping global temperatures to a 2°C change by 2050 [15] and many doubt that even this goal can be met.

Because water is scarce in the MENA region, and becoming more scarce, it is important to view the issue as a core concern of human and state security, and to manage development patterns and the long-term impact of population growth and land use decisions with a strategic approach to water. By framing water as a security issue, governance institutions can more effectively coordinate with critical constituencies, including populations and communities, civil society organizations, private sector institutions, and scientific and technical experts. At the heart of meaningful progress is coordination among these stakeholders and governance institutions to address the environmental, economic, and human aspects of water so that potential threats may be better analysed and priority concerns may be addressed. Above all, progress depends on strengthening the role of water governance frameworks – policies, laws, and institutions – and engaging key stakeholders.

2.2 Defining Environmental and Water Security

A meaningful definition of environmental security starts with the concept of state security, which has evolved in the years following the Cold War. From a more traditional view, state security is concerned with maintaining territorial integrity and domestic peace. The state is the point of reference. From a more modern perspective, state security also values economic prosperity, stability and the health and well-being of populations [2–4, 6, 7, 12, 13]. Human populations become a central concern and point of reference.

As security is defined more broadly, the importance of environmental challenges becomes immediately apparent. It is well documented that environmental stresses have both direct and indirect impacts on populations and economies under a range of circumstances. Human health is at risk from practices ranging from poor sanitation to mis-managed industrial wastes. The economic costs of environmental stress are potentially substantial. The human cost is often incalculable.

Environmental security builds on and focuses human security concerns through a strategic, risk-based analysis of non-traditional security priorities. Just as defense analysts target resources at what they discern is the greater threat (in terms of immediacy or magnitude), policy analysts must target resources at the greater threats to populations and economies. The threats posed, for example, by water scarcity, unsafe drinking water, unstable development patterns and resource scarcity – analyzed from a security perspective – can be more strategically addressed through policies and institutions aimed at early intervention and minimized escalation. Risk assessment has been used as a tool in both the defense and environmental disciplines for many years – and an environmental security framework calls for these disciplines to be integrated to assess a growing range of concerns.

2.3 Causes of Environmental Stress – Anthropogenic Drivers

In order to assess risks and set policy priorities, it is important to concentrate on the human element in environmental stress. While natural events such as earthquakes and drought can be anticipated, policy planning can be more effective when it is aimed at human behaviour – arguably more easily targeted by policy makers. Thus even where natural events are a concern, event forecasting must be combined with meaningful long-term planning for behavioural change (such as increased resource conservation or more efficient resource use and distribution) to assure that human disaster is mitigated or avoided. More importantly, the full range of environmental stresses, including those that directly affect populations, can be addressed through long-term planning.

In general, the anthropogenic drivers of environmental stress may be described as they relate to resource claims, resource depletion, and resource degradation, as detailed below.

2.3.1 Resource Demand

The effort to access natural resources and claims of right to those resources are important historical causes of environmental stress. In the MENA region, many historical disputes between states can be attributed to resource claims – including water. Egypt and Sudan, for example, have long disputed control of the Nile River, which was allocated through a colonial-era agreement that many believe is inequitable and ill-suited to current conditions in the Nile River Basin [9, 10]. Former United Nations Secretary-General Boutros Boutros-Ghali has predicted that water will be a more likely source of future international conflict in the MENA region than politics [11].

2.3.2 Resource Depletion

Resource depletion leads to increased scarcity that can have destabilizing economic or social impacts, particularly where communities are highly dependent on specific resources. Water scarcity in the MENA region, for example, is driven by increased demand from an increasing population against a background of reduced aquifer recharge. Even as technical solutions and new sources through processes like desalination are developed, conservation of use and more efficient forms of delivery are necessary. Where resource depletion becomes extreme, drought and famine may result – with devastating impacts on life as well as territorial integrity.

2.3.3 Resource Degradation

Human activity can degrade the environment through the introduction of contaminants, the destruction and dislocation of species, and the destruction of habitat. Of the three categories of anthropogenic drivers, this is the one that arises most often from modern industrial practices that are not well regulated, and thus presents the most unique set of challenges for traditional security analysts. It also has the most potential for solution from modern technologies and sound environmental management practices. To some extent, degradation is a by-product of industrial development that is an important regional goal. In many cases, the impact of resource degradation is direct and the cost in human health and economic terms is high. But degradation can lead indirectly to harm where natural conditions (such as background water quality) or events (such as drought) intervene.

2.3.4 Summary of Anthropogenic Drivers

The following table summarizes the causal factors described in Sects. 2.3.1–2.3.3 above.

Table 2.1 Summary of anthropogenic drivers

Causal factor	Description
Resource demand	Intra- and inter-state claims of right to (or efforts to access) territory or resources through the operation of informal or formal systems (such as land tenure rules), or the denial of access through economic or social structures that place resources beyond the reach of demandeurs
Resource depletion	The reduction of available resources through agricultural, industrial or domestic activity leading to increased scarcity of resources or commodities (including food and water), or in extreme cases the destruction of resources
Resource degradation	The degradation of resources through agricultural, industrial, or domestic activity leading to direct or indirect human health threat or ecosystem stress (such as the discharge of untreated human or industrial waste to air, water, or terrestrial systems)

Note that any of these three causal factors can be gradual or acute, and can be aggravated by natural conditions and/or events, as well as by social, economic and political effects

2.3.5 New Urgency

Some have argued that environmental security concerns are not new, and reflect traditional security problems that have been faced by governments and defense officials for years. To be sure, resource issues have historically sparked cross-border and internal conflict as states have looked outside their borders for new resources and populations have been driven from lands or livelihoods by scarcity. Yet environmental security takes on a new and broader dimension in a post-industrial context. The frequency and magnitude of resource claims and resource degradation have increased as growing populations and industrial demand have placed increasing pressure on available resources. New technologies allow more rapid depletion of some resources, and often allow the exploitation of resources that were beyond the reach of commercial ventures in the past. Growing populations put increasing demands on already-scarce resources. Thus, environmental security bears a relation to traditional state security concerns, but it has critical new dimensions that call for new thinking.

2.4 Environmental Security Dynamics

The link between human activity, environmental stress and security is complex, and the dynamics of the inter-relationship are difficult to describe. Fig. 2.1, below, is a model that represents one way of thinking about the interaction of human activity (anthropogenic causes) and natural events or conditions as "environmental stressors" within an economic, social and political context.

Under this model, anthropogenic drivers are represented by the "policy challenges" on the left side of the diagram. These challenges are resource demand, depletion, and degradation (described more fully above in part 3 and highlighted in Table 2.1).

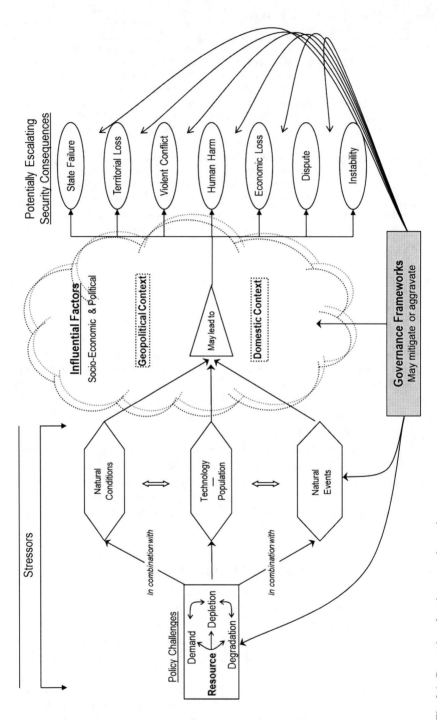

Fig. 2.1 Dynamics of environmental security

They are referred to as policy challenges because they are the aspect of environmental security most directly related to human behaviour and thus most directly the concern of policymakers. These policy challenges are understood to act directly or in combination with specific natural conditions (e.g., water scarcity, groundwater salinity, etc.) or events (e.g., drought) to threaten security through a range of consequences. Technology and population (demographics and size) may also be influential factors. Policy challenges in combination with natural conditions, natural events, and technology/population factors are collectively called "stressors" because they exert stress on natural resources and natural systems.

Stressors may result in security consequences – which are outlined at the right side of the diagram, but the stressors are always influenced by socio-economic and political factors and mediated through geopolitical and domestic (including cultural) contexts. These factors and contexts may reduce or increase the likelihood of a security impact.

The "security consequences" noted on the right side of the diagram are provided in a hierarchy that is consistent with traditional notions of state security (ranging from "mere" instability at the bottom to state failure at the top) but differing moral, ethical, political, cultural, and other concerns could cause one to view the potential consequences as being ordered in a different priority. These security consequences may also escalate ("escalation" being relative to initial ordering) or interact with each other and may be driven by social, political or economic factors – often in a nonlinear fashion.

The model shows that governance frameworks (policies, laws and institutions – represented at the bottom of the diagram) may intervene within the broader economic, political and social context to minimize the cause, avoid the consequence, or prevent escalation. This intervention can occur, in theory, anywhere within the model, but earlier (to the left side of the diagram) is preferable. The magnitude of any specific security threat depends on the ability of frameworks to respond, and states with adequate frameworks are more likely to recognize environmental stresses and minimize their impact.

2.5 Policy Challenges

Any effort to address environmental security in the MENA region will face certain conceptual and political challenges that should be addressed to assure progress both regionally and domestically. Two of these are outlined below.

2.5.1 *Sustainability, Prosperity, Stability and Environmental Security*

Those seeking to address environmental security concerns must recognize that sustainability, prosperity, stability, and environmental security are not independent or collateral goals, but must be viewed integrally. Prosperity cannot be sustained, or broadly attained, where economic stability and the well-being of populations cannot be assured. Environmental security should be seen as a baseline for stability because it focuses on protecting the health and well-being of populations and the stability of resource-based economies. Focusing on environmental security allows a pragmatic and strategic approach to building national stability and wealth. It does not seek decisions based on competing abstract ideals, but instead takes an approach to assessing development choices through a quantitative, risk-based analysis.

2.5.2 Disparate Institutions and Bureaucratic Competence

A second challenge in addressing environmental security concerns is that institutions dealing with the range of issues that affect environmental security often fail to interact, much less be integrated, with those dealing with security. Recent efforts in the MENA region to coordinate among institutions at a domestic and regional level, such as the efforts of the Arab Water Council, are important mechanisms for addressing this challenge.

Coordination will be critical, and states must identify institutions best suited to manage environmental security concerns and then give them a meaningful leadership role. This challenge will ultimately be met by states on the basis of their national priorities and traditions, but a careful analysis of environmental security concerns will help to discern and coordinate the proper role of agencies that have not been traditional partners.

2.6 The Governance Response

As the dynamic model at Fig. 2.1 suggests, governance frameworks form part of the context within which causal factors arise and lead to the range of possible consequences. But they are more than just contextual – they are also potential management tools. These frameworks can be deliberately conceived and brought to bear to assess, avert, or eliminate environmental stress and conflict and thus reduce security threats.

National and local environmental authorities have a particularly important role in dealing with environmental security concerns, as they are the stewards of policy and legal frameworks that promote sound environmental management, protection and conservation. They may be in the best position (working with relevant sectoral ministries) to minimize environmental stresses, and to intervene – even after stress has led to some degree of instability – by preventing escalation to higher-order conflict and human harm. While some level of conflict is inevitable, environmental authorities, operating within nationally-relevant conflict resolution paradigms, can offer a conflict resolution process to defuse threats of instability and minimize further environmental, human, or economic harm.

Governance frameworks should also create a basis for affected and interested parties to become engaged in the process of making and implementing policies relating to the environment. A recent Arab Water Council workshop in Cairo engaged experts in discussing the role of the public in water management and water governance – and represented part of a broader Council project on Public Engagement in Water Management [8]. Such a recognition of the public role is critical. At a domestic level, public participation can help shape policies to address and reduce resource demand, and to minimize degradation and depletion. The public is integral to efforts to equitably distribute water – particularly during times of increasing scarcity. Citizens can also serve an important role in monitoring resource availability and providing early warning of potential scarcity. They can monitor compliance with applicable norms designed to minimize environmental degradation, and may even be given the right to seek redress where the rules have been violated. Participatory compliance and dispute resolution systems provide a socially-acceptable way to minimize environmental harm and to manage conflict even where some dispute is inevitable.

2.7 Conclusion

Environmental and water security concerns warrant the attention of policymakers at the regional and domestic levels throughout the MENA region. National and local governments, as well as regional institutions, have a role in managing environmental issues and addressing concerns of resource demand, depletion and degradation. While the link between environmental stresses and security may be complex, and the search for appropriate responses may be analytically challenging, it is important to embrace the challenge. Governance frameworks should be adapted to assess environmental and water security risks, establish policy priorities, and take action. These frameworks should engage government leaders with critical constituencies, including populations and communities, civil society organizations, private sector institutions, and scientific and technical experts, to assure a broad perspective and to strengthen support for policies and measures that are adopted to increase security.

Acknowledgements This paper is drawn in part from research conducted by the author as part of a project funded by the US Agency for International Development (USAID) and in connection with the work of the Foundation for Environmental Security and Sustainability (FESS). Much of the background material herein was first published in Eric Dannenmaier, "Environmental Security and Governance in the Americas," Canadian Foundation for the Americas (FOCAL), Policy Paper No 01-4 (March 2001). The model at Fig. 2.1 and outline in Table 2.1 were designed by the author and early versions were first published in the FOCAL paper.

Input on the model at Fig. 2.1 was provided by participants in The Hague Conference on Environment, Security and Sustainable Development, at The Peace Palace, The Hague, The Netherlands May 9–12, 2004. The model was also discussed in relation to the Middle East and North Africa region at the NATO Advanced Research Workshop "Environmental Security: Water Security, Management and Control", in Marrakech, Morocco, May 31–June 2, 2010. The author wishes to thank those who have commented on and helped to improve the ideas described in this chapter through each of these processes and dialogues. The author is also indebted to those who offered insights on this work during subsequent discussions with the Policy, Economics, and Law Working Group of the Richard G. Lugar Center on Renewable Energy at Indiana University/Purdue University in Indianapolis. He is also greatly indebted to the organizers of the Marrakech conference, and in particular Dr. Andrea Scozzari, for their insights and support. He also wishes to thank Stephanie Boxell and Melissa Buckley for their research and editorial assistance.

References

1. Arab Water Council (2009) MENA/Arab countries regional document. 5th World Water Forum, Istanbul
2. Axworthy L (1999, Apr) Human security: safety for people in a changing world. Department of Foreign Affairs and International Trade, Ottawa
3. Brown L (1977) Redefining security, worldwatch Paper No. 14. Worldwatch Institute, Washington, DC
4. Homer-Dixon TF (1999) Environment, scarcity, and violence. Princeton University Press, Princeton, NJ, pp 53
5. LeRoy P (Summer 1995) Troubled waters: population and water scarcity. Colo J Int Law Policy 6:299
6. Masters SB (2000) Environmentally induced migration: beyond a culture of reaction. Georget Immigr Law J 14:855
7. Mathews JT (1989) Redefining security. Foreign Aff 68(2):166
8. Mollinga PP (2010, Jan) Public engagement in water management project report launch workshop: designing project action plan, coordination, key indicators and information portals
9. Pflanz M (2010) Egypt, Sudan lock horns with lower Africa over control of Nile River. Christian Science Monitor, 4 June 2010
10. Rotberg RI (2010) The threat of a water war: Egypt and Sudan draw battle lines with upstream nations over access to the Nile. Boston Globe, 2 July 2010. Available at http://www.boston.com/bostonglobe/editorial_opinion/oped/articles/2010/07/02/the_threat_of_a_water_war/
11. Thomson M (2005) Ex-UN chief warns of water wars. BBC News, 2 Feb 2005. Available at http://news.bbc.co.uk/2/hi/africa/4227869.stm
12. Ullman R (1983) Redefining security. Int Secur 8(1):153
13. United Nations Development Programme (1994) Human development report 1994: new dimensions of human security. UNDP, New York
14. United Nations Development Programme (2008) Human development report 2007/2008: fighting climate change: human solidarity in a divided world. UNDP, New York
15. United Nations Framework Convention on Climate Change (2010) United Nations Climate Change Conference in Cancun, COP 16/CMP 6, 29 Nov–10 Dec 2010, Draft decision [–/CP.16]
16. The International Bank for Reconstruction and Development/The World Bank (2007) Making the most of scarcity: accountability for better water management results in the Middle East and North Africa. World Bank, Washington, DC

Chapter 3
European and International Funding Programmes for Environmental Research with African Countries

Thomas Ammerl

Abstract In the light of global environmental and societal challenges there is a huge demand for international cooperation in research and development. The following article shows different funding possibilities of international organisations (e.g. NATO, United Nations and the World Bank) and has its main focus on the European programmes and strategies (Europe 2020, EU-Africa). The further development of the European Research Area (ERA) and its focus especially on Africa is discussed together with the structure and implementation of the Seventh Framework Programme for Research and Technological Development (FP7). The possible international dimension of the FP7 for Northern African countries is demonstrated via an explanation of the different FP7-funding schemes and the environmental topics in the cross-thematic FP7-call for Africa in the year 2010. At the end the active participation of African institutions in the FP7-projects CLIMB, WASSERMed and CLICO and their integration in the research cluster CLIWASEC is highlighted.

Keywords FP7 • Environmental research • African countries

3.1 Programs for Research and Development of International Organisations

There are different research and development programs from international institutions with a focus on environmental challenges and respective solutions for which African partners are eligible.

T. Ammerl (✉)
Bavarian Research Alliance, Prinzregentenstr. 52, D-80538, Munich, Germany
e-mail: ammerl@bayfor.org

The Science for peace and security program (SPS) of the North Atlantic Treaty Organization (NATO) supports practical cooperation in civil science and innovation (www.nato.int/science). Its main focus is on security and environmental sustainability, and this program aims to create links from science to society by collaboration, networking and capacity building activities. Since 2008 there has been an emphasis on strategically important Mediterranean Dialogue countries (Algeria, Egypt, Israel, Jordan, Mauritania, Morocco and Tunisia) with the main objective of fostering regional cooperation [15].

The International Strategy for Disaster Reduction (ISDR) of the United Nations (UN) aims at building disaster resilient communities by promoting increased awareness of the importance of disaster reduction as an integral component of sustainable development [11]. Its main goal is the reduction of human, social, economic and environmental losses due to natural hazards and related technological and environmental disasters (www.unisdr.org). Due to an increase in the incidence of disasters on the continent over the last decade, an African regional strategy for disaster risk reduction was also developed. Within the New Partnership for Africa's Development (NEPAD), Regional Economic Communities (RECs) are expected to make a contribution to disaster management.

As a main source of financial and technical assistance to developing countries, the World Bank also offers different programs for providing resources, sharing knowledge, forging partnerships and building capacity (www.worldbank.org). For Africa, there are regional coordination partnerships and strategic alliances with different partners, e.g. African Development Bank, African Partnership Forum, African Union, European Commission, and United Nations Development Program etc. Within these strategic alliances the World Bank also built up Thematic Sectoral Partnerships with a link to environmental issues. In 2005, an African Action Plan (AAP) was designed to make progress on accelerating shared growth, building capable states, sharpening the focus on results, and strengthening the development partnership [17]. The AAP consists of eight different focus areas, of which three (agriculture, clean energy, and water & sanitation) have an increasing focus on environmental management.

3.2 Strategies "Europe 2020" and "EU-Africa"

The new developed Europe 2020 strategy has also a main focus on research cooperation with international partners [7]. Following its "smart growth"- priority and the measurable target of "3% of Europe's GDP for research and innovation" there is the flagship initiative "innovation union" to ensure that innovative ideas can be turned into products and services that create growth and jobs (http://ec.europa.eu/eu2020). The goal for the European Union is to become a knowledge-based, resource-efficient economy. The longstanding tradition in international science and technology cooperation (S&T) with non-European partners will be fostered in the future. Its main objectives are to support Europe as an effective player in the world scene with a more coordinated and consistent approach. On an institutional level, the Strategic Forum for International S&T cooperation (SFIC) was developed to "facilitate the further development, implementation and monitoring of the international dimension of ERA by the sharing of information and consultation between the partners with a view to identifying common priorities which could lead to coordinated or joint initiatives, and coordinating activities and positions vis-à-vis third countries and within international fora" [2]. In 2010, the former European Union Scientific and Technical Research Committee (CREST) was renamed as the European Research Area Committee (ERAC) to better reflect the shared competence between the Member States and the EU and its strategic policy mission [10].

Special attention is also placed on S&T-cooperation between the European Union (EU) and the African Union (AU) which has rapidly developed during the last years. The new EU-Africa-strategy, adopted at the Lisbon EU-Africa Summit in December 2007, fosters the scientific-political and economic relationship between the relevant stakeholders of both continents (www.africa-eu-partnership.org). Cooperation is based on the recognition that science, technology and innovation are essential to eradicating poverty, combating disease and malnutrition, stopping environmental degradation and building sustainable agriculture and economic growth in Africa. It builds upon different strategic priorities, which will be implemented through partnerships (peace and security, democratic governance and human rights, trade, regional integration and infrastructure, millennium development goals, energy, climate change, migration, mobility and employment, science, information society and space). These eight partnerships will focus on concrete actions at the global, continental or regional level.

Environmental issues are covered by the partnerships for security, infrastructure, energy, and climate change. The platform on climate change strengthens the dialogue and cooperation for adaptation and mitigation measures and has two priority actions: the common agenda on climate change policies and cooperation on land degradation and increased aridity, including the "Green Wall for the Sahara Initiative". It will also make contributions to the Global Climate Change Alliance (GCCA), whose main objective is raising awareness about the impacts of climate change for the most vulnerable developing countries [3]. These countries will be affected in terms of food security, sustainable water supply and extreme weather phenomena (e.g. floods, droughts, land degradation). Within the Seventh Framework Programme for Research and Technological Development of the European Union (FP7) the EU-Africa partnership will be implemented through projects for research and development (R&D).

3.3 The European Research Area and Its Focus on Africa

In order to establish the European Union as the most dynamic and competitive knowledge economy in the world with the target of 3% of the Gross Domestic Product (GDP) by 2010, science, technology and innovation were designated as the crucial elements of the European society of the twenty-first century [1]. To achieve the goal of becoming a worldwide leading arena for research and innovation, the concept of the European Research Area (ERA) was created in the year 2000. Its long-term goal is to secure economic competiveness by means of the use of scientific resources, to respond to global challenges, to meet the demographic and educational challenge of human resources and to promote political cooperation, dialogue and trust. Also the consensual awareness regarding national and European research activities should be improved. Furthermore, mobility of scientists and already existing research infrastructures should be advanced [14].

The ERA must not only deepen the integration within the European Union, but also interact with other parts of the world [4]. Therefore, in 2008 the European Commission proposed a new strategic framework to strengthen science and technology cooperation with non-EU countries to address global challenges (e.g. climate change, water supply, food and energy security) in a better coordinated way [5]. Uncoordinated national policies cannot provide socially and environmentally sustainable solutions to respond to global challenges. Therefore, this new and more internationally oriented strategy is coordinated by the Strategic Forum for International Science and Technology Cooperation (SFIC).

3.4 The Seventh Framework Programme of the European Commission and Its International Dimension in Northern Africa

The key tools for building the ERA with its international dimension are the European Research Community Framework Programmes (FP) for research and technological development (RTD). These programs were established in 1984 with the aim of promoting world-renowned and state-of-the-art research, with scientific excellence as the crucial criterion. The scientific topics for the public calls for proposals of the Framework Program are developed by the Directorate-Generals of the European Commission (e.g. DG Research, DG Environment, and DG Energy etc.) with the decisions of the European Parliament and the European Council. That means that the Brussels Administration provides a "top-down" list for the scientific topics, prioritising according to the demands of European-financed research. Scientific consortia are then requested through so-called "calls for proposals" to develop high-potential approaches to solutions for the drafted research problems.

The current 7th Research Framework Programme (FP7), with a running period from 2007 until 2013 and a budget of 54 billion €, has different eligibility criteria for participation. Eligible in the FP7 are institutions from the 27 EU-Member States (e.g. universities, universities of applied sciences, research institutions, Small and Medium-Sized Enterprises, SMEs) as well as European Union Candidate countries for future accession (Croatia, Iceland, Former Yugoslav Republic of Macedonia, Serbia, Turkey). The so-called "Associated Countries" are also eligible partners in FP7 projects. These countries (Albania and Montenegro, Bosnia & Herzegovina, Israel, Lichtenstein, Norway, Switzerland, Faroe Islands) have an agreement on science and technology or signed a memorandum of understanding for RTD-cooperation with the European Commission (www.bayfor.org).

The FP7 also offers schemes for enhancing and emphasizing the importance of international RTD cooperation. "Third country" institutions from International Cooperation Partner Countries (ICPCs) are encouraged to participate in the specific FP7-programmes "Cooperation", "Ideas", "People" and "Capacities" (www.cordis.eu). Also African researchers are enabled in these EU-funded programmes to make a contribution to the resolution of local, regional and global problems. Meanwhile, the "Ideas" Specific Programme supports excellent individual researcher-driven frontier research from all over the world, the "People" Specific Programme offers a lot of funding schemes for career and mobility development, where international participation is a must (e.g. International Incoming and Outgoing Fellowship Programmes, Initial Training Networks, Industry-Academia Partnerships and Pathways, International Reintegration Grants and International Research Staff Exchange Scheme).

The "Capacities" Specific Programme also has a strong international dimension implemented via various funding schemes (Specific Activities of International Cooperation, bi-regional activities "INCO-NET", bilateral activities BILAT and the international ERA-NET and ERA-NET PLUS actions). An important contribution with regard to the joint Europe-Africa S&T-dialogue is the Mediterranean Innovation and research coordination action MIRA (www.miraproject.eu) and the CAAST-NET (A Network for the Coordination and Advancement of Sub-Saharan Africa-EU Science & Technology Cooperation; www.caast-net.org). One of the main objectives of international cooperation under the "Capacities" Specific Programme is to address the problems that third countries face, or problems that have a global character. In the respective Capacities Work Programme 2011 there is also a special focus on the improvement of closer scientific cooperation with the European Neighbourhood Policy (ENP). This will be implemented through an ERA-WIDE support action for the Mediterranean countries (Algeria, Egypt, Jordan, Lebanon, Libya, Morocco, Palestinian-administered areas, and Syria).

Most of the FP7 funding for international cooperation is available under the Specific Programme "Cooperation" where excellent and innovative researchers are funded in high-quality peer reviewed collaborative research projects (small to medium scale focused research projects, large scale integrating projects) or other different funding schemes like Coordination and Support Actions (CSAs), Joint Technology Initiatives (JTIs), Joint Programming Initiatives (JPIs) and the coordination of non-community research programmes (ERA-NETs, Article 169/185, EUREKA, COST). All ten different themes under the "Cooperation programme" (health, food, agriculture and fisheries, biotechnology, information & communication technologies, nanosciences, nanotechnologies, materials & new production technologies, energy, environment (including climate change), transport (including aeronautics), socio-economic sciences and the humanities, space, security) put special emphasis on international cooperation.

The participation of third countries from outside Europe helps to enhance European scientific excellence and to address problems of a global character. There are also Specific International Cooperation Actions (SICAs) for the dedicated involvement of third countries such as developing and emerging countries. In this case efforts have to be made to align these activities to development goals (including the Millennium Development Goals, MDGs). SICAs are also focused on non-associated candidate countries of the European Union or target countries of the European Neighbourhood Policy like Egypt, Morocco and Tunisia. The main focus of the ENP, created in 2004, are European Neighbourhood and Partnership Instruments (ENPI) and bilateral S&T-Action Plans between the EU and each partner, e.g. the integration of African researchers into the ERA and the FP7, the development of scientific and technological capacity and scientific high-level exchanges (www.ec.europa.eu/world/enp). From the year 2007 until now the international research community was able to integrate a huge amount of institutions from third partner countries to the FP7 [8].

Within the 6th theme "Environment (including climate change)" there are five central activities with corresponding sub-activities. Concrete topics are proposed under these sub-activities (see Table 3.1).

Table 3.1 Activities and sub-activities of the 6th thematic priority, environment (including climate change) within the FP7-cooperation-programme (www.cordis.eu)

Activities	Sub-activities
Climate change, pollution and risks	Pressures on environment and climate
	Environment and health
	Natural Hazards
Sustainable management of resources	Conservation and sustainable management of natural and man-made resources and biodiversity
	Management of marine environments
Environmental technologies	Environmental technologies for observation, simulation, prevention, mitigation and adaptation, remediation and restoration of the natural and man-made environment
	Protection, conservation and enhancement of cultural heritage, including human habitat
	Technology assessment, verification and testing
Earth observation and assessment tools for sustainable development	Earth and ocean observation systems and monitoring methods for the environment and sustainable development
	Forecasting methods and assessment tools for sustainable development taking into account differing scales of observation
Horizontal activities	Dissemination and horizontal activities

3.5 African Participation in the FP7-Projects CLIMB, WASSERMed and CLICO, and the Research Cluster CLIWASEC

The FP7-projects CLIMB (Climate Induced Changes on the Hydrology of Mediterranean Basins - Reducing Uncertainty and Quantifying Risk; www.climb-fp7.eu), WASSERMed (Water Availability and Security in Southern EuRope and the Mediterranean; www.wassermed.eu) and CLICO (Climate Change, Water Conflict and Human Security; www.clico.org) contribute to the integration of African scientific partners with Europe's neighbouring regions. In order to better assess the consequences and uncertainties of climate impact on man-environment systems, a coordinated topic was established in 2009 between the FP7-themes "Environment (including climate change)" and the "Socio-Economic Sciences and the Humanities". All three projects granted so far began in early 2010, and focus on climate induced changes in water resources as a threat to security. To foster scientific synergies and to improve policy outreach, the projects are joining together to establish the research cluster CLIWASEC (Climate induced changes on water and security in southern Europe and neighbouring regions; www.cliwasec.eu). The cluster comprises of a critical mass of scientists from 44 partners (29 institutions from the EU, 5 institutions from S&T-countries and 10 international institutions). CLIWASEC runs 10 African study sites for field campaigns in different countries (Egypt, Ethiopia, Morocco, Niger, Tunisia, and Sudan) and intends to develop a close collaboration with regional stakeholders (e.g. from ecology, socio-culture, politics, and industries) to communicate the project findings in an uncomplicated and efficient manner that puts research results into immediate practice [12].

3.6 The "Call for Africa" of the FP7-2010-Programme

In the "Cooperation" Work Programme 2010, the European Commission gave special priority to S&T cooperation with Africa, following the conclusions of the European Union (EU) and the Africa College of 2008 [6]. With an overall budget of 63 m €, the multi-disciplinary "Call for Africa" was implemented through cross-thematic collaborations on various themes (health, food, agriculture and fisheries, biotechnology and environment, including climate change). Its main focus is on water and food security as well as on health research. There are different considerations within the integrated approach: interaction with climate change, demographic changes, globalization processes, sustainability, various geographical, sectoral and cultural differences, migration and resettlements, urbanisation, health care systems and programme interventions, destabilisation of national food reserves, variations of food and oil prices [6]. Within the theme Environment (including climate change) five individual topics were published, all of them focusing on climate change and/or environmental change and its water-related impacts (see Table 3.2).

Table 3.2 Call for Africa – FP7 Cooperation Work Programme 2010: environment (including climate change); [6]

Topics of the Call for Africa (FP7-Africa-2010-Environment)	Expected impacts by the European Commission
ENV.2010.1.2.1-1 The effect of environmental change on the occurrence and distribution of water related vector-borne diseases in Africa	More accurate and reliable predictions for the distribution of water related vectors and vector-borne diseases in Africa and Europe. Strengthening of the early warning, surveillance and monitoring systems for vector-borne diseases. Support to policies on climate change and health
ENV.2010.1.3.3-1 Early warning and forecasting systems to predict climate related drought vulnerability and risks in Africa	Increase our knowledge on the relation between drought and climate change and provide contribution to improved early warning and forecasting systems. Help to better identify vulnerable regions and to further strengthen preparedness and planning capacities in Africa. Contribute to capacity building
ENV.2010.2.1.1-1 Integrated management of water and other natural resources in Africa	Since the outcome of the project should be a tool-box for both integrated natural resources management that could be used in a variety of environmental and socioeconomic conditions in Africa and assess potential future scenarios as well as proposed policies and programs, the expected impact is a long-term integrated management of natural resources in line with sustainable development principles and a better capacity for assuring the economic and social well being at local and regional levels
ENV.2010.3.1.1-3 Decentralised water supply and sanitation technologies and systems for small communities and peri-urban areas	Development of criteria with regard to the adoption of particular technological approaches/solutions. Bridging the water and sanitation gaps, thus supporting the achievement of Millennium Development Goals. To tailor water resources management to local conditions, capacities and institutional settings and help African countries to cope with water adaptation to climate changes
ENV.2010.3.1.1-4 Water harvesting technologies in Africa	Strengthening the potential and sustainability of rainfed agriculture in Africa and increasing food production and security. Improving the livelihoods of rural communities, using innovative appropriate water management techniques

In addition to the Framework Programmes there is also support for the research community via the European Development Fund (EDF) and the Development Cooperation Instruments (DCIs) with six domains (food security, investing in people, environmental and sustainable management of natural resources, non-state actors and local authorities in development, human rights and democracy, migration and asylum). The European Neighbourhood Instruments (ENPI) also offers support to the African research community, e.g. through the ENPI's capacity building initiative [13].

All North African countries (Algeria, Egypt, Libya, Morocco and Tunisia) are included in the ENP with Egypt, Morocco, and Tunisia as ENP-target countries and so-called "Mediterranean partner countries". All northern African countries are also in the Union for the Mediterranean (UfM) where the cooperation between the European Union and its Southern neighbouring countries is fostered, e.g. on environmental foci (depollution of the Mediterranean, Mediterranean Solar Plan for alternative energies etc.). The Monitoring Committee for Euro-Mediterranean Cooperation in RTD (MoCo) has the objective to stimulate the Euro-Mediterranean S&T-cooperation [9].

3.7 Conclusions

The European Research Area (ERA) offers different tools for developing Europe into the worldwide leading arena for research and innovation. Within the Seventh European Framework Programme for Research and Technological Development (FP7), the cooperation with international partners from non-European countries (e.g. African countries) is a key feature for innovation and the transfer of knowledge. In the light of the society's global environmental changes, the international, intersectoral and interdisciplinary FP7-research consortia have to response to these challenges. By bridging the gap between the scientific and policy community, the research consortia should also make a contribution to guide policy makers in investment decisions and the sustainable use of the research findings [16].

References

1. Archibugi D, Coco A (2005) Is Europe becoming the most dynamic knowledge economy in the world? J Common Mark Stud 43:433–59, Blackwell Publishing Ltd 2005
2. CREST-SFIC (2010) First report of Activities of the Strategic Forum for International S&T Cooperation (SFIC. CREST-SFIC, Brussels
3. European Commission (2007) Limiting global climate change to 2 degrees celsius: the way ahead for 2020 and beyond. European Commission, Luxembourg
4. European Commission (2008a) Opening to the world: international cooperation in science and technology. Report of the ERA expert group. European Commission, Luxembourg
5. European Commission (2008b) A strategic European framework for international science and technology cooperation. European Commission, Brussels
6. European Commission (2009) Work Programme 2010 – Theme 6 Environment (including climate change). European Commission C (2009) 5893 of 29 July 2009, Brussels
7. European Commission (2010a) Europe 2020 – a strategy for smart, sustainable and inclusive growth. European Commission, Brussels
8. European Commission (2010b) Catalogue of FP7 projects 2007–2010. COOPERATION Theme 6 – Environment (including climate change). DG Research, Environment Directorate, Brussels
9. European Commission (2010c) The changing face of EU-African cooperation in science and technology. European Commission, Luxembourg, p 68. ISBN 978-92-79-15658-8

10. European Council (2010) Resolution on the developments in the governance of the European Research Area. 3016th Competitiveness Council meeting. European Council, Brussels
11. Innocenti D (2010) International workshop on climate change impacts and adaptation: reducing water-related risks in Europe, University Club Foundation, Brussels, 6–7 July 2010
12. Ludwig R, Ammerl T (2010) The water's edge. In: Public service review: science and technology, issue 6. Publicservice.co.uk Ltd, Staffordshire, pp 164–165
13. Makhan VS (2009) Making regional integration work in Africa: a reflection on strategies and institutional requirements. African Capacity Building Foundation, Harare. ISBN 978-1-77937-013-6
14. Muldur U, Corvers F, Delanghe H, Dratwa J, Heimberger D, Sloan B, Vanslembrouck S (2007) A new deal for an effective European Research Policy. Springer, Dordrecht, p 289. ISBN 1402055501
15. Pedrazzini F (2010) Environmental security – water security, management and control. In: Advanced Research Workshop, Marrakech, 31 May–2 June 2010
16. Quevauviller P (2010) Climate change impacts and adaptation: reducing water-related risks in Europe. In: International science-policy interfacing workshop, Brussels, 6–7 July 2010
17. World Bank (2007) Accelerating development outcomes in Africa – progress and change in the Africa action plan. Development Committee, Washington

Selected Web Resources

A Network for the Coordination and Advancement of Sub-Saharan Africa-EU Science & Technology Cooperation (CAAST-NET). www.caast-net.org
Bavarian Research Alliance (BayFOR). www.bayfor.org
Climate change, water conflict and human security (CLICO). www.clico.org
Climate induced changes on the hydrology of Mediterranean Basins – reducing uncertainty and quantifying risk (CLIMB). www.climb-fp7.eu
Climate induced changes on water and security in southern Europe and neighbouring regions (CLIWASEC). www.cliwasec.eu
Community Research & Development Information Service (CORDIS). www.cordis.eu/fp7
European Commission on International Cooperation and the international dimension of the European Research Area (ERA). www.ec.europa.eu/research/iscp/index.cfm
European Commission on its European Neighbourhood Policy (ENP). www.ec.europa.eu/world/enp
Mediterranean Innovation and research coordination action (MIRA). www.miraproject.eu
Water Availability and Security in Southern Europe and the Mediterranean (WASSERMed). www.wassermed.eu

Chapter 4
Space Technology for Global Water Resources Observations

Jérôme Benveniste

Abstract Since the launch of the Radar Altimeters on-board ERS-1 and TOPEX/POSEIDON 19 years ago, significant advances in all facets of Radar Altimetry have resulted in a high accuracy over the open ocean to the centimeter level. Over inland water bodies such as rivers, lakes and reservoirs, the measurements are tenfold less precise, however it is recognized by the scientific community that, with special processing algorithms which are constantly being improved, useable results are obtained to support observations of watershed systems. This paper reports upon the strategic outlook for exploiting the current and future potential of Radar Altimetry missions and other space-borne data products to monitor surface water storage. The scope of this technique is to monitor thousands of river and lake heights worldwide, with the access to almost two decades of historical data now permitting analysis of trends and identification of climate signatures.

The results illustrate the current capability and future potential of these approaches to derive a global picture of the Earth's inland water resources, depicting each catchment basin, worldwide, and to identify both climate signatures and regions where human usage is depleting the resource beyond its capacity to recharge.

Important progress has been made recently in using Satellite data for hydrology, juxtaposed with in-situ data and the modelling effort. Particular attention is paid to the support of management and preservation of water resources, a key aspect of environmental security, by supplying to both scientists and managers accurate global datasets, for large scale, trans-boundary and local scale monitoring frameworks, particularly where and when there are no in-situ hydrological measurements available. Space borne measurements are global in nature, usually with a free and open access data policy, which is the case for the future operational observing

---

J. Benveniste (✉)
European Space Agency, Via Galileo Galilei, Frascati (Roma) 00044, Italy
e-mail: Jerome.Benveniste@esa.int

"Sentinel" satellites of the European Global Monitoring for Environment and Security Programme, and as such their worldwide scientific exploitation contributes to peace and security.

Keywords Satellite altimetry • Hydrology • Global water resources

4.1 Introduction

The effective management of the Earth's inland water is a major challenge facing scientists and governments worldwide. However, whilst demand for this often scarce resource continues to grow, the number and distribution of in-situ hydrological gauge stations is steadily falling and many catchments basins in the developing world are now entirely ungauged. The same goes for water quality measurements. There is a visionary strategic outlook for exploiting the current and future potential of Radar Altimetry missions and other space-borne data products to monitor surface water storage and quality. The scope of this technique is to monitor thousands of river and lake heights worldwide, with the access to almost two decades of historical data now permitting analysis of trends and identification of climate signatures. Satellite radar altimeters have been collecting exploitable echo series over inland water since 1992, but only a tiny fraction of these data has been successfully mined for information on river and lake heights. The European Space Agency has launched a research and development activity 6 years ago to extract the river and lake level information from the complex radar echoes over the continental waters, with the idea to eventually provide this information to scientists and water managers in near real time. The ESA River and Lake Project [8] is aimed at developing, demonstrating and assessing an information service based on inland water altimetry, globally and both in near real time and for analysing decadal time series. In particular to Africa, the data retrieved over this continent is supporting ESA's TIGER initiative. Following the 2002 Johannesburg World Summit on Sustainable Development, the European Space Agency has launched the TIGER Initiative – focusing on the use of space technology for water resource management in Africa and providing concrete actions to match the Resolutions. The objective of the River and Lake research and development project is to build and validate a system that retrieves inland water heights from EnviSat and Jason satellites, globally, in near real time, i.e., without time for *a posteriori* verification by scientists, for a substantial set of lakes and rivers, which is the main challenge of the development [1, 4]. Data are also provided from the historical archive amassed by ERS-2. The ultimate goal is to drastically increase the number of measured inland water bodies, by improving the data processing, meanwhile demonstrating the accuracy and robustness of the approach through a pilot demonstration. When it is confirmed as mature and when the users are satisfied with the data products, this processing should be sustained as an operational service, relying on data from an operational space-borne mission, such as the series of Sentinel-3 satellites. An important aspect for this demonstration is the involvement of users in the local verification of the accuracy of the river stage derived from the radar altimetry data.

Furthermore, the quality of an operational service relies on timeliness, frequency of repetition and location, in other terms, usefulness, for which feedback shall also be gathered from users.

It might not be useless at this point to recall the basic principle of Radar Altimetry and its suite of applications.

4.2 Radar Altimetry Principle

Satellite altimetry measures the altitude of the target vertically below (e.g., the river or lake surface), firing 1800 low-energy microwave pulses every second, averaged onboard by groups of one hundred. Each averaged 18 Hz waveform corresponds to 350 m of progression along the orbit track. This means that a great number of radar echoes bounce off the water surface, which in turn can be changed into virtual limnograph readings of the stage [4]. The measurement principle relies on measuring the radar range between the target and the spacecraft and computing the altitude of the latter, the difference being the altitude of the target (Fig. 4.1).

The full construction of the radar target altitude involves a suite of corrections due to propagation in the atmosphere and reflexion from the surface. The equation is:

$$H_i = H_{orbit} - H_{range} + E_r = S_g + S_s + S_v + S_t + E_o + E_r \tag{4.1}$$

where:

S_g is the Geoid signal;
S_s is the Stationary signal;
S_v is the Variability;
S_t is the tides signal;
E_o is the Orbital error;
E_r is the cumulated sum of other errors and corrections (wet and dry tropospheric corrections, solid tides, loading effect, inverse barometer effect, etc.).

A complete tutorial on Radar Altimetry can be found in the Radar Altimetry Tutorial, as part of the Basic Radar Altimetry Toolbox, accessible on the web at earth.esa.int/brat [11]. This a priori simple equation leads to different types of ocean and land applications, such as:

- Oceanographic analysis

  - Meso-scale dynamic topography (currents, eddies, kinetic energy, heat transfer)
  - Large scale topography/large scale variability (basin gyres, strong currents, mean sea level, global and regional mean sea level trend monitoring)

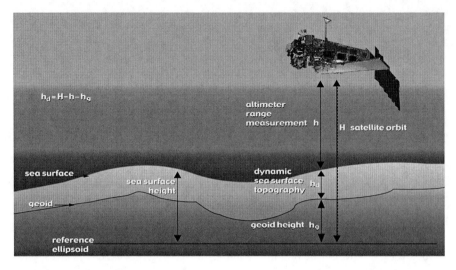

Fig. 4.1 The radar altimetry principle

- stationary signal (mean reference surface, estimation of the stationary dynamic topography)
- tides study (hydrodynamic models constrained by altimetric data)
- Assimilation in dynamic models of the oceanic circulation.

- Glaciology
  - Digital Elevation Models of the ice caps and ice sheets
  - Long term monitoring of the topography for seasonal or secular variations.
  - Input data for forcing, initialisation or test of ice flow dynamic models
  - Sea-ice extent and thickness.

- Land topography
  - Global Digital Elevation Models obtained from the full 336 days of the ERS-1 geodetic phase, e.g. the most accurate ACE2 Global DEM [5].

- Rivers and Lakes level
  - Long term, global, surface water monitoring
  - Study of the response of lakes to climate for water resources management, fisheries, water quality and conservation.

4.2.1 Return Power Waveform

Over the surface of the ocean or a large lake, the radar return power waveform, also called radar echo waveform, has a characteristic shape that can be described analytically, known as the Brown model [6]. The reflected pulse received by the radar

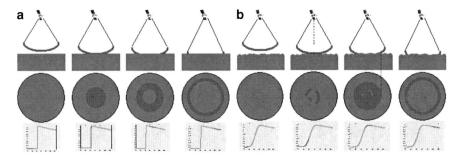

Fig. 4.2 The radar echoes as a function of time: (**a**) low wave height, (**b**) high wave height. The *second row* shows the instantaneous footprint corresponding to a given time interval; the *third row* of the plots shows the time series of the echo signal, where the portion corresponding to each time interval is highlighted

altimeter varies in intensity over time. Figure 4.2 illustrates the building up of the echo power in the receiving window. Where the sea surface is flat (a), the reflected wave's amplitude increases sharply from the moment the leading edge of the radar signal bounces off the surface. However, in sea swell or rough seas (b), the radar pulse strikes the crest of one wave and then a series of other crests, which cause the reflected wave's amplitude to increase more gradually. Ocean wave height is derived from the information in this reflected wave, since the slope of the curve representing its amplitude over time is proportional to wave height.

From the radar echo waveform shape (timeseries of the echo signal), six parameters can be deduced, by matching the received waveform with the theoretical curve (see Fig. 4.3):

- Epoch (discretised time measurement) at mid-height: this gives the time delay of the expected return of the radar pulse (estimated by the tracker algorithm) and thus the time the radar pulse took to travel the satellite-surface distance (or 'range') and back again
- P: the amplitude of the received signal. This amplitude with respect to the emission amplitude gives the backscatter coefficient, sigma0, which is related to the roughness of the target observed at nadir, for instance the effect of the wind on the water
- Po: thermal noise
- Leading edge slope: this can be related to the significant wave height
- Skewness: the leading edge curvature
- Trailing edge slope: this is linked to any mispointing of the radar antenna (i.e. any deviation from nadir of the radar pointing).

In the case of echoes returning from the coastal zone, lakes reservoirs and rivers, the echoes from the land nearby, within the field of view of the antenna (15–20 km) will superimpose on the echoes stemming from the target to be measured. This makes the retracking process to derive the radar range, therefore the water body's altitude much more complicate than over the ocean.

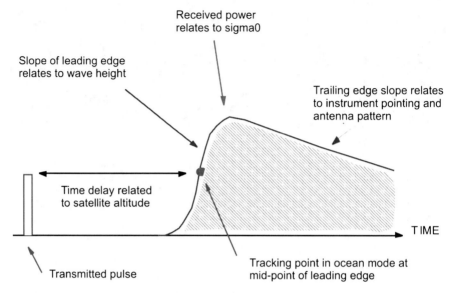

Fig. 4.3 Radar altimeter received echo waveform shape and the parameters that are estimated by the processing called "retracking"

4.3 Current Coverage Capabilities

By applying a special processing based on the analysis of the echo waveform and selecting the most appropriate retracker among a bank of 10–15, the water level can be retrieved, with an accuracy, which depends on the target and its surroundings rather than on the instrument itself. So let see what can the current generation of retracking systems recover over inland water, globally. A huge global analysis carried out of waveform recovery over inland water from ERS-2, TOPEX Jason-1 and EnviSat was performed. Every location where at least 80% of cycles have valid waveforms over the targets, which can be processed in to water heights, was identified and flagged. Figure 4.4 (top) shows the typical coverage for TOPEX and Fig. 4.4 (bottom) shows EnviSat targets, with one dot for each crossing flagged. The increase in global target coverage is only partly due to the 35- vs 10-day orbit but more to the EnviSat RA-2 ability to maintain lock by autonomously switching back and forth from the highest resolution mode to the wider medium resolution mode range receiving window.

Data over inland water are selected using a mask, which is set conservatively so that only those targets known to give reliable height estimates are included. The echoes are then analysed and those with simple shapes known to be associated with inland water are retained. These echoes are then 'retracked' using one of a set of algorithms configured for each echo shape, which determines which part of the returned echo corresponds to the 'mean surface' directly below the altimeter. The range to surface is then adjusted to use this value, and the data are combined with Geoid

4 Space Technology for Global Water Resources

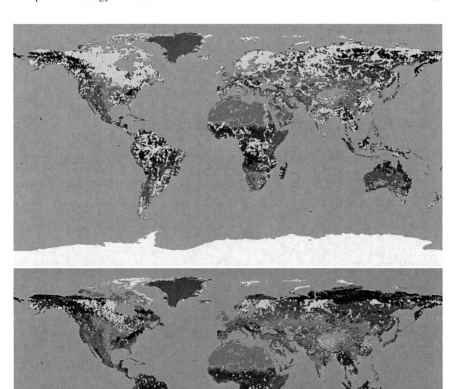

Fig. 4.4 Global water targets for TOPEX (*top*) and EnviSat (*bottom*)

model EGM96 [10] to calculate a height above mean sea level [3]. Two important questions are: (1) How many virtual limnographs can satellite altimetry generate? And: (2) Can these virtual limnographs supplement the existing but decreasing worldwide limnograph network? Taking Central Africa as an example, we see on Fig. 4.5 that there are approximately 30 EnviSat river and lake crossings that can potentially be processed into virtual limnographs. However, not all readings go through quality threshold tests all year round, due to environmental variations, from season to season, perturbing the radar echoes. As an indication, the accuracy of the best readings is nearing the accuracy of ocean measurements, around 5 cm and degrades depending on the water body dimension and environment. When the measurement is too convolved with returns from surfaces other than the targeted river,

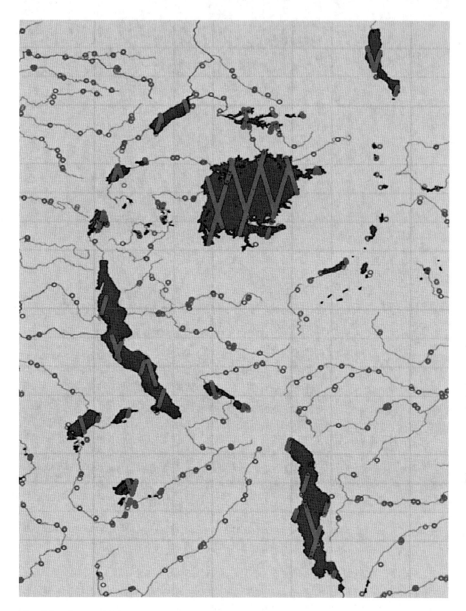

Fig. 4.5 River and lake crossings (*circles*) in Central Africa of the Radar altimeter supplying measurements than can be turned into river stage or lake level

then the data are not supplied in near real time, as it will require a more sophisticated processing which can be done off-line to accurately retrieve the water component within the echoes and convert it to river stage and ensuring a proper validation of the result. It is worth noting that the primary limitation is now not the retracking but the atmospheric corrections error, particularly the wet tropospheric correction.

4 Space Technology for Global Water Resources   39

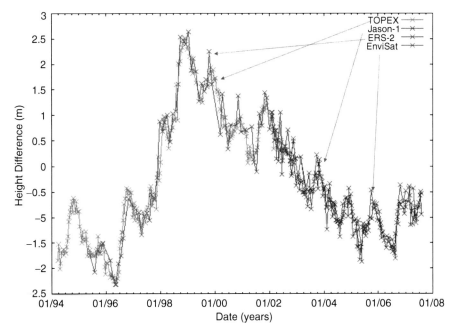

Fig. 4.6 14-year series of Lake Turkana (*upper right* of Fig. 4.5) level versus time, from 4 altimeters. TOPEX and ERS-2 are monitoring until 2002, when they are replaced with Jason-1 and EnviSat

An example of a time series that can be generated on one of these targets is shown in Fig. 4.6. This is a 14 year series of Lake Turkana (upper right of Fig. 4.5) level versus time, from 4 altimeters.

Figure 4.7 shows the same plot for the River Congo. One can appreciate the performance of the "River and Lake" system over large rivers, which provides stage data as well as lake levels. The data can be accessed by scientists and water managers on the "River and Lake" web site [8]. Similar data with a different processing system can be accessed at the "Hydroweb" web site [9].

4.4 Validation Over Amazonia with River Gauge Station

To assess the current accuracy of altimeter derived height time series, a validation exercise has been undertaken in the Amazon Basin, using the measurements from the available river gauge network. A typical example of the comparison of the "River and Lake" output product using ERS-2 before mid-2003 and EnviSat afterwards is shown here (Fig. 4.8). The comparison from 1995 to mid-2003 for ERS-2 yields a correlation coefficient of 0.98 and rms of 63 cm and afterwards until 2007 EnviSat measurements yield a correlation coefficient of 0.99 and rms of 47 cm. Figure 4.9 is similar to Fig. 4.8 but for TOPEX and Jason-1. The comparison from

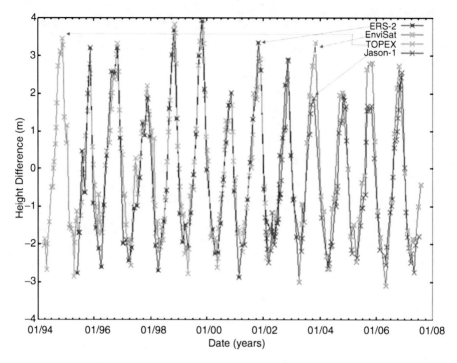

Fig. 4.7 Same as Fig. 4.6 for a river crossing over the River Congo

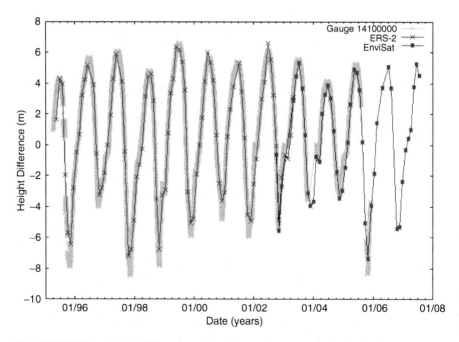

Fig. 4.8 ERS-2 (until mid-2003): rms against Gauge: 0.63 m, Correlation: 0.98. EnviSat: (after mid-2003) rms against Gauge: 0.47 m, Correlation: 0.99. The Gauge data are plotted with crosses very high time sampling appearing as a *fat line*

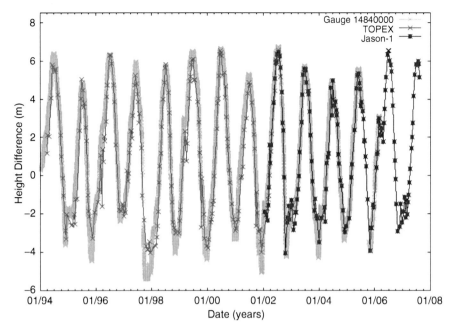

Fig. 4.9 TOPEX (until end-2003): rms against Gauge: 1.83 m, Correlation: 0.86. Jason-1: (after Feb-2002) rms against Gauge: 1.22 m, Correlation: 0.93. The Gauge data, plotted with crosses at very high time sampling, appear as a *fat line*

1995 to end-2003 for TOPEX- yields a correlation coefficient of 0.86 and rms of 183 cm and from Feb-2002 until 2007 Jason-1 measurements yield a correlation coefficient of 0.93 and rms of 122 cm. The rms value is not to be taken as the error of the river level measured by radar altimetry as it includes a difference due to the distance between the satellite river crossing and the location of the river gauge. The very high correlation coefficient shows that the altimeter is retrieving the annual cycle with high fidelity. The probable reason for higher RMS values for TOPEX and Jason-1 is a higher distance to the gauge.

4.5 Lake Volume Variation

By combining knowledge of the area of the lake, and if accessible the variation of this area with lake level, a lake volume variation can be estimated. Using the estimated lake level time series from the "River and Lake" system, series of lake volume are computed. An example is given in Fig. 4.10 for six lakes in the East African region: Lake Turkana, Lake Mweru, Lake Bangweulu, Lake Rukua, Lake Mweru-Wantipa and Lake Pangani. Some lakes are quite stable, but two of them, Lake Turkana and Lake Mweru present important volume variation over this 8-year time series.

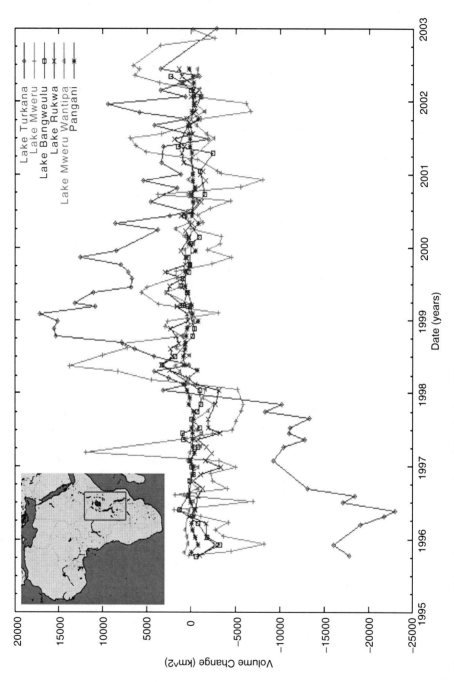

Fig. 4.10 Time series of lake volume variation (in km³) for six East African lakes

By using another source of remote sensing data, in particular optical or radar (to avoid cloud coverage) this surface estimation can be improved, therefore, the lake volume variation can also be estimated more precisely.

4.6 Near-Real Time Products Pilot Demonstration

In order to demonstrate and qualify the usefulness of this novel data flow, a pilot experiment was set up to supply river and lake stage data in near real time. The main steps to be performed are summarized in the following:

- Radar Altimeter waveforms and geophysical corrections are fetched directly from the EnviSat ground segment; the level 1B data [2] which contain the waveforms are available within 1–2 days of measurement, and fusing these data with the IGDR product [2], which provides accurate orbit data and the series of atmospheric corrections required to compute the range from the satellite to the surface enables generation of the height measurements.
- An ad hoc Ultra Stable Oscillator correction is computed and applied.
- The near real time River&Lake processor is run as soon as the data is received and corrected.
- The "River&Lake for Hydrology" (RLH) output product is stored on the web site.

The longest lag is the availability of the DORIS orbit (3–4 days), but this could improve to a few hours as the DORIS Navigator orbit is fully processed in real time, at the cost of a slightly lower accuracy in the orbit solution with the future generations of DORIS on Jason-2, CryoSat and Sentinel-3. A comparison between the near real-time and off-line SGDR product-derived river level shows very good agreement to the centimeter level. This means that the loss of accuracy for the near real time product derived river and lake heights is within a few cm, compared to the final precision product (Fig. 4.11).

The near real time system has been demonstrated to be nearly as precise as the delayed processing, but it is worth noting the limitation of using just one altimeter to supply data over an ungauged basin. One altimeter will supply 6 measurements per day over Africa, only. It is clear that a constellation of traditional altimeters or/and a two-dimensional swath altimeter is required to obtain an appropriate space-time sampling to mitigate for ungauged basins.

4.7 Conclusion and Outlook

The "River and Lake" processing system has now matured to a point where river and lake levels can be supplied in near real time, meaning that it produces and disseminates the high quality output products, without human intervention.

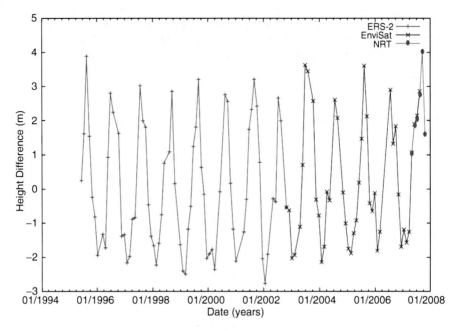

Fig. 4.11 Comparison between the near real-time and off-line SGDR product-derived river level shows very good agreement to the centimeter level

However, two problems remain to be solved: today, only 25% of the potential data can be supplied accurately in NRT and the data from one single mission are too sparse in space and time to be used directly for assimilation in river basin models. These two problems cannot be resolved without an evolution of the "River and Lake" System. The System will need to be further developed (1) to increase the number of rivers that can be measured from space, processed and delivered automatically without validation and (2) to include the data coming from Jason-2.

Concerning the historical time series, a huge amount of waveforms are already gathered over inland water targets globally. Processing these complex echoes to retrieve decadal time-series of height changes has already recovered information over hundreds of targets worldwide. The use of a specific retracking system is essential, as none of the existing retrackers, such as the four retrackers running in parallel in the EnviSat ground processing are suitable for estimating river stage.

The applications of radar altimetry range from near-real-time monitoring for water resource management to decadal climate change indicators; its spatial scales allow correlations with monthly gravity variation data from GRACE, as well as they permit the monitoring of hundreds of river systems. Thus, the unique contribution of satellite radar altimetry to global inland surface water monitoring and the importance of continued measurements is evident. As the existing network of in-situ gauges falls out of maintenance, more and more catchments are becoming ungauged, whilst the demand on this increasingly scarce resource continues to escalate. Using the remote measurement capability of altimetry, particularly the near-real-time capability, is now beginning to allow water resource managers access to both the NRT data and its context, in the form of decadal historical information. The global monitoring capability now being achieved using multi-mission satellite radar altimetry reveals changing patterns of use, as stress on water resources increasingly depletes drainage basin resources beyond their capability to recharge.

The scientific challenge is to fully extend to the global inland water bodies (including the coastal ocean) the success of altimetry in monitoring the global open ocean.

To satisfy hydrologists' requirements we need:

- Better techniques to process current instruments' data
- Better tracking for the future instruments
- Better spatial/temporal sampling - this will require new technology, such as interferometric altimetry, or constellations of traditional altimeters as proposed by the CEOS Strategic Implementation Team
- Integration of measurement and forecast systems - satellites, river gauges, discharge and current meters, tide gauges, hydrographic measurements, models.

In hydrology (including the coastal zone), altimetry cooperation is essential at European and international level, as the problems are global and the expertise needed is interdisciplinary and geographically distributed. The best excellences need to be networked and complementarities exploited. This is actually happening with project funding and regular thematic workshops (e.g. the ESA, 2nd Space for Hydrology Workshop, [7]) and needs to be sustained with further adequate funding. Concerning instrument data processing and auxiliary corrections, an active network is required for gathering local data, both used for altimeter measurement corrections and validation, to be patch-worked into a global product, both for inland water, estuaries and oceanic coastal zone.

Once this is done, the objective is to run another pilot demonstration phase, and support the effort to assimilate the data produced into river catchment models. Current efforts funded by ESA are on-going to assimilate the river stage data into several catchment basin models. This is a necessary step to mitigate for the low time sampling, using high density of satellite limnographs. The objective is to estimate river discharge in all parts of the catchment basin. So far, we have demonstrated the compatibility of the data with different models and preliminary results of assimilation show the value of the water level data from space. A preliminary report on model performance, accuracy and capability to predict water storage and fluxes was provided at this NATO-Advanced Research Workshop, but it will be the subject of a future article.

Acknowledgements The lecture "Space Technology for Global Water Resources Observations" delivered at the NATO-ARW workshop by the author was put together with contributions from: Philippa Berry and team at De Montfort University, Ole Anderson, Danish Nat. Space Center, Peter Bauer-Gottwein, Danish Technical University, Anny Cazenave & Jean-François Crétaux, LEGOS (France). They are thanked with gratitude for their contributed material for the lecture. Most of the work and results reported here was produced under the ESA-ESRIN contract 21092/07/I-LG awarded to De Montfort University, Earth and Planetary Remote Sensing (EAPRS) Laboratory and coordinated by the author.

References

1. Benveniste J, Berry PAM (Feb 2004) Monitoring river and lakes from space. ESA Bulletin (117) – February 2004
2. Benveniste J et al. (2002) EnviSat RA-2/MWR Product Handbook, Issue 1.2, PO-TN-ESR-RA-0050. European Space Agency, Frascati
3. Berry PAM, Garlick JD, Freeman JA, Mathers EL (2005) Global inland water monitoring from multi-mission altimetry. Geophys Res Lett 32(16):L16401. doi:10.1029/2005GL022814
4. Berry PAM, Freeman JA, Smith RG, Benveniste J (2007) Near real time global lake and river monitoring using the EnviSat RA-2. EnviSat Symposium 2007, ESA Pub. SP-636 2007
5. Berry PAM, Smith RG, Benveniste J (2010) *ACE2*: the new *global* digital elevation model. In: Gravity, geoid and earth observation, vol 135, part 3, International Association of Geodesy Symposia. Springer, Berlin/Heidelberg, pp 231–237. doi: 10.1007/978-3-642-10634-7_30
6. Brown GS (1977) The average impulse response of a rough surface and its applications. IEEE Trans Antennas Propag 25(1):67–74, doi: 10.1109/TAP.1977.1141536
7. ESA (2007) 2nd space for hydrology workshop, Geneva (Switzerland), Nov 2007. http://earth.esa.int/workshops/hydrospace07/
8. ESA (2011) River&Lake, Feb. 2011, [Online]. http://earth.esa.int/riverandlake
9. LEGOS (2011) Hydroweb system, Feb. 2011, [Online]. http://www.legos.obs-mip.fr/fr/soa/hydrologie/hydroweb/
10. Lemoine FG, Kenyon SC, Factor JK et al (1998) The Development of the Joint NASA GSFC and the National Imagery and Mapping Agency (NIMA) Geopotential Model EGM96, NASA/TP-1998-206861
11. Rosmorduc V, Benveniste J, Bronner E, Dinardo S, Lauret O, Maheu C, Milagro M, Picot N (2011) In Benveniste J, Picot N (eds.) Radar Altimetry Tutorial. http://earth.esa.int/brat

Chapter 5
Remote Sensing for Environmental Monitoring: Forest Fire Monitoring in Real Time

Abel Calle, Julia Sanz, and José-Luis Casanova

Abstract The use of remote sensing techniques for the study of forest fires is a subject that started already several years ago and whose possibilities have been increasing as new sensors were incorporated into earth observation international programmes and new goals were reached based on the improved techniques that have been introduced. In this way, three main lines of work can be distinguished in which remote sensing provides results that can be applied directly to the subject of forest fires: risk of fire spreading, detection of hot-spots and establishment of fire parameters.

Keywords Remote sensing • Forest fires risk • Fire detection

5.1 Introduction

Forest fires are one of the major natural risks and the Earth observation by means of satellites is a very useful tool for the main stages of a fire, allowing locating the areas of maximum risk, "hot spots" location and, during the fire evolution, it takes information about its characteristics: temperature, released power, burning area etc. Before the fire, spatial observation allows us to know the status of the vegetation cover, both forest and otherwise, including savannas. This makes it possible to determine which areas have lower photosynthetic activity, and therefore are more likely to burn. Also, through a multi-temporal analysis, it is possible to determine the rate of decay of vegetation and estimating the risk and the areas with the highest

A. Calle • J. Sanz • J-L. Casanova (✉)
Remote Sensing Laboratory. LaTUV, University of Valladolid, Valladolid, Spain
e-mail: jois@latuv.uva.es

amount of stored fuel. Other factors such as vegetation moisture, soil temperature and evapotranspiration, can ultimately establish the risk levels of a given area.

The use of remote sensing techniques for the study of forest fires is a subject that started already several years ago and whose possibilities have been increasing as new sensors were incorporated into earth observation international programmes and new goals were reached based on the improved implemented techniques. In this way, three main lines of work can be distinguished in which remote sensing provides results that can be applied directly to the subject of forest fires: risk of fire spreading, detection of hot-spots and establishment of fire parameters and, finally, cartography of affected areas.

With respect to the risk of fires, remote sensing has provided very valuable results in real time, which was the required aim. However, in order to be able to predict the existence of fires, it is necessary to incorporate indicators of very heterogeneous types which sometimes fall out of the field of earth observation studies. Indicators related to economy, social and human activities or historical statistics among others, should, for example, be taken into account. That's why remote sensing must be restricted to a very limited aspect which makes it only suitable for the estimation of the spreading risk related to the vegetation dryness and surface temperature values. The main magnitude used as an indicator is the vegetation index, above all, the Normalized Difference Vegetation Index (NDVI). In Spain, the first results in the estimation of the fire risk, although not in real time, were obtained through analyses on the Advanced Very High Resolution Radiometer (AVHRR) sensor, belonging to the National Oceanic and Atmospheric Administration (NOAA) satellites, through accumulated decreases in the NDVI analyzed on 10×10 km^2 cells [18]. Other potential indexes have also been used for the characterization of the forest cover [2]. Later on, further indicators coming from the same sensors were incorporated so as to improve the algorithms and include the information relative to meteorological conditions. The first one was the surface temperature obtained through satellites with very high accuracy. Thus, the combination of the NDVI with the surface temperature has given place to a mixed index in which the lineal regression slope in both magnitudes established on 20×20 km^2 cells presents a good correlation with the vegetation evapotranspiration [22]. The use of the slope in this relation has been incorporated through different algorithms by different authors in order to establish another risk indicator; thus, Casanova et al. [6] introduced it to work in real time within the operation in Mediterranean countries and Prosper-Laget et al. [26] used it in France after the establishment of clusters that characterized levels according to slope values. The possibility of using the spectral information in the middle infrared, in the 1.6 µm region, has given place to the introduction of other indicators related to the fuel's moisture given that the vegetation's reflectivity in this wavelength interval is strongly influenced by the water contained in it. Hunt and Rock [12] suggested a new vegetation index similar in the equation to the NDVI but including the reflectance in the near infrared and the reflectance in the 1.6 µm region, an index indicating the fuel's moisture. At first, this index could only be applied to the Landsat-TM sensor for the creation of fuel maps [7]. Today, it can be used in real time on the AVHRR and the Moderate Resolution Imaging

Spectroradiometer (MODIS) sensors to be incorporated to the risk maps as a new indicator.

The fire detection can be done by polar and geostationary satellites. The polar satellites are more sensitive and can detect fires small or less active, but given its temporal periodicity useful as locators fire is not very high. By contrast geostationary satellites send an image every 15 min, allowing for near real-time monitoring, but given its distance from Earth its detection limit is lower and only detects fires very intense or very large, exceeding several tens of hectares. When the fire has already been developed, it is possible to know the burning area, the emitted power or temperature of the fire. These data are important for fire fighters and to know the destructive effect of the flames, as well as their potential impact on human settlements, roads or any inhabited place.

Despite its limitations the NOAA-AVHRR sensor has been the most important for fire detection and has provided a benchmark for subsequent sensors. An excellent revision of the algorithms used on AVHRR can be found in Li et al. [17]. The case of the European sensor Advanced Along Track Scanning Radiometer (AATSR) and the World Fire Atlas from 1997, published by the ESA and performed by data coming from the first and second European Remote-Sensing Satellites (ERS-1 and ERS-2) [1] has been used to demonstrate its availability to fire detection and assessment of vegetation fire emissions [27]. The appearance of the MODIS sensor heralded a significant step forward in the observation of forest fires [11, 14]. At this moment, the MODIS fire product is a consolidated product and a reference for global Earth observation [13]. Fire product has been identified by the International Geosphere and Biosphere Programme (IGBP) as an important input for global change analysis [19]. However, although the radiometric availability is satisfactory, the main problem is the time resolution to operate in real time.

Detection of High Temperature Events (HTE) through geostationary satellites has been taken into account with the different perspective. The improvements introduced in the sensors have allowed us to use geostationary satellites beyond their meteorological capabilities, adapting them to Earth observation. This is the response to the need for series of stable fire activity observations for the analysis of global change, changes in land use and risk monitoring [13]. The Geostationary Operational Environmental Satellite (GOES) has been the worldwide reference for fire monitoring through geostationary platforms. Since 2000, the Geostationary Wildfire Automated Biomass Burning Algorithm (WF-ABBA) has been generating products for the western hemisphere in real-time with a resolution of 30 min [23, 24]. This detection system has been operational within the NOAA National Environmental Satellite, Data, and Information Service (NESDIS) programme since 2002. The GOES-East and GOES-West spacecrafts are located in the Equator at 75°W and 135°W, providing diurnal coverage of North, Central and South America and data based on fire and smoke detection. The results provided by the GOES programme have been the starting point of a global geostationary system for fire monitoring, initially comprising four geostationary satellites that were already operational: two GOES platforms, from the USA, the European Meteosat Second Generation (MSG), situated at longitude 0°, and the Japanese Multifunctional Transport Satellite (MTSAT) situated at 140°E.

The minimum fire detection sizes of GOES, MSG and MTSAT, with time resolution less than 30 min has allowed the international community to think that the global observation network in real time may become a reality. The implementation of this network is the aim of the Global Observations of Forest Cover and Land Cover Dynamics (GOFC/GOLD) FIRE Mapping and Monitoring program, internationally focusing on decision-taking concerning research into Global Change. The GOFC/GOLD FIRE programme and the Committee on Earth Observation Satellites (CEOS) Land Product validation held a workshop dedicated to the applications of the geostationary satellites for forest fire monitoring. This workshop was hosted by the European Organization for the Exploitation of Meteorological Satellites (EUMETSAT), whose most relevant conclusions can be seen in Prins et al. [25].

Throughout this paper, we will describe the details of a methodology aimed at making operative the generation of fire risk index maps, the hot spots detection and the fire monitoring

5.2 Degradation Vegetation Indicator, DVI

The vegetation state must be obtained through a vegetation index. The use of indexes based in the ground line demands the introduction of coefficients which are dependent on the zone analysed. That's why, due to the global character of the product searched, the NDVI will be used. Its value will provide the information from the photosynthetic state of vegetation.

The study of the degradation implies the comparison between the values in the current day and the historical values. In this way, the main idea for risk determination is that the NDVI value on a concrete day is lower than the expected value considering the normal evolution during the former days. In order to carry out the analysis, a time series must be used throughout a period previous to a given day. We propose a period of 60 days previous to that day. These historical images are represented through NDVI Maximum Value Composite (NDVI-MVC) images.

The degradation vegetation indicator will be applied to clearly differentiate fast response processes. As was shown in the review of the specialised literature, algorithms of accumulation of differences in the NDVI have been proposed. After studying these techniques, we have decided to look for an expected NDVI through an adjustment equation and compare this value with the real one obtained. Thus, the NDVI behaviour for a pixel "i" is characterized from the equation:

$$NDVI_i = a_i t^2 + b_i t + c_i$$

In which the datum from the real day, $NDVI_{real}$ has not been used in order not to introduce this information in the search of the tendency. In the equation, t is the historical day variable, the middle day of each composite. The coefficients a_i, b_i and c_i are the specific regression coefficients of that pixel. In this way, the pixel's

5 Forest Fire Monitoring in Real Time 51

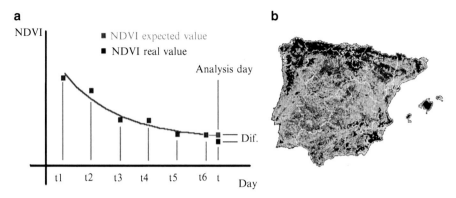

Fig. 5.1 (a) Extrapolation used to obtain the expected NDVI value. (b) Image of the Iberian Peninsula showing the normalized residual from AVHRR sensor

expected vegetation evolution in the current day, $NDVI_{i,expected}$, will be given by the extrapolation of that curve to the instant t_{day}. In other terms:

$$NDVI_{i,expected} = a_i \cdot t_{day}^2 + b_i \cdot t_{day} + c_i.$$

According to this reasoning, the difference between values ($NDVI_{i,expected} - NDVI_{real}$), which from now on will be called residue, should already be a degradation vegetation indicator: negative values indicate good behaviour, positive values indicate that the vegetation is even worse than the value foreseen by the tendency and consequently there is a risk situation. Figure 5.1a represents a scheme of the algorithm shown.

However, it is still a slanted indicator, since a concrete residue should be more serious in a pixel with a value lower than the NDVI than with a high value. That's why, the final indicator proposed will be a residue normalised in this way:

$$R_{i,normalized} = \frac{NDVI_{i,expected} - NDVI_{i,real}}{NDVI_{i,real}}$$

The negative values of R are used as DVI's. Figure 5.1b represents an image of the Iberian Peninsula of normalised residues for any given day generated through NOAA-AVHRR sensor, and consequently, with a resolution of 1×1 km². It must be pointed out that the difference introduced by MODIS sensor with respect to the AVHRR sensor is that the former's NDVI values will be dispersed in a slightly higher interval. That is, there will be lower minimums and higher maximums caused by the higher spatial resolution and, consequently, by the capacity of observing a more homogeneous land.

Our aim is to distinguish those zones whose behaviour has "peaks": characterising the whole territory as high risk can be real but it is not very useful information for the analyst. That's why we suggest applying the classification

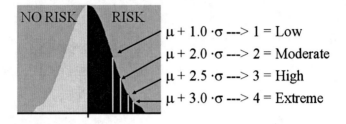

Fig. 5.2 Distribution of differences between expected and current NDVI values

on provincial forest values. For it, the statistic distribution is established in the province analysed. Those situations which go over the mean value of such distribution will be considered risk situations (see Fig. 5.2). Also, the indicator values will be stratified through increasing factors of the standard deviation in order to quantify 4 different risk classes.

5.3 Evapotranspiration Indicator: EPTI

In order to characterise the vegetation water stress, some authors have suggested establishing the lineal relation existing between the surface temperature and the NDVI value, as has been commented in the literature review section. Besides, this lineal relation between these magnitudes is different for the different crops. For this reason, it is a good indicator for the establishment of fire risk levels.

The main idea is that under moisture conditions (for example, due to rain), the soil behaves similarly to the vegetation covers and the relation surface temperature (T_s) vs. NDVI is represented by a horizontal alignment of points. Under drought conditions, the surface temperature rises above the air temperature and there is a very steep slope in the straight line of the former relation in a negative way. The straight line's slope thus generated is an indicator of the relation between the potential and the real evapotranspiration, and consequently, of the vegetation water stress [22].

In order to establish the relation between the NDVI and the surface temperature, all the satellite's image coverage has been divided in 20×20 km cells. A lineal regression adjustment between the points in each cell has been calculated in order to find the relation:

$$T_s = m \cdot NDVI + n$$

Figure 5.3a shows the relation T_s vs. NDVI in a 20×20 km-sized generic cell.

The parameter that characterises the relation T_s vs. NDVI is the regression's slope, m, which has negative values in practically all situations. Positive values have

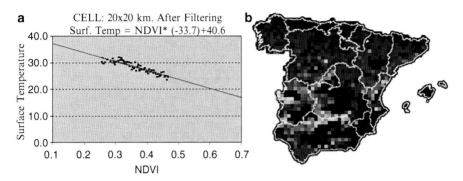

Fig. 5.3 (**a**) Linear regression Ts vs. NDVI. (**b**) Display of the Ts *vs* NDVI slope on the Iberian Peninsula over 20×20 km cells

been found, although close to zero, in situations of very high moisture content and for zones with a high percentage of vegetation cover, such is the case in the north of the Iberian Peninsula at the beginning of spring. The more the vegetation water stress increases in each cell, the more negative the slope values are.

The generation of a new image from the slope values of the T_s vs. NDVI regression line in 20×20 km² cells gives place to the image shown in Fig. 5.3b, in which the zones with no data due to the cloud cover appear in black.

Finally, the quantification of the risk indexes provided by this indicator will be established according to: $m<M_1 \rightarrow$ Low; $m<M_2 \rightarrow$ Moderate; $m<M_3 \rightarrow$ High; $m<M_4 \rightarrow$ Extreme.

5.4 Wind Indicator: RI

In the specialised literature, the wind does not appear as an indicator used in the methodologies that use exclusively remote sensing, even though it is known as one of the main factors in the spreading of the seats produced. However, it is true that the LATUV has included it within the indexes presently at work.

In the first place, we will use the speed as a first indicator associated to the ground's topography in order to establish the risk indicator. Such speed cannot be obtained directly through the MODIS sensor, although we can obtain the pressure as it is described in the Algorithm Theoretical Basis Document [28] which shows the determination of atmospheric profiles (the MOD07 product). This MODIS's product includes the pressure fields in the levels 5, 10, 20, 30, 50, 70, 100, 150, 200, 250, 300, 400, 500, 620, 700, 780, 850, 920, 950 and 1,000 mbar. We suggest to use the 1,000 mbar level. For it, it is necessary to establish the spatial scale first, in which we intend to provide the final data. We suggest using the same one as the evapotranspiration indicator, of 20×20 km². The generation of pressure fields will be carried out through the spatial interpolation algorithms with the inverse of the square of the distance. So, the pressure in each point is:

$$P(x,y) = \frac{\sum_{i=1}^{N} P_i \cdot \frac{1}{d_i^2}}{\sum_{i=1}^{N} \frac{1}{d_i^2}}$$

where P_i is the pressure obtained by the MODIS and d_i the Euclidean distance from the MODIS points to the interpolation point (x,y). For the final determination of the field of winds, the geostrophic wind equations, whose main analysis factor is the pressure's gradient, will be applied.

5.5 Risk Index

The fusion of all the indicators proposed is a very relevant aspect since it determines and provides the final risk value. Evidently, it is not an instantaneous task, for two reasons: in the first place, each indicator is obtained at a very different spatial resolution; secondly, because they have a completely different nature.

On the other hand, and due to the second aspect, the mean of all indicators cannot be a reliable measure: if one of them "reports" an extreme situation it should be noticed, even if the rest had low values.

In the first place, the spatial resolution in which we'll provide the final index should be established. We suggest the resolution of 10×10 km² cells. We cannot expect either that all the pixels in a cell have an extreme situation in order to consider the cell as such. That's why we establish a threshold of N = 10% of its size to analyse a concrete level in the cell of study. Logically, the highest risks will be considered first. Thus, for each of the indicators established [6] (with the exception of the evapotranspiration indicator in which the cell has already provided a calculated value) we will obtain the risk value in the whole cell, as follows:

$$\Sigma N_4 \geq N \Rightarrow \text{Cell Risk} = 4$$
$$\Sigma N_4 + \Sigma N_3 \geq N \Rightarrow \text{Cell Risk} = 3$$
$$\Sigma N_4 + \Sigma N_3 + \Sigma N_2 \geq N \Rightarrow \text{Cell Risk} = 2$$
$$\Sigma N_4 + \Sigma N_3 + \Sigma N_2 + \Sigma N_1 \geq N \Rightarrow \text{Cell Risk} = 1$$
$$\text{In other case} \Rightarrow \text{Cell Risk} = 0$$

The final risk for a 10×10 km cell will come from the weighted mean of all the indicators. As follows:

$$\text{Risk\_Index} = F_1 \cdot DVI + F_2 \cdot EPTI + F_3 \cdot RI$$

where the weight factors, F_i, are normalized.

5.6 The VERDOR GIS as Tool for Drought Vegetation Warning

AVHRR and MODIS images from the NOAA & NASA satellites are especially suitable due to the large area of coverage of the sensor as well as the excellent temporal resolution as daily images are available. Indicators of the state of the vegetation applicable to these images are normally derived from the NDVI. With the aim of having a software available which allows us to obtain vegetation state from these images and thus be able to compare the vegetation evolution in certain areas of interest as well as handle the large amount of accumulated information in the shortest possible time, the LATUV has developed a software for the treatment of these data which also includes additional information for referencing the vegetation processes.

The Fig. 5.4 shows the flow chart diagram of the software, which explains the general structure of the Geographical Information System (GIS). The permanent inputs are files as the digital terrain model, land cover map and processed images which represent the main information manage by the GIS.

As interactive inputs the GIS uses the maps and database of the regions to be studied. These regions could be brought in or edited by the user, following his own requirements. The GIS outputs are very numerous, including DVI and EPTI risk indicators, but we can stand out the graphic outputs, made by processed images and

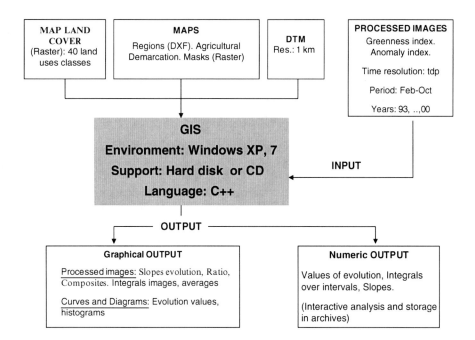

Fig. 5.4 VERDOR flow chart

evolution curves, and the histograms for different areas and land cover classes. Other possible output is made by numeric data coming from evolution analysis and integration. The software has been written in C++ for Windows XP and Windows-7 visual environment

The data dealing with the land uses have two possibilities: firs option is the CORINE LC map (Coordination of Information of the Environment Land Cover, funded by the ministers of UE, CE/338/85), containing 40 different classes. Second possibility is the Land uses map provided by NASA USGS (Geological Survey of United States). The GIS VERDOR (the name of Geographical Information System is VERDOR, that is, Greenness in Spanish) can, however, operate on all the types grouped in the main groups. The reference which uses the digital terrain model is of special interest. The DTM (Digital Terrain Model) treated corresponds, as in the previous case, to a spatial resolution of 1 km per pixel.

The input data from remote sensing are the NDVI values of each pixel. These NDVI values can be referred to the historical maximum and minimum values of each pixel in such a way that the RGI (Relative Greenness Index) is obtained as [3]:

$$RGI = 100 \cdot \frac{NDVI - NDVI_{min}}{NDVI_{max} - NDVI_{min}}$$

Also images of NDVI anomaly could be obtained for each ten days period of analysis, the ratio between NDVI value of this year (for this ten days period) and the average NDVI for all years, without the year to be analyzed:

$$Anomaly_i = \frac{NDVI_i}{Avg(NDVI_j)}$$

VERDOR permits simple or multiple visualization of any area in any 10-day period in any month of any year (including maximum, minimum and average years). These visualizations allow a comparison of the evolutive state of the vegetation in different areas. Visualization may be carried out using different color palettes chosen by the user.

When more detailed analysis is required, there is an important function which is able to discriminate more "precisely" the changes which have occurred in the vegetation: ratio, which consists of carrying out a division of the greenness indices for a specific area between different periods. Although these periods may be chosen at will, a great deal of information may be obtained when the ratio is carried out for the same 10-day period of the same month between different years.

VERDOR performs a graphic representation of the images of the integrals of the greenness indices, carried out over the whole period under analysis for each year. These representations, also providing numerical values, are very important when it comes to making rapid and easily interpretable comparisons of two specific years which is extremely useful to determine the higher or lower fire risk in comparison with previous years. In this way, a "medium year" with a medium amount of fires could be used as reference.

Evolution curves of greenness values could be carried out by VERDOR. This modeling can be selected between (a) Evolution in a linear regression and (b) Evolution in a parabolic regression. User can obtain these results selecting years of analysis and land uses of terrain, which allow analyzing different fuel classes: forest, savannah, etc.

The setting of different time intervals is available, in order to perform integration under the evolution curve (intervals which can be overlapped in time). This is important as it permits a total integral under the evolution curve of a specific fuel class to be compared simultaneously with other integrals partially distributed on both sides of the obtained maximum. The usefulness of semi-integrals will serve to determine the temporal displacement undergone by specific species in a specific year. With this knowledge the high or low incidence of forest fires can be analyzed. As previously mentioned, the analyses carried out can be performed globally or discriminating certain fuel classes. In this case, not only an individual choice can be performed by consulting the corresponding legend of the 40 total types but work can be done with different fuel types.

5.7 Hot Spot Detection in Real Time

Detection and parameterization of forest fires is a task traditionally performed by polar-orbiting sensors, AVHRR, AATSR, The Bispectral InfraRed Detection (BIRD) and MODIS. However, their time resolution is a problem to operate in real time. New geostationary sensors have proven their capacity for Earth observation. GOES and MSG are already operative with time resolutions below 30 min [5]. The international community feels that a real-time global observation network may become a reality, which is the aim of the GOFC/GOLD programme, focusing internationally on decision-making concerning research into global change. This paragraph shows the operation in real time by the MSG SEVIRI (Spinning Enhanced Visible and Infrared Imager) sensor over the Iberian Peninsula. For fire detection, a temporal gradient of temperature 3.9 mm is used, which is more efficient at eliminating incoming false alarms from solar reflection. Capacity to detect hot forest fires below 0.3 ha in Mediterranean latitudes has been analyzed along with the conditions in which it is possible to apply the MODIS methodology to establish the Fire Radiative Energy (FRE) by means of experimental relation. We found that in this case, the fire size can introduce maximum differences of 40% in FRE for small-scale fires. The capacity of this sensor is shown in different cases that have greatly impacted Spain and Portugal because of the loss of human life.

One of the most important tasks for fire detection in real time by means of polar satellites, NOAA and MODIS, is the generation of hot-spot maps and their cartography. Two different types of algorithms that use the information in the Middle Infrared (MIR) and Thermal Infrared (TIR) spectral bands jointly could be used for the detection: an algorithm based on adaptive thresholds, and a contextual algorithm whenever the environment's conditions made it possible.

The use of thresholds has been established in the easiest way possible in order to respond to the requirement of real-time processing. The detection test analyses the temperature values in the MIR and TIR bands and the difference (MIR–TIR). This detection scheme is found in the literature applied to different sensors among which the NOAA-AVHRR stands out remarkably.

In a schematic way, a pixel could be classified as hot-spot if it fulfilled the following triple test:

- Brightness temperature MIR > T_{MIR}
- Brightness temperature TIR > T_{TIR}
- Brightness temperature (MIR-TIR) > T_{DIF}

The detection algorithms based on the contextual analysis screen each pixel's environment and establish thresholds according to the statistical values found. Thus, if the brightness temperature value in the analysed band is higher than the mean value found in the environment in a certain value of the standard deviation, the result of the test in this band is positive in the detection. In the literature, different contextual algorithms can be found which have in common the same structure but differ in the values of the parameters applied. One of the most important parameters is the area of the environment analysed and the multiplying factor of the standard deviation found in the analysis.

On the other hand, the tests found in the literature applied to other sensors such as MODIS or AVHRR include the contextual analysis of the reflectance in the MIR band. The test gives a positive result when the pixel's reflectance is lower than the environment's statistic. For this reason, the contextual analysis will be only applied to completely cloud-free images in order to avoid the appearance of false alarms produced on their edges. In this way, the test applied to the band NIR does not take away efficiency to the results.

Schematically, a pixel is classified as a hot-spot if the following tests were fulfilled:

- Brightness temperature MIR $> (\mu_{MIR} + f_{MIR} \cdot \sigma_{MIR})$
- Brightness temperature $(MIR - TIR) > (\mu_{DIF} + f_{DIF} \cdot \sigma_{DIF})$

Usually algorithm applied over MODIS follows the methodology showed on Giglio et al. [11]. This MODIS algorithm has been checked on different environments and always has given very good results. Also, at Fire Information for Resource Management System (FIRMS, reference in http://maps.geog.umd.edu/firms/; [8]) hot spot detection around the entire globe by using this MODIS algorithm could be found.

The European MSG satellite, a geostationary satellite operated by the ESA, is a spin-stabilized satellite. The main sensor and the most important for the purpose of hot spot detection is the SEVIRI, comprising 11 spectral bands and a visible broadband, the High Resolution in the Visible (HRV). The spatial resolution of the HRV band is 1 km and that from remaining bands is 3 km in the nadir and the images are sent every 15 min.

5 Forest Fire Monitoring in Real Time

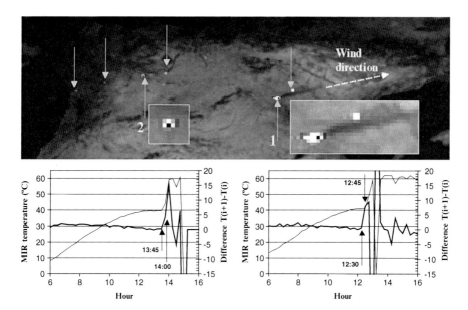

Fig. 5.5 Methodology to detect the start of a fire. The *upper part* of the figure shows the 3.9 μm band, highlighting several fires as well as wind direction. The *second part* shows the thermal evolution, in the *left scale*, and the time thermal gradient, in the *right scale*, in °C/15 min, for the two selected cases #1 and #2

The strong influence of the latitude on the SEVIRI images, which modify the observation angle, sun angle and pixel dimension, makes the methodology proposed to detect fire outbreak no longer valid as the fire keeps developing since the temperature differences between the different scenes undergo strong variations. Even the frequent appearance of saturated pixels causes sharp changes that cannot be analysed. Further, for the subsequent monitoring of the fire a new methodology for detecting hot spots is required.

Establishing the outbreak of a fire as accurately as possible is crucial to alerting fire-fighting teams as quickly as possible. If the detection process takes into account the comparison with the previous image the delay can be up to 30 min in the worst cases. To establish some representative results we have analyzed the day on which Spain's worst fire in the previous decades in terms of human losses occurred. This fire, which started between 12:30 and 12:45 on 16th July 2005, when it was detected by the LATUV, spreads for over five consecutive days and devastated around 13,000 ha. Figure 5.5 shows the image of the 3.9 μm band corresponding to a few hours after the fire. The visual analysis of the image shows the existence of many fires in Spain and Portugal. Given their importance, two have been highlighted and shown. Number 1 is the fire in Guadalajara (Spain) that claimed the lives of 11 people and number 2, one of the fires that affected the natural park of Lago de Sanabria (Zamora, Spain) during the summer of 2005, whose initial characteristics, as will be seen, differ from the first. In the figure, we have indicated the wind direction in fire #1 from the smoke plume, which is perfectly visible and which will be useful later to analyze the spread of the fire. Below in the same figure are the two thermal evolution diagrams corresponding to these fires. The diagram shows the temperature evolution of band 3.9 μm, in °C, in the primary axis of the ordinate according to the time of the day, between 06:00 and 16:00 GMT.

The secondary axis of the ordinate shows the evolution of the time thermal gradient of the same band, in °C/15 min. If we compare both temperature evolution curves, we can see that they are practically identical on the primary axis up to the moment at which the fire starts, at 12:30 in #1 and at 13:45 in #2 despite being different vegetation covers with different fuel moisture content since they occur in different climate zones. The analysis of the curve of the time thermal gradient is much more conclusive. The change in the temperature value is 1.5°C/15 min in both curves prior to the outbreak of the fire reaching a maximum of 2.3 in #1 and 1.8 in #2, which are exceptional considering the rest of the values. Standard deviation was approximately 0.5 in both cases. These results are coherent with the theoretical analysis in the previous paragraph. Case #2 was a fire that started with a time thermal gradient of 4.2°C/15 min in the first scene at 14:00 GMT, immediately jumping to 15°C/15 min in the following scene at 14:15 GMT. It is clear that it began between 13:45 and 14:00 as the figure shows. The case of fire #1 presents a much more abrupt beginning, with a time thermal gradient of 8°C/15 min in the first scene at 12.45 GMT. In this case, the fire broke out between 12:30 and 12:45 GMT. These results entirely agree with the comments in the methodology paragraph, where the value 4°C/15 min was presented as the initial detection threshold. Apart from its initial causes, the characteristics of a fire at its onset depend on the combustible material and moisture. In this comparison, it is not surprising that the outbreak was slower in case #2, whose gradient was below #1, as this was a climate zone with higher moisture content. Other fires have been analyzed and the same qualitative characteristics have been found [5].

5.8 Fire Monitoring

For the knowledge of fire parameters, we need first an analysis at a sub-pixel level through the application of Dozier's [9] methodology. This methodology allows us to establish both the fire temperature and the fraction of the area that is burning simultaneously. This procedure can be applied to any sensor and it is based on the solution of the following system of equations: given a pixel affected by a fire at a temperature T_f that occupies the fraction of the pixel p, and it is surrounded by a surface at a temperature T_{surf}, then the radiances detected in the MIR and TIR bands will be given by the expressions:

$$\begin{cases} L_{MIR} = pB(\lambda_{MIR}, T_f) + (1-p)B(\lambda_{MIR}, T_{surf}) \\ L_{TIR} = pB(\lambda_{TIR}, T_f) + (1-p)B(\lambda_{TIR}, T_{surf}) \end{cases} \quad 0 < p < 1$$

where L_{MIR} and L_{TIR} are the radiances observed by the sensor in the spectral regions of 3.7 and 11 μm respectively and B(λ,T) is the function of Plank. This system of equations provides the fire temperature value and the fraction of the pixel that is burning.

Before analyzing some approximations taken in this methodology, we must point out two very important restrictions concerning its operating capacity. In the first place, it must be said that the equations are based on the establishment of the radiance emitted by the thermal spectrum. The 11 μm region has no other nature, but the radiance observed in the MIR region has a reflection component that has been analyzed in the false alarms section. That's why, the application of these equations to day images should include an additional solar term. Otherwise, they would only be valid for night images. On the other hand, in order to obtain reliable results, it is necessary to avoid saturation as much as possible. This is not so obvious if we take into account that this methodology was originally applied to the AVHRR sensor [15, 20, 21] with the saturation problems in the MIR band exposed in the detection algorithms section.

Dozier's system of equations is very simple to understand although many of the approximations it takes are not realistic and should be analyzed. In the first place, the pixel observed is divided into two parts, fire and surface, those are considered homogenous, but this is not the case, especially because of the surface's heterogeneity. On the other hand, the atmospheric effects have been neglected in this scheme. The most serious approximation with respect to the error magnitude is probably found in the establishment of a surface temperature value. Dozier suggested for this value the mean value of the pixels surrounding the fire but not affected by it. It must be highlighted that the results obtained depend to a great extent on this parameter. Simulations carried out on a real fire changing the surface temperature value [4] show that the error in the surface temperature affects the fire temperature with a value multiplied by 10. Finally, another approximation taken is not to include in the equations the emissivity of the radiance received by the sensor. Although it is true that the fire performance is very similar to that of a blackbody, the same does not happen with the non-affected surface, which seems to have variable values. The deduction of the emissivity is justified in the fact that the zones observed for fire purposes are always forest zones and the emissivity values in this kind of environment are comprised in the interval [0.983–0.995] for the TIR band [29]. A more realistic scheme derived from Dozier's methodology is the one suggested by Giglio and Kendall [10]. This scheme modifies the former one by including terms of emissivity, atmospheric effects and sun reflection in the radiance equation of the MIR band. The following are the modified equations of Dozier:

$$\begin{cases} L_{MIR} = \tau_{MIR} pB(\lambda_{MIR}, T_f) + (1-p) L_{surf,MIR} + pL_{atm,MIR} \\ L_{TIR} = \tau_{TIR} pB(\lambda_{TIR}, T_f) + (1-p) L_{surf,TIR} + pL_{atm,TIR} \end{cases} \quad 0 < p < 1$$

where $L_{atm,MIR}$ and $L_{atm,TIR}$ are the radiances emitted by the atmosphere to the sensor in the MIR and TIR bands respectively. These terms are worthless with respect to the radiances emitted by the surface, $L_{surf,MIR}$ and $L_{surf,TIR}$, and can be neglected.

τ is the atmosphere's spectral transmittance. The difference in these equations with respect to the original ones lies in the intervention of the radiances of the surrounding pixels instead of the temperature and finally, although they are taken into account, the surface's emissivity and temperature are not usually known explicitly. The techniques mentioned for the obtaining of fire parameters imply some difficulties related to the errors that are made. In the first place, they are not analytic equations so that their solution must be found by means of numerical calculation techniques. However, it must be said that their solution comes, in the end, from a convergent system. Other important sources of errors have their origin in different magnitudes that have been analyzed by Giglio and Kendall [10] and that will be mentioned here next.

First, a source of error in the results is the error in the calculation of the surface's radiance introduced in the equations. The values for the fire temperature and size are more sensitive to errors in the radiance of 11.0 μm than in the 3.7 μm. At low temperatures, this is not a big error, but, with a high fire temperature, the error increases noticeably both in the fraction of the pixel affected and in the fire temperature itself. Another source of error to consider is the one corresponding to the atmospheric transmittance. However, in this case, the errors, made in the temperature and fraction of the pixel affected, are compensated in the MIR and TIR bands as long as such errors are caused by either an underestimation or an overestimation in both cases. Otherwise, the errors in the results will add up. Thus, an overestimation in the MIR transmittance overestimates the temperature calculated whereas an overestimation in the TIR transmittance produces the opposite effect. A third source of error in the calculations is due to the instrument's noise, although in this case it introduces an accidental systematic error. Finally, the omission of the atmospheric radiance that reaches the sensor is less important than the causes considered formerly, so that in no case does the temperature go over 1.5 K or the area over 2%.

A very interesting aspect in the theory developed is the one that refers to the fire's emissivity. A fire has always been considered as a blackbody. In fact, and strictly speaking, this is only true when the length of the flame seen from the sensor is larger than 6 m [16]. This would make us reconsider this aspect in the case of smaller fires so that in these cases we should consider the fire as a grey body. In these cases that separate from the characteristics of a blackbody, the errors made for considering that the fire has an emissivity one, result in an underestimation of both the fire temperature and the fire area, and they are independent from the fraction of the pixel that is affected and the fire temperature.

For the MODIS sensor case, Kaufman and Justice [14], have suggested an empiric expression in order to fix the intensity, this magnitude is the Fire Radiative Power (FRP), that is, the power of fire in MWatts, from the brightness temperature of the pixel affected by the fire. This expression corresponds to:

$$FRP = 4.34 \cdot 10^{-19} \left(T_{MIR}^8 - T_{MIR,background}^8 \right)$$

where T_{MIR} is the brightness temperature of the band of the 4 μm of the pixel affected by the fire and $T_{MIR,background}$ is the same temperature in the adjacent pixels.

5.9 Conclusions

Earth observation from the space has shown to be a very good tool to establish a fast and reliable fire risk index. Fire detection and fire monitoring could be accomplished from the space with high accuracy, allowing a fast answer to prevent human or environmental damages. Both polar and geostationary satellites could work on a combined way to add its possibilities and mutually improving them.

Concerning fire cartography, today the good spatial resolution of sensors is suitable in order to provide a quality product, once the time resolution is not a requirement for this purpose.

The international community feels that a real-time global observation network may become a reality, by means of geostationary sensors such as GOES, MSG once they are improving its technical features (more sensitive sensors in thermal spectrum and finer spatial resolution, mainly). This is the aim of the GOFC/GOLD programme, focusing internationally on decision making concerning research into global change. Major efforts are also being made by ESA EUMETSAT to increase the use of MSG in environment observation tasks as the Research Announcement of Opportunity, to which this paper belongs, proves.

Acknowledgements This paper has been funded by the CGL2004-00173 project, from the Spanish Science Ministry and the European FEDER Funds.

References

1. Arino O, Paganini M, Simon M, Ferruchi F, Barro E, Benvenuti M, Chuvieco E, Dwyer E, Fierens F (2001) Current and future activities of ESA related to forest fires. In: Proceedings of the EARSeL 3rd international workshop, Paris, 17–18 May 2001
2. Boyd DS, Ripple WJ (1997) Potential vegetation index for determining global forest cover. Int J Remote Sens 18:1395–1401
3. Burgan RE (1993) Monitoring vegetation greenness with satellite data. In: USDA forest service, INT-297, Ogden
4. Calle A, Romo A, Sanz J, Casanova JL (2004) Analysis of forest fire parametres using BIRD, MODIS and MSG-SEVIRI sensors. In: EARSeL symposium, Dubrovnik (Croatia), May 2004
5. Calle A, Casanova J-L, Romo A (2006) Fire detection and monitoring using MSG Spinning Enhanced Visible and Infrared Imager (SEVIRI) data. J Geophys Res 11, G04S06. doi:10.1029/2005JG000116
6. Casanova JL, Calle A, González-Alonso F (1998) A forest fire risk assessment obtained in real time by means of NOAA satellite images. Forest fire research. In: III International conference on forest fire research and 14th conference on fire and forest meteorology, vol I, Luso, pp 1169–1179
7. Chuvieco E, Riaño D, Aguado I, Cocero D (2002) Estimation of fuel moisture content from multitemporal análisis of Landsat Thematic Mapper reflectance data: applications in fire danger assessment. Int J Remote Sens 23(11):2145–2162
8. Davies DK, Ilavajhala S, Wong MM, Justice CO (2009) Fire information for resource management system: archiving and distributing MODIS active fire data. IEEE Trans Geosci Remote Sens 47(1):72–79

9. Dozier J (1981) A method for satellite identification of surface temperature fields of subpixel resolution. Remote Sens Environ 11:221–229
10. Giglio L, Kendall JD (2001) Application of the Dozier retrieval to wildfire characterization. A sensitivity analysis. Remote Sens Environ 77:34–49
11. Giglio L, Descloitres J, Justice CO, Kaufman YJ (2003) An enhanced contextual fire detection algorithm for MODIS. Remote Sens Environ 87:273–282
12. Hunt ER, Rock CR (1989) Detection of changes in leaf water content using near and medium infrared reflectances. Remote Sens Environ 30:43–54
13. Justice CO, Korontzi S (2001) A review of satellite fire monitoring and requirements for global environmental change research. In: Ahern FJ, Goldammer JG, Justice CO (eds.) Global and regional wildfire monitoring: current status and future plans. SPB Academic Publishing, The Hague, pp 1–18
14. Kaufman Y, Justice C (1998) MODIS fire products, algorithm theoretical basis document. MODIS science team. EOS ID#2741
15. Langaas S (1993) A parametrised bispectral model for savana fire detection using AVHRR night images. Int J Remote Sens 14:2245–2262
16. Langaas S (1995) A critical review of sub-resolution fire detection techniques and principles using thermal satellite data. PhD thesis, Department of Geography, University of Oslo, Oslo
17. Li Z, Nadon S, Chilar J, Stocks B (2000) Satellite mapping of Canadian boreal forest fires: evaluation and comparison of algorithms. Int J Remote Sens 21:3071–3082
18. López S, González F, Llop RA, Cuevas M (1991) An evaluation of the utility of NOAA-AVHRR images for monitoring forest fire risk in Spain. Int J Remote Sens 12:1841–1851
19. Malingreau JP (1990) The contribution of Remote Sensing to the global monitoring of fires in tropical and subtropical ecosystems. In: Goldammer JG (ed.) Fire in the tropical biota, ecosystem processes and global challenges. Springer, Berlin, pp 337–399
20. Matson M, Dozier J (1981) Identification of sub-resolution high temperatures sources using a thermal IR sensor. Photo Eng Remote Sens 47(9):1311–1318
21. Matson M, Holben B (1987) Satellite detection of tropical burning in Brazil. Int J Remote Sens 8:509–516
22. Nemani RR, Running SW (1989) Estimation of regional surface resistance to evapotranspiration from NDVI and thermal IR AVHRR data. J Appl Meteorol 28(4):276–274
23. Prins EM, Menzel WP (1992) Geostationary satellite detection of biomass burning in South America. Int J Remote Sens 13:2783–2799
24. Prins EM, Menzel WP (1994) Trends in South American burning detected with the GOES VAS from 1983-1991. J Geophys Res 99(D8):16719–16735
25. Prins E, Govaerts Y, Justice CO (2004) Report on the joint GOFC/GOLD fire and CEOS LPV working group workshop on global geostationary fire monitoring applications, GOFC/GOLD report no. 19, EUMETSAT, Darmstadt, 23–25 Mar 2004
26. Prosper-Laget V, Douguedroit A, Guinot JP (1995) Mapping the risk of forest fire occurrence using NOAA satellite information. EARSeL Adv Remote Sens 4(3-XII):30–38
27. Schultz M (2002) On the use of ATSR fire count data to estimate the seasonal and inter-annual variability of vegetation fire emissions. Atmos Chem Phys 2:387–395
28. Seemann SW, Borbas EE, Li J, Menzel WP, Gumley LE (2006) MODIS atmospheric profile retrieval algorithm theoretical basis document (ATBD07). Version 6, Oct 2006
29. Wan Z (1999) MODIS land-surface temperature. Algorithm theoretical basis document. MODIS land team

Chapter 6
Electromagnetic Methods and Sensors for Water Monitoring

Francesco Soldovieri, Vincenzo Lapenna, and Massimo Bavusi

Abstract This paper deals with a brief presentation of non or minimally invasive techniques of interest for water security and monitoring applications, based on electromagnetic sensing. The techniques are presented according to the frequency range and to the degrees of novelty for the specific applicative context. For each technique, we briefly sketch the basic theory and the general and specific interest for the water monitoring. Besides well assessed techniques such as Electrical Resistivity, Electromagnetic Induction, Ground Penetrating Radar and Time Domain Reflectometry, Self Potential and Magnetometry, we give a short presentation of techniques of recent interest, some of them still in course of development, such as: Hyperspectral and TeraHertz Imaging.

Keywords Electromagnetic methods and sensors • Water monitoring • Non invasive diagnostics and monitoring

6.1 Introduction

The use of electromagnetic methods in water monitoring for security and resource management is gaining increasing interest also for the improved awareness that the stakeholders are achieving about this thematic.

F. Soldovieri (✉)
Institute for Electromagnetic Sensing of the Environment, CNR, Via Diocleziano 328, Naples 80124, Italy
e-mail: soldovieri.f@irea.cnr.It

V. Lapenna • M. Bavusi
Institute of Methodologies for Environmental Analysis, CNR, Contrada Santa Loja, Tito Scalo 85100, Italy

In fact, many applications in water management can be favourably affected by the use of the non- or low-invasive monitoring techniques based on electromagnetic sensing such as: groundwater flow monitoring; identification of water flow paths; water content determination, pollution and contaminant detection and characterization; soil salinity and texture; soil drainage classes; water recharge and, very recently, agriculture of precision.

In fact, the electromagnetic sensing methods permit the indirect measure of the water content and quality-related parameters through the determination of electromagnetic parameters in static or low frequency regime and microwave/higher frequencies. In particular, the determination of electromagnetic quantities such as electrical conductivity and dielectric permittivity as well as transmission and reflection coefficients of the probing radiation at the investigated medium, permits to achieve information in a remote and minimally invasive way about the final parameters of interest. As an example of this interrelation we recall the fact that the dielectric permittivity of the water is very high (about 80 for the relative dielectric permittivity), so that the increase in the soil water content can be detected by means of the increase in the soil dielectric permittivity.

As said above, the main advantage of the electromagnetic sensing techniques resides in their non or minimal invasivity; in fact, they do not require the extraction of samples as in the case of gravimetric measurements. In addition, the measurement of the electromagnetic static fields (in term of currents and voltages) and waves allows to give a distributed and global vision of the monitored area, also thanks to the remote sensing observations. Finally, the integration of different techniques can allow a water monitoring that is multi-sensing, at different spatial scales (in terms of extent of the area and resolution), multi-depth and multi-temporal (from continuous to real-time).

In particular, here we will describe the potentiality of several electromagnetic sensing techniques that are classified according to the working frequency band and related sensing mechanism.

First, the chapter gives an outline of the two main and usually exploited static (or low frequency) active techniques as Electrical Resistivity (ER) and Electromagnetic Induction (EMI). These techniques are able to detect changes in the electrical conductivity of the subsurface that are useful to gain direct and indirect information about quantities relevant to the water security and management, such as: water content; ground water recharge; soil salinity assessment; soil drainage classes; herbicide and pollutant diffusion due to the groundwater flow.

Besides these active techniques, at a low frequency regime other passive techniques are relevant as the Self Potential (SP) and the magnetometry methods. These methods are concerned with the evaluation of the spatial and temporal variations of the "natural" electric and magnetic field that can be related to anomalies present in the subsurface. The SP technique consists in the passive measurements of the electrical potential distribution at the ground surface associated with natural polarization mechanisms occurring at depth. The main contribution to these self-potential signals is usually associated with groundwater flow mainly through the electrokinetic (hydro-electric) coupling, also known as the "streaming potential effect".

The SP method is a fast, inexpensive, and is very simple to be implemented in field conditions also due to the fact that the sensors (non-polarisable electrodes) are cheap.

Magnetometry is a passive remote sensing technique that records the Earth's local magnetic field at the sensor location and has found application in agriculture for the detection and location of the drainage pipes when the Ground Penetrating Radar (GPR) is not able to work, due to the high attenuation of the soil.

By turning to higher frequencies from tens of MHz to some GHz (Radiofrequency and microwaves regime), the technique usually exploited in water monitoring is based on Ground Penetrating Radar, which is sensitive to changes of the soil dielectric permittivity and electrical conductivity. This makes possible to determine with high spatial resolution and accuracy the water content distribution through semi-empirical laws that relate the water content with the dielectric permittivity values retrieved by the GPR measurement. In this way, one can think to evaluate the effectiveness of the irrigation operations in agriculture of relevant interest in water scarcity conditions, and to prevent the soil from pollution due to herbicide and pesticide diffusion. In addition, the flexibility of the GPR allows to tackle other applications as: to detect buried drainage pipes; to detect depth of the water table; to determine the shallower layering of the soil.

In the work frequency band similar to that of GPR, we stress the usefulness of the Time Domain Reflectometry (TDR), which is now a standard tool to determine volumetric soil water content. In particular, TDR is relatively insensitive to soil composition and texture, and it is thus a good method for 'liquid' water measurement (content and salinity) in soils, taking into account that the reflected wave takes information about soil salinity as well. In fact, the most powerful aspect of TDR is the ability to monitor the temporal development of soil water content and bulk soil electrical conductivity in the same volume at a selected number of locations with a high temporal resolution.

Besides the above said techniques, which exhibit a rather good degree of maturity, other techniques, covering the highest part of the electromagnetic frequency spectrum, are gaining interest as Tera-Hertz technology and Hyperspectral imaging/spectroscopy.

As new application of THz waves, we point out the remote sensing of the hydration state of soil and vegetation that can provide significant information in the sustainable development of ecosystems, as well as in an optimal exploitation of agriculture resources.

Also, the effectiveness of reflectance spectroscopy, in the Visible, Near-InfraRed and ShortWave InfraRed spectral ranges (400–2500 nm) is now under investigation for its capability to perform fast and non-destructive surveys, which provide potentially useful alternatives to time-consuming chemical methods of soil analysis. In fact, the characteristics of soil reflectance spectra are controlled by mineral composition, organic matter, water (hydration, hygroscopic, and free pore water), iron form and amount, salinity, and particle size distribution. These attributes of soil, basically determine their capacity to perform production and environmental functions.

In the following, we describe briefly each of the said above techniques by grouping them for frequency band. It must be noted that due to the brevity of the work, we aim principally at providing very basic information about the sensing mechanisms and the applicative potentialities of the techniques, whereas a more detailed study of each technique can be made using as starting point the referenced papers.

6.2 Active Low Frequency Techniques: ERT and EMI

This Section is devoted at presenting the two main low frequency active electromagnetic methods as the Electrical Resistivity Tomography (ERT) and the Electromagnetic Induction (EMI). The two methods are both sensitive to the electrical conductivity but through different sensing mechanisms. The wide range of spatial and temporal variability of the electrical conductivity of the rocks is the key of success of the electromagnetic sensing technologies ([46] and reference therein). The soil electrical conductivity is dependent on a combination of physical-chemical properties including soil water content together with soluble salts, clay content and mineralogy, bulk density, organic matter, and soil temperature. Accordingly, measurements of electrical conductivity have been exploited at field scales to map the spatial variation of soil salinity, clay content or depth to clay-rich layers, soil water content, the depth of flood deposited sands, and organic matter [15]. In addition, apparent electrical soil conductivity (EC_a) has been used at field scales to determine a variety of anthropogenic properties: leaching fraction, irrigation and drainage patterns, and compaction patterns due to farm machinery of relevant interest in the framework of the agriculture precision [11].

The electrical method is based on the measurement of the electrical resistivity (the inverse of the electrical conductivity) of the soil; in its basic configuration (of principle) the system is made up of four electrodes penetrating or in contact with the soil surface [8, 9, 48]. As shown in Fig. 6.1, two electrodes (current electrodes) inject a DC current I in the soil whereas the other two electrodes are used to measure the corresponding induced Voltage V; then by applying the well known Ohm's law in an homogenous half-space, the electrical resistivity is achieved as $\rho = k(V/I)$, where k is a geometrical coefficient related to the distance of the electrodes.

Different electrode configurations can be adopted for the field survey. The above described electrode configuration is referred to as a Wenner array when the four electrodes are equidistantly spaced in a straight line at the soil surface with the two outer electrodes serving as the current or transmission electrodes and the two inner electrodes serving as the potential or receiving electrodes as it is shown in Fig. 6.1 [9]. The depth of penetration of the electrical current and the volume of measurement increase when the inter-electrode spacing, a increases. For a homogeneous soil, the volume interested by the measurement is roughly πa^3. A good flexibility of the electrical method is possible since the depth and the extent of the investigation region can be easily changed by changing the spacing between the electrodes.

6 Electromagnetic Methods for Water Monitoring

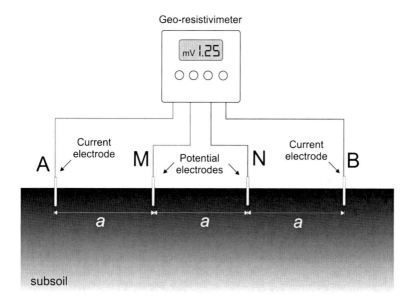

Fig. 6.1 The basic configuration of the electrical resistivity measurement

Accordingly, the electrical method is a powerful tool for mapping the resistivity pattern of subsoil at different scales and with different resolution. However, the resistivity measured in the field with any electrode layouts coincides with the true resistivity of subsoil only in the case of an homogenous half-space. In the geological applications where the subsoil is extremely complex and the investigated volume is interested by many electrical discontinuities the concept of "*apparent resistivity*" has been introduced. The transformation of the apparent resistivity values, representing the experimental data measured during the geophysical field surveys, in the "*true resistivity*" values of the subsoil is the key challenge of the electrical method.

For tackling this basic problem different strategies for data acquisition and advanced mathematical inversion algorithms have been adopted. The Vertical Electrical Sounding (VES) is a method to estimate the variation of electrical resistivity with the depth. A sequence of apparent resistivity values is obtained for different increasing values of the distance between the energizing and receiving electrodes. Under the assumptions of a subsoil modeled by horizontal-plane layers and using direct or inverse algorithms for data interpretation, the resistivity vertical distribution is obtained. The VES technique is suitable for geoelectrical investigations in alluvial deposits, for identifying the depth of the groundwater and for investigating the geometry of geological layers.

In presence of lateral variations of the electrical resistivity, the Electrical Profiling (EP) technique is applied. The apparent resistivity measurements are performed moving the electrode array along a profile/direction on the earth surface without any changes of the mutual distance between the electrodes. By combining parallel profiles it is easy to obtain resistivity maps. Profiles and maps are largely applied to localize buried objects in subsoil (cavities, groundwater channels, pollutant plumes etc.).

In the past two decades, the electrical resistivity surveys have progressed from the conventional vertical soundings and profiles to advanced techniques that provide two-dimensional and even three-dimensional high-resolution electrical images of the subsurface (Electrical Resistivity Tomography). Technically, an ERT survey can be carried out using different electrode configurations (dipole–dipole, Wenner, etc.) that are dislocated on earth surface to inject into the ground the electric currents and to measure the generated voltage signals.

For a dipole–dipole configuration, the electric current (I) is sent into the ground via two contiguous electrodes x meters apart, and the potential drop (V) is measured between two other electrodes x meters apart in line with the current electrodes. The spacing between the nearest current and potential probes is an integer n times the basic distance x and the maximum number of measurements n depending on the signal-to-noise ratio of the voltage recordings [27].

The values of the apparent resistivity ($\rho_a = [\pi n x(n+1)(n+2)] \cdot V/I$) for each traverse are assigned, along a horizontal axis, at the intersections of two converging lines at 45° from the center of the current dipole and the center of the measuring dipole. All values of the apparent resistivity, choosing significant variable contour intervals, make a first tomographic image of the electrical subsurface structure, the so-called "pseudo-section" (Fig. 6.2).

In a second step, it is necessary to transform the apparent resistivity values obtained during the field survey into real resistivity values of the subsoil. Recently, many methods have been proposed to solve this problem. In a seminal paper Loke and Barker [30] propose a new method for the automatic 2D inversion of apparent resistivity data. The inversion routine is based on the smoothness constrained least-squares inversion [45] implemented by a quasi-Newton optimization technique. The optimization adjusts the 2D resistivity model trying to reduce iteratively the difference between the calculated and the measured apparent resistivity values. The root mean square (RMS) error gives a measure of this difference. After this paper other algorithms have been proposed and to-date the ERT method is a powerful tool for the near-surface investigations and finds a wide range of interesting applications in applied geology, geomorphology, environmental geology, hydrogeology, geotechnical engineering etc. [47].

Nowadays, the electrical methods are currently applied for tackling new scientific problems in hydrogeology. One of the most interesting application is the use of the electrical resistivity to monitor the influence of climate and vegetation on root-zone moisture, bridging the gap between remotely-sensed and in-situ point measurements. The time-lapse electrical imaging is the key to follow the time-dependent changes of hydrological parameters and to quantify large seasonal differences in root-zone moisture dynamics [2, 5]. The near surface hydrological parameters are pivotal to better understand the changes in global climate and land use, affecting important processes from evapotranspiration to the groundwater recharge.

Electromagnetic induction (EMI) performs electrical conductivity measurements through a different sensing mechanism. EM induction is a noninvasive technique that measures a depth-weighted average of EC, termed the apparent electrical conductivity, EC_a [1].

6 Electromagnetic Methods for Water Monitoring

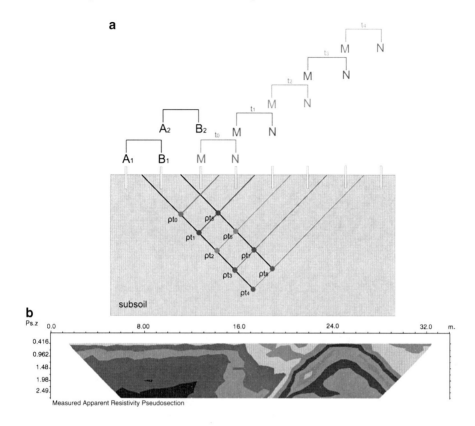

Fig. 6.2 Measurement configuration and pseudo section for the apparent resistivity

The EMI instrument (Fig. 6.3) normally operates at a fixed frequency (frequency domain, FEM). The alternating current that passes through the transmitter coil generates a primary magnetic field. This primary magnetic field H_i induces small alternating current loops in the soil. Each current loop generates a magnetic field H_p that is proportional to the value of the current flowing within the loop. The secondary magnetic field, a combination of the primary and the induced magnetic fields ($H_s = H_i + H_p$), induces a small alternating current in a receiver coil.

The receiver coil measures the amplitude and phase of the secondary magnetic field, which consists partly of signals from the soil layers at different depths corresponding to the different loops.

All of the measured signals are amplified and summed into an output voltage, which is directly related to a depth-weighted average EC_a calculated from [33]

$$EC_a = \frac{4}{2\pi f \mu_0 s^2} \left(\frac{H_s}{H_p} \right) \quad (6.1)$$

where f is the frequency, μ_0 is the magnetic permeability and s is the spacing between the two coils.

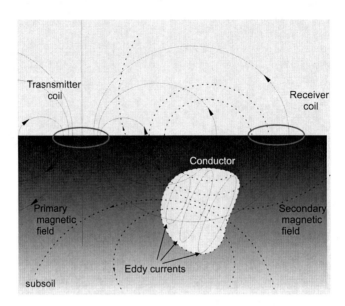

Fig. 6.3 The physical phenomenon at the basis of the Electromagnetic Induction

For such a sensing method, the most commonly used EM conductivity meters in soil science and in vadose zone hydrology are the Geonics EM-38, whose depth of measurement can reach 1.5 m in the vertical coil configuration [20, 33], and the Geosonics EM-31 whose investigation depth can be of 6 m. Very small – milli Siemens/meter (mS/m) changes in soil electrical conductivity/resistivity can be detected and recorded. The data is gathered by simply moving the equipment over the site under investigation on a predetermined grid and no direct contact with the ground is necessary. However, the main difficulty is in the calibration phase that allows to pass from the electrical conductivity to the measured field.

The other type of EMI instrument in common use is time-domain EMI (TEM) or "transient EM" (see Fig. 6.4). A short low energy electromagnetic pulse from the transmitter coil couples with the ground and a receiver coil measures the decaying signal induced into the ground with respect to time. This technique allows very sensitive detection of shallow and deep buried metal objects. The depth of measurement of the EM-61 is approximately 3 m.

6.3 Passive Low Frequency Techniques: Self Potential and Magnetometry

The Self-Potential method (SP) is among the oldest methods in geophysics; it consists in measuring natural electrical fields that arise in the subsurface due to several mechanisms: electrokinetic coupling (streaming potential); thermoelectric coupling; electrochemical effects; cultural activity [10].

6 Electromagnetic Methods for Water Monitoring

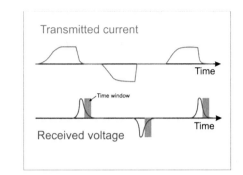

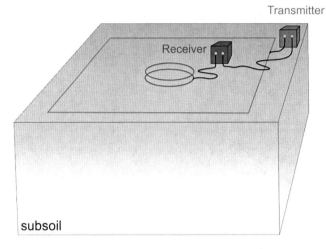

Fig. 6.4 The working principle of the Time-Domain EMI

Self-potential field surveys are performed by measuring electrical potential differences between pairs of electrodes that are inserted or placed in contact with the surface of the earth and located at a number of survey stations in the area of interest (see Fig. 6.5). In particular, the measurement of potential is usually performed along a transect according to two different modes (fixed offset or multi-offset) and then represented as a profile along the transect. SP gradients are generally measured along both closed loops and linked traverses by alternating the leading and following electrodes (leap-frog technique), in order to reduce cumulative errors caused by electrode polarization. The distance between the measuring electrodes and the length of the profile dictate the resolution and the investigation depth of the SP survey ([46] and reference therein). Grid contouring and visualization tools are finally used for representing SP profile and maps.

The SP method has been applied in a wide class of geological problems: mineral prospecting, detection and delineation of thermal sources in geothermal and volcanic areas, and groundwater investigations [6, 41].

In groundwater studies the SP anomalies are mainly due to the streaming potentials generated by fluid flow in porous rocks, because of a pore pressure gradient.

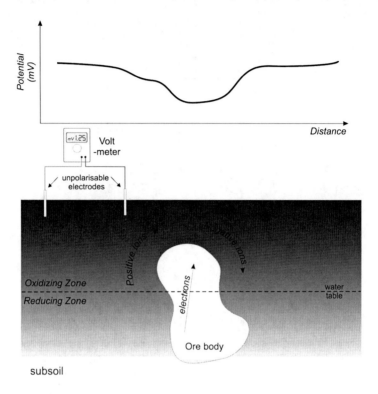

Fig. 6.5 The self potential method

The electrokinetic effect is controlled by the relative motion occurring along a shear plane of the electrical double layer at the mineral water interface and associated with an electrical diffuse layer in the pore fluid. The ζ potential is the electrical potential on the shear plane [40] and is related to the streaming potential V as follows:

$$V = \frac{\rho \varepsilon \zeta}{\eta} \Delta P \qquad (6.2)$$

where ρ, ε and η are the electrical resistivity, dielectric permittivity and dynamic viscosity of the pore fluid, respectively, and ΔP is the drop in pressure along the flow path. In this scheme, positive charges are carried in the direction of the fluid flow, producing positive SP anomalies on the surface, where water discharge is located, and negative in the sites of infiltration [46].

Recently, new methodologies for SP data inversion and novel hydrogeophysical models have been developed to better describe the water flow movements in porous media. To-date, the SP method is one of most promising and challenging geophysical methods for the hydrogeology. In a pioneer paper Patella [36] introduced a tomographic procedure able to depict the most probable distribution of electrical charge accumulations in the subsoil, responsible for the measured SP field. The 2D SP

Tomography (SPT) is based on a cross-correlation algorithm between a theoretical scanner function and the observed SP electrical field. In the 2D case, a Charge Occurrence Probability (COP) function is defined as the probability to find, in a point of the cross section through the profile, a positive or negative electric charge accumulation, which is responsible for the whole field observed.

Recently, new geophysical models have been studied in order to describe the generating mechanism of SP anomalous patterns [41]. These recent developments provide more quantitative results in hydrogeological studies, such as those regarding the fluid flow pattern in landslide bodies or the contaminant diffusion in the water table.

The self-potential method can be considered as a powerful tool for mapping groundwater flow in near surface. The underlying physics describing the coupling between the flow of water through a porous continuum and the generation of an electrical field is well understood at the scale of the representative elementary volume of the porous continuum and described by the electrokinetic theory (e.g., [3, 42]). Accordingly, the self-potential method can be applied for mapping the groundwater flow and to estimate hydraulic conductivity as demonstrated during pumping tests [43]. Another important application is the use of the self-potential surveys for monitoring the contaminant plume pollutant in groundwater. In fact, self-potential and electrical resistivity methods have been successfully applied to identify the presence of organic and inorganic contaminants close to the landfills [35].

In addition, Self-Potential is suitable for groundwater investigations and in geotechnical engineering applications for seepage studies. Self potential measurements (mainly associated to the electrokinetic mechanism) have been exploited to identify leaks and determine functioning condition in drainage pipes for irrigation; self potential methods can be suitable to provide information about the extent and the magnitude of the water table extension due for example to the well irrigation; other possible application is the determination of the spatial profile of the soil salinity.

To complete this short review, we mention some studies regarding the use of electrical methods to identify the groundwater pressure changes induced by earthquake activities. The study of the fluid migration processes in tectonic active areas is one of the most promising and challenging problems in seismology. The integration of the electrical methods can give information about fluid movements in the focal areas and to identify electrokinetic phenomena triggered by local and/or distant earthquakes [7, 29].

Finally, we emphasize the scientific relevance of a recent paper in which the authors introduce seminal concepts to jointly invert the active (resistivity) and passive (self-potential) electrical measurements [24]. The 3D jointly inversion of the self-potential and electrical measurements is the key challenge for the future and could disclose new and more interesting applications of the electrical methods in hydrogeology.

The magnetic mapping is a passive low frequency method exploiting natural geomagnetic field as source; in particular it measures the magnitude of the Earth's local magnetic field at the sensor location. Secondary regional and local variation of the earth's magnetic field are produced by objects locate over or beneath the soil surface exhibiting magnetic properties. The sensor used for the measurement, a magnetometer, can be located at the ground or on an airborne and satellite based platform or in the soil through borehole configuration. The measurement are performed along transects. One of the applications concerned with the magnetometry is the location of drainage pipes and pollution flow in the soil.

6.4 Active Mid-frequency Techniques Ground Penetrating Radar and TDR

The possibilities offered by a Ground Penetrating Radar (GPR) system in water monitoring, hydrology and underground stratigraphy have been gaining interest among the scientific community in the last few years [13, 34]. In particular, there is a strong applicative interest in the fields of the agricultural geophysics [22, 23] and in monitoring subsoil pollution.

Related to these points, the Ground Penetrating Radar (GPR) results a fast and simple tool for the determination of the water content of the soil [23, 26]. In particular, it allows to achieve spatial maps of the water content at an intermediate scale (and with an intermediate resolution) between the scale provided by satellite systems and the scale provided by Time Domain Reflectometry (TDR) systems [32].

GPR (Fig. 6.6) is based on the same operating principles of classical radars [12]. In fact, it works by emitting an electromagnetic signal (generally modulated pulses or continuous harmonic waves) into the ground; the electromagnetic wave propagates through the opaque medium and when it impinges on a non-homogeneity of the electromagnetic properties, in terms of *dielectric permittivity and electrical conductivity*, a backscattered electromagnetic field arises. Such a backscattered field is then collected by the receiving antenna located at the air/opaque medium interface and undergoes a subsequent processing and visualization, usually as a 2D image.

With reference to the specific application of the water monitoring, GPRs have been used in a cross-hole configuration where the antennae are located in boreholes. This allows to gain information about the dielectric permittivity of the soil up to the depth of ten meters and more [4]. However, performing borehole measurements requires the necessity of vertical holes in the soil and thus this configuration is more difficult to be implemented compared to the GPR prospecting at soil-air interface.

The other main measurement method is based on the GPR measurements at the air/soil interface and two different configurations are customarily exploited [23]: the first one exploits a single measurement spatial point and the electromagnetic velocity (related to the inverse of the square root of the searched for dielectric permittivity) is achieved from the knowledge of the depth of a buried reflecting target. This configuration exploits a one-dimensional model of the electromagnetic scattering, and thus problems can arise in data interpretation if there is some spurious buried object in the neighbors of the known target. Moreover, under this configuration, the resolution available in the determination of the object's depth is quite poor, and this affects the precision in the determination of the Two Travel Time. Still, in many

6 Electromagnetic Methods for Water Monitoring

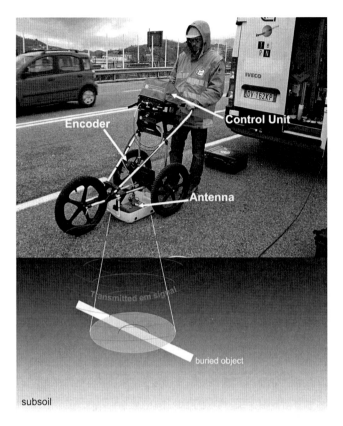

Fig. 6.6 GPR measurement

cases it is not reliable to assume the depth of the reflector as a perfectly known quantity: in particular, this happens if the same reflector is exploited for months or years to monitor the time behaviour of the soil characteristics.

A second configuration exploits jointly several measurement points. Within this framework, further possibilities of choice are offered by common offset, common mid point or WARR configurations [12, 16]. The characteristics of the soil, in this case, can be worked out from the comprehensive image of the buried object obtained by several GPR traces. In particular, several procedures have been proposed to retrieve the EM propagation velocity in the soil from a common offset configuration or B-scan "radargram" [23, 49].

Independently from the adopted measurement method, GPR data are processed in two steps in order to provide information about the water content of the soil. The first step consists in the determination of the dielectric permittivity of the soil, obtained from the Two-Way-Time (TWT) of the EM wave or/and the strength of the reflected EM energy. The second step consists in the determination of the water content in the soil starting from the estimated dielectric permittivity and in this framework empirical models can be adopted [21, 50].

Another way in which the GPR can be used in studying the soil moisture is exploiting 3D GPR images so to determine not the value of the water content but its spatial distribution in a volume at sub-meter resolution. A recent study has pointed out a correlation between the variation of the GPR signal and the water content [25] at the spatial scale typical of GPR surveys.

For each of the previous configurations, the result is often provided by the direct interpretation of the collected data. However, even if the interpretation of radar signals might be a relatively simple task in the case of a single target buried in a homogeneous half space, it can become quite more difficult if several targets are present (including among them also clutter and layers). This entails that a high degree of user knowledge is necessary in order to achieve a good quality of data and reliable interpretation. This consideration is exemplified by the Fig. 6.7 (upper and middle panel) where the true scene is compared with its GPR image (raw data).

Very recently, a new approach based on the microwave tomography has been developed [28, 37]. However, in order to exploit a tomographic approach, it is required to cope with an inverse scattering problem, which entails some mathematical difficulty due to its inherent non-linearity and ill-posedness. In particular, ill-posedness entails to adopt a regularization technique to make the solution more robust (even if somehow less accurate) against the effects of the noise and the parametric uncertainties affecting the model. In addition, non-linearity of the data-unknown relationship imposes attention to the problem of false solutions that can make the inversion results quite different from the ground truth.

These difficulties can be coped with by making use of a simplified model of the electromagnetic scattering. In this paper, in particular, we will adopt the linearized model funded on the Born Approximation (BA) [28, 37]. Under BA, a stable solution of the problem can be easily achieved and the problem of the false solutions is automatically avoided because the adopted model is linear. Moreover, the computational burden is considerably reduced with respect to that customarily needed within different (non-linear) models, and this makes it possible to investigate large areas. The reliability of the method is pointed out by Fig. 6.7 where the lower panel depicts the tomographic reconstruction of the scene shown in the upper panel; the result points out the high quality interpretability of the tomographic image compared to the raw data (middle panel).

Time domain reflectometry (TDR) represents an useful tool for measuring the velocity of an electromagnetic wave propagating in the material under test. Thus, the major application of such technique is the determination of the soil water content, via the bulk soil permittivity and conductivity [44, 50]. In addition, recently the TDR method has been used to estimate propagation velocity and attenuation of electromagnetic waves in order to support GPR data inversion and interpretation [31, 32, 39, 44].

In its basic configuration, a TDR system is based on an open circuit transmission line filled with (or embedded in) the test material (hereafter referred to as probe line), excited by a stepwise signal [38, 44]. The probe line is designed to have a characteristic impedance quite different from that of the generator feeder line

6 Electromagnetic Methods for Water Monitoring

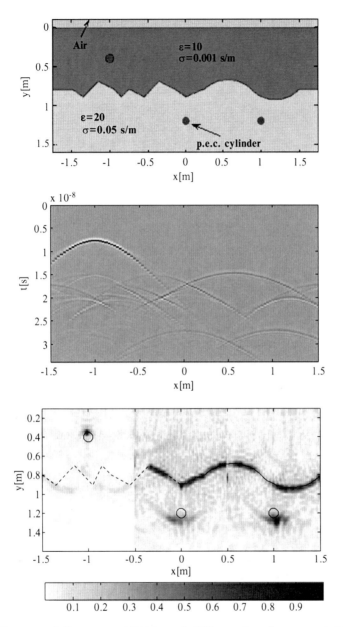

Fig. 6.7 *Upper panel*: True scene; *Middel panel*: GPR raw data; *Lower panel*: Microwave Tomographic reconstruction

(100–200 Ω against the usual 50 Ω of the feeder line) so as signal reflection signatures arise and allow to measure the wave velocity; these measured electromagnetic quantities permit to determine the dielectric permittivity and the conductivity of the material under test.

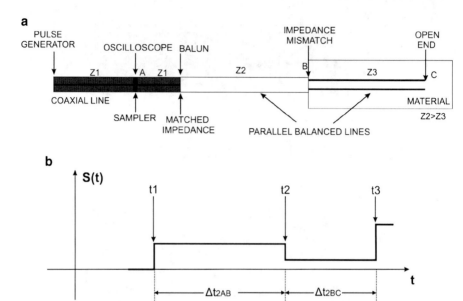

Fig. 6.8 (a) Schematic drawing of a basic TDR system, consisting of a pulse generator, a sampler, an oscilloscope, a coaxial cable, a balun, a shielded two-wire transmission cable, and two parallel metal rods which are inserted in the material to be tested. (b) Ideal wave form as a function of time detectable with a TDR equipment

In particular, a TDR system generally consists of a pulse generator, a sampler, an oscilloscope, a coaxial cable, a shielded two-wire transmission cable, and two parallel metal rods which are inserted in the material to be tested; the scheme is shown in Fig. 6.8a [38].

The pulse generator applies a fast rise-time voltage step (typically 200 ps) to a 50 V coaxial cable and triggers a sampler. The step pulse travels down the coaxial cable and, through an impedance matching transformer (balun), into a shielded two-wire transmission cable until it reaches the probe's metal rods inserted in the material being tested. At this point part of the signal is reflected back toward the cable tester due to the mismatch in the impedance (due to the different materials) and part of the signal is transmitted in the material; this transmitted part travels and reaches the end of the rods where another reflection arises and propagates back towards the oscilloscope.

The wave form displayed on the oscilloscope represents the superposition of incident waves (the step signal emitted by the pulse generator) and the reflected waves "phase" or "in counter phase" arisen at each impedance mismatch in the system. Figure 6.8b shows an ideal TDR wave form related to the characteristic impedances Z_i of the different sections. All measured times are relative to t_1 (i.e., point A, where the generated step passes the sampler/ oscilloscope). The two-way travel time of the electromagnetic pulse in the cable between points A and B is Δt_{2AB}, corresponding to $t_2 - t_1$. The two-way travel time related to the propagation of the signal between the start point B and the end point C of the two parallel rods is Δt_{2BC}, corresponding to $t_3 - t_2$. This value can be used to calculate the propagation velocity of the signal in the material. In fact, for nonmagnetic and nonconductive materials, the travel time Δt_{2BC} is given by $\Delta t = 2L\sqrt{\varepsilon_{mat}}/c$ where L is the length of the probe line, c the light speed in a vacuum, and ε_{mat} is the apparent relative permittivity given by:

$$\varepsilon_{mat}(f) = \varepsilon'(f)\left(1 + \sqrt{1 + \varepsilon''^2(f)/\varepsilon'^2(f)}\right)/2 \qquad (6.3)$$

where f is the frequency, ε' and ε'' (accounting for losses in the material) are the real and imaginary parts of the dielectric permittivity of the material, respectively [32]. In the case of loss-less material $\varepsilon'' = 0$, we have the usual expression $\varepsilon_{mat} = \varepsilon'$ where no dispersion effects are present.

The main advantage offered by TDR concerns the possibility to monitor the temporal behaviour of soil water content and bulk soil electrical conductivity in the same soil volume at a selected number of locations (punctual measurement) with a high temporal resolution. Some disadvantages are the small sampling volume, the high costs of a complete TDR system and the inability to automatically monitor soil water content of larger areas with a single TDR system.

6.5 High Frequency Methods: Hyperspectral and TeraHertz

Hyperspectral images are produced by instruments called imaging spectrometers. Spectroscopy is the study of light that is emitted by or reflected from materials and its variation in energy with the wavelength (Fig. 6.9). As applied to the field of optical remote sensing, spectroscopy deals with the spectrum of sunlight that is diffusely reflected (scattered) by materials at the Earth's surface. Instruments called spectrometers (or spectroradiometers) are used to make ground-based or laboratory measurements of the light reflected from a test material, over a wind range of wavelength ranging from 0.4 to 2.4 µm (visible through middle infrared wavelength ranges).

Remote imagers are designed to focus and measure the light reflected from many adjacent areas on the Earth's surface. In many digital imagers, sequential measurements of small areas are made in a consistent geometric pattern as the sensor platform moves and subsequent processing is required to assemble them into a global image. Until recently, imagers were restricted to one or a few relatively broad wavelength bands by limitations of detector designs and the requirements of data storage, transmission, and processing. Recent advances in these areas have allowed the design of imagers that have spectral ranges and resolutions comparable to ground-based spectrometers.

In reflected-light spectroscopy the fundamental property that we want to obtain is spectral reflectance: the ratio of reflected energy to incident energy as a function of wavelength. Reflectance varies with wavelength for most materials because the

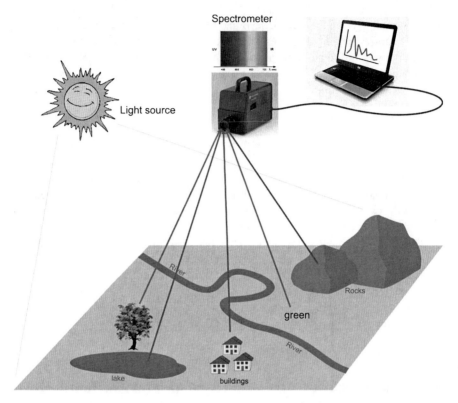

Fig. 6.9 The working principle of the spectrometer

electromagnetic energy is scattered and/or absorbed in dependence of its wavelength. These reflectance variations are evident when we compare spectral reflectance curves (plots of reflectance versus wavelength) for different materials, as shown in Fig. 6.10. Pronounced downward deflections of the spectral curves mark the wavelength ranges for which the material selectively absorbs the incident energy. These features are commonly called absorption bands. The overall shape of a spectral curve and the position and strength of absorption bands in many cases permit to identify and discriminate different materials. For example, vegetation has higher reflectance in the near infrared range and lower reflectance of red light than soils.

The spectral reflectance curves of healthy green plants also have a characteristic shape that is dictated by various plant attributes. In the visible portion of the spectrum, the curve shape is governed by absorption effects from chlorophyll and other leaf pigments. Chlorophyll absorbs visible light very effectively but absorbs blue and red wavelengths more strongly than green, producing a characteristic small reflectance peak within the green wavelength range. As a consequence, healthy plants appear to us as green in color. Reflectance rises sharply across the boundary between red and near infrared wavelengths (sometimes referred to as the red edge) to values of around 40–50% for most plants.

6 Electromagnetic Methods for Water Monitoring

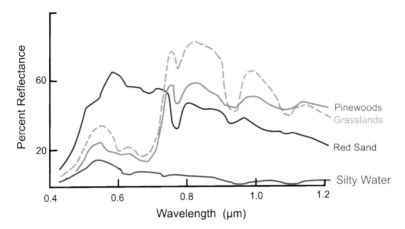

Fig. 6.10 Behavior of the reflectance of different materials versus wavelength

This high near-infrared reflectance is primarily due to interactions with the internal cellular structure of leaves. Most of the remaining energy is transmitted, and can interact with other leaves lower in the canopy. Leaf structure varies significantly between plant species, and can also change as a result of plant stress. Thus species type, plant stress, and canopy state all can affect near infrared reflectance measurements. Beyond 1.3 μm reflectance decreases with increasing wavelength, except for two pronounced water absorption bands near 1.4 and 1.9 μm.

At the end of the growing season leaves lose water and chlorophyll. Near infrared reflectance decreases and red reflectance increases, creating the familiar yellow, brown, and red leaf colors of autumn.

The quantitative determination of water content in leaf tissues is of great concern to most physicochemical and environmental plant physiologists, as well as plant breeders, biotechnologists, and genetic engineers. When plant cells are not more or less saturated with water, they cease to function normally and the plant is said to be subject to water stress. This results in, at best, reversible growth inhibition, and can lead to irreversible cell damage.

A possible method for the water content estimation in the plants is based on the use of the THz radiation (from hundreds of GHz to some THz) in a transmission configuration [18]; under this configuration a THz source radiates the EM wave that passes through the target (the leaf) and the EM wave is then collected by a receiving antenna (detector). In this case, the sensing mechanism exploits the property that THz-frequency electromagnetic radiation is strongly absorbed by water. Therefore, the measured transmission of the EM wave through the leaf provides information about the water content. On the basis of the measured water content it is possible discriminate between a plant subject to water stress and an unaffected plant. THz technique offers important number of advantages such as: the method is non-invasive and works in remote sensing (contactless); minor angular misalignments between the leaf surface and the incident beam have a fairly limited effect; the terahertz beam can have a cross section which averages over a large section in the leaf structure.

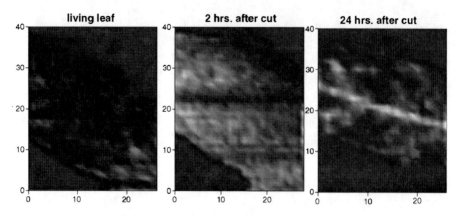

Fig. 6.11 THZ images of the time-behaviour hydration stress of a tomato leaf

Due to the very small wavelength, leaf surface irregularities do not affect the measurements through induction of excessive scattering. Furthermore, transmission measurements at microwave frequencies are impractical because the diffraction limited minimum spot size for a free-space beam is too large to avoid beam spillover around most leaves, while in the infrared and optical parts of the spectrum, strong chlorophyll absorption masks that due to water. Therefore, the THz frequency band is, therefore, a unique window in which this type of measurement may be made. At the actual state, THz experiments are performed in laboratory controlled conditions, as shown in Fig. 6.11 [14, 17, 19] and the research aim is to develop instrumentation for remote sensing imaging in realistic conditions. Figure 6.11 shows THz images depicting the time-behaviour of the tomato leaf after its cutting; the images have been collected by the reflective THz imaging system that has been developed at ENEA-Frascati [14, 17].

Acknowledgments The authors would like to thank Dr. G.P. Gallerano (ENEA- Advanced Physics Technologies) for providing them Fig. 6.11 of this Chapter.

References

1. Abdu H, Robison DA, Jones SB (2007) Comparing bulk soil electrical conductivity determination using the DUALEM-1S and EM38-DD electromagnetic induction instruments. SSSAJ 71:189–196
2. Al Hagrey SA (2007) Geophysical imaging of root-zone, trunk, and moisture heterogeneity. J Exp Bot 58:839–854
3. Bernabe´ Y (1998) Streaming potential in heterogeneous networks. JGR 103:20827–20841
4. Binley A, Winship P, Middleton R, Pokar M, West J (2001) High-resolution characterization of vadose zone dynamics using cross-borehole radar. Water Resour Res 37:2639–2652
5. Binley A, Winship P, West LJ, Pokar M, Middleton R (2002) Seasonal variation of moisture content in unsaturated sandstone inferred from borehole radar and resistivity profiles. J Hydrol 267:160–172

6. Bogoslovsky VA, Ogilvy AA (1977) Geophysical methods in the investigation of landslides. Geophysics 42:562–571
7. Brodsky EE, Roeloffs E, Woodcock D, Gall I, Manga M (2003) A mechanism for sustained groundwater pressure changes induced by distant earthquakes. JGR 108(B8):2390
8. Burger HR (1992) Exploration geophysics of the shallow subsurface. Prentice Hall PTR, Upper Saddle River
9. Corwin DL, Hendrickx JMH (2002) Solute content and concentration – indirect measurement of solute concentration–electrical resistivity: Wenner array. In: Dane JH, Topp GC (eds.) Methods of soil analysis, Part 4 – Physical methods. Soil Science Society of America Book Series 5. Soil Science Society of America, Madison, pp 1282–1287
10. Corwin RF, Hoover DB (1979) The self-potential method in geothermal exploration. Geophysics 44:226–245
11. Corwin DL, Lesch SM (2005) Apparent soil electrical conductivity measurements in agriculture. Comput Electron Agric 46:11–43
12. Daniels D (2004) Ground penetrating Radar, 2nd edn. IEE Press, London
13. Daniels JJ, Roberts R, Vendl M (1995) Ground penetrating radar for the detection of liquid contaminants. J Appl Geophys 33:195–207
14. Doria A, Gallerano GP, Germini M, Giovenale E, Lai A, Messina G, Spassovsky I, d'Aquino L (2006) Imaging in the frequency range between 100 GHz and 1 THz using compact free electron lasers. In: Infrared millimeter waves and 14th international conference on teraherz electronics, 2006. IRMMW-THz 2006. Joint 31st international conference on, Shanghai, pp 161–162
15. Friedman SP (2005) Soil properties influencing apparent electrical conductivity: a review. Comput Electron Agric 46:45–70
16. Galagedara LW, Parkin GW, Redman JD (2003) An analysis of the ground-penetrating radar direct ground wave method for soil water content measurement. Hydrological Process 17:3615–3628
17. Gallerano GP, Doria A, Germini M, Giovenale E, Messina G, Spassovsky IP (2009) Phase-Sensitive reflective imaging device in the mm-wave and Terahertz regions. J Infrared Milli Terahz Waves 30:1351–1361
18. Hadjiloucas S, Karatzas LS, Bowen JW (1999) Measurements of leaf water content using terahertz radiation. IEEE Trans Microwave Theory Tech 47:142–149
19. Hellicar A, Li L, Greene K (2007) Design and implementation of a THz imaging system. In: Proceedings of the 10th Australian symposium on antennas, Sydney
20. Hendrickx JMH, Kachanoski RG (2002) Nonintrusive electromagnetic induction. In: Dane JH, Topp GC (eds.) Methods of soil analysis. Part 4. Physical methods. SSSA Book Series 5. SSSA, Madison, pp 1297–1306
21. Herkelrath WN, Hamburg SP, Murphy F (1991) Automatic, real time monitoring of soil moisture in a remote field area with time domain reflectometry. Water Resour Res 22:857–864
22. Hubbard S, Grote K, Rubin Y (2002) Mapping the volumetric soil water content of a California vineyard using high-frequency GPR ground wave data. The Leading Edge, pp 552–559
23. Huisman J, Hubbard S, Redman J, Annan A (2003) Measuring soil water content with ground penetrating radar: a review. Available at www.vadosezonejournal.org. Vadose Zone J 2:476–491
24. Jardani A, Revil A, Boleve A, Dupont JP (2008) Three-dimensional inversion of self-potential data used to constrain the pattern of groundwater flow in geothermal fields. JGR 113:B09204
25. Knight R, Irving J, Tercier P, Freeman G, Murray C, Rockhold M (2007) A comparison of the use of radar images and neutron probe data to determine the horizontal correlation length of water content. In: Hyndman DW, Day-Lewis FD, Singha K (eds.) Subsurface hydrology: data integration for properties and processes, AGU Geophysical Monograph Series vol 171. doi:10.1029/170GM01
26. Lambot S, Rhebergen J, Van den Bosch I, Slob EC, Vanclooster M (2004) Measuring the soil water content profile of a sandy soil with an Off-Ground monostatic ground penetrating radar. Vadose Zone J 3:1063–1071
27. Lapenna V, Macchiato M, Patella D, Satriano C, Serio C, Tramutoli V (1994) Statistical analysis of non-stationary voltage recordings in geoelectrical prospecting. Geophys Prospect 42(N.8):917–952

28. Leone G, Soldovieri F (2003) Analysis of the distorted Born approximation for subsurface reconstruction: truncation and uncertainties effects. IEEE Trans Geosci Remote Sens 41:66–74
29. Linde A, Sacks I, Johnston M, Hill D, Bilham R (1994) Increased pressure from rising bubbles as a mechanism for remotely triggered seismicity. Nature 371:408–410
30. Loke MH, Barker RD (1996) Rapid least-squares inversion of apparent resistivity pseudo-sections by a quasi-Newton method. Geophys Prospect 44:131–152
31. Mattei E, De Santis A, Di Matteo A, Pettinelli E, Vannaroni G (2005) Time domain reflectometry of glass beads/magnetite mixtures: A time and frequency domain study, Appl Phys Lett 86:224102, 1–3. doi:10.1063/1.1935029
32. Mattei E, Di Matteo A, De Santis A, Pettinelli E, Vannaroni G (2006) Role of dispersive effects in determining probe and electromagnetic parameters by time domain reflectometry. Water Resour Res 42:W08408. doi:10.1029/2005WR004728
33. McNeill JD (1980) Electromagnetic terrain conductivity measurement at low induction numbers. Tech. Note TN-6. Geonics Ltd, Mississauga
34. Nakashima Y, Zhou H, Sato M (2001) Estimation of groundwater level by GPR in an area with multiple ambiguous reflections. J Appl Geophys 47:241–249
35. Naudet V, Revil A, Rizzo E, Bottero JY, Begassat P (2004) Groundwater redox conditions and conductivity in a contaminant plume from geoelectrical investigations. Hydrol Hearth Syst Sci 8(1):8–22
36. Patella D (1997) Introduction to ground surface self-potential tomography. Geophys Prospect 45:653–681
37. Persico R, Bernini R, Soldovieri F (2005) The role of the measurement configuration in inverse scattering from buried objects under the Born approximation. IEEE Trans Antennas Propag 53:1875–1887
38. Pettinelli E, Cereti A, Galli A, Bella F (2002) Time domain reflectometry: Calibration techniques for accurate measurement of the dielectric properties of various materials. Rev Sci Instrum 73:3553–3562
39. Pettinelli E et al (2006) Electromagnetic propagation features of ground penetrating radars for the exploration of Martian subsurface. Near Surf Geophysics 4:5–11
40. Revil A, Pezard PA, Glover PWJ (1999) Streaming potential in porous media. 1. Theory of the zeta potential. J Geophys Res 104:20021–20031
41. Revil A, Hermitte D, Voltz M, Moussa R, Lacas JG, Bourrié G, Trolard F (2002) Self-Potential signals associated with variations of the hydraulic head during an infiltration experiment. Geophys Res Lett 29(7):1106. doi:1029/2001 GL014294
42. Revil A, Naudet V, Nouzaret J, Pessel M (2003) Principles of electrography applied to self-potential electrokinetic sources and hydrogeological applications. Water Resour Res 39(5):1114
43. Rizzo E, Suski B, Revil A, Straface S, Troisi S (2004) Self-potential signals associated with pumping tests experiments. JGR 109:B10203. doi:10.1029/2004JB003049
44. Robinson DA, Jones SB, Wraith JM, Or D, Friedman SP (2003) A review of advances in dielectric and electrical conductivity measurements in soils using time domain reflectometry. Vadose Zone J 2:444–475
45. Sasaki Y (1992) Resolution of resistivity tomography inferred from numerical simulation. Geophys Prospect 54:453–463
46. Sharma PS (1997) Environmental and engineering geophysics. Cambridge University Press, Cambridge
47. Steeples DW (2001) Engineering and environmental geophysics at the millenium. Geophysics 66(1):31–35
48. Telford WM, Gledart LP, Sheriff RE (1990) Applied geophysics, 2nd edn. Cambridge University Press, Cambridge
49. Tillard S, Dubois JC (1995) Analysis of GPR data: wave propagation velocity determination. J Appl Geophys 33:77–91
50. Topp G, Davis J, Annan AP (1980) Electromagnetic determination of soil water content: Measurements in coaxial transmission lines. Water Resour Res 16:574–582

Chapter 7
Subsurface Permeability for Groundwater Study Using Electrokinetic Phenomenon

Alexander K. Manstein and Mikhail I. Epov

Abstract Electric parameters of rocks and soils are valuable and important information for the purpose of groundwater management. The existing methods for the study of electric parameters are able to give exact information about the structure of any media saturated with a fluid, even in noisy urban and industrial areas. The presented work is devoted to the interaction between a geological structure saturated with groundwater and its electrical parameters. The Near-surface Electro-Magnetic Frequency Induction Sensor (NEMFIS) has a narrower sensing volume with respect to Electric Resistance Tomography (ERT), does not require a direct contact with the ground and exhibits an high immunity to noise. In addition, NEMFIS is able to work with 14 frequencies in every station, providing a vertical resolution of 0.5–0.7 m. This sensor has been successfully used for groundwater contamination mapping in the past. The electroosmosis is a well known phenomenon used for dehydration of porous media and for impregnation. Such phenomenon has been used to study the electrokinetic features of rock samples. The sustainable filtration process of fluid through sandstone was observed, and the volumetric velocity of the fluid current was measured. The interaction between direct electric current and specific electric resistance was studied experimentally. The presented results show the possibility to create the electroosmosis process and control the transport on a large volume of groundwater. The analysis of electric conductivity changes measured by NEMFIS, allows us to assess the relative water permeability of the media.

Keywords Electrokinetics • Frequency induction

A.K. Manstein (✉) • M.I. Epov
Trofimuk Institute of Petroleum Geology and Geophysics, Siberian Branch of Russian Academy of Science, Bld. 3, Koptyuga street, Novosibirsk 630090, Russia
e-mail: MansteinYA@ipgg.nsc.ru

7.1 Introduction

Development of new methods for the improvement of soils and rocks features for various tasks, e.g. building construction and exploitation, is one of the task of geophysical engineering. This work proposes the use of mass transport phenomena due to the injection of a direct electric current where the transport velocity can be measured using geoelectric methods. The measurement of rocks and soils electric conductivity is very important in view of it. Such a parameter is very sensitive to the water saturation and is applicable for a media soaked in electrolytes of any kind.

Features used to characterize solutions in physical chemistry can be divided in two groups of parameters: (i) parameters that can be studied by equilibrium system analysis; (ii) parameters that can be studied by non-equilibrium system analysis. The second group is formed by those parameters that can be assessed by diffusion, electric conductivity and viscosity [3]. Porous geological media soaked with fluids, mainly with electrolytes, are divided into sandy and argillaceous rocks [5]. Electrokinetic potential in water saturated clay layers is higher than in sands. The range of potential values is 10–80 mV for any alluvial sediments [1].

The electric field affects the electric conductivity of solutions, as it is well known. For instance, M. Win (Materials of 90-th convention of German Medical Doctors and Naturalists, 1928 [7]) has discovered that the electric conductivity of electrolytes is not a constant parameter, and increases with the intensity and frequency of an electric field imposed. He also showed that the increase is a function of concentration, valency and origin of the solution.

The electric conductivity of a porous geological media is determined by the presence of attenuated electrolyte ions in the rock matrix. The electric potential for such current carriers under the hypothesis of constant concentration (c) is:

$$\varphi = (\frac{d\Phi}{dn})c \qquad (7.1)$$

where: Φ is the thermodynamic potential; n is the amount of ion charges in a unit of solution volume (see [4], pp. 151).

Considering the chemical potential gradient ($\nabla \xi$) the current density should include the additional part:

$$j = \sigma(E - \beta \cdot \nabla \xi). \qquad (7.2)$$

Here β is the additional electric parameter of the media, which is describing the diffusion and electric processes. For the electroosmosis it is the electrokinetic potential [2]. The chemical potential gradient exists close to metal electrodes of grounded power lines. Chemical potential ξ is the derivative of the thermodynamic potential of a mass unit of solution by its concentration, at given P and T conditions. Concentration here is the ratio between the electrolyte mass and the total fluid mass in a given volume. Constancy of the chemical potential, as well as pressure and temperature constancy are the conditions for thermodynamic equilibrium.

7 Electrokinetics for Groundwater Study

Direct electric current in porous geologic media saturated by fluid at constant temperature causes a flux of electrolyte mass. Flux density of electrolyte mass, transported by macroscopic fluid movement is equal to the product of speed, solution density and concentration. Another component of the total flux density is the electrolyte transport by molecular diffusion. The above mentioned flux is proportional to the product of the electric current density and an additional electric parameter of the media. Let us designate the flux density of mass of electrolytes with the fluid in general as:

$$\rho c v \qquad (7.3)$$

here v is the speed and ρ is the density of the solution. Given I as the diffusion flux density, the full electrolyte mass flux density is:

$$\rho c v + i \qquad (7.4)$$

Irreversible diffusion leads to the increase of entropy. The entropy change speed is:

$$\frac{dL}{dt} = \int \frac{E \cdot j}{T} dV - \int \frac{i \cdot \nabla \xi}{T} dV \qquad (7.5)$$

The diffusion flux, as well as the electric current density, can be expressed as a linear combination of j and $\tilde{N}\xi$. The diffusion flux density is:

$$i = -\frac{\rho D}{\left(\frac{d\xi}{dc}\right)_{P,T}} \nabla \xi + \beta j \qquad (7.6)$$

here D is the electric field induction. When there's no electric current, and pressure and temperature are constant, the flux is due only to diffusion:

$$i = -\rho D \nabla c \qquad (7.7)$$

All the electric-diffusion processes of porous media saturated with electro-lyte under thermodynamic equilibrium are described in the Eqs. 7.2 and 7.6.

The experimentally measured electric conductivity, diffusion and viscosity values that are needed to describe the non-equilibrium state of open system will be obviously depend on time. The instantaneous values of electric conductivity with gradient of electrolyte concentration are not able to describe the system. The speed of values change must be measured too. Equations 7.3 and 7.6 show that the diffusion flux density is small in comparison to macroscopic fluid flux in the media. Under the hypothesis of constant viscosity in the porous media under a low density of electric current, the electric conductivity and electroosmosis fluid volume flux speed are the values to be measured to describe the porous media.

The experiments described were performed to record the change of electric conductivity in time under direct electric current.

Taking into account that the total electrolyte mass flux density is proportional to the amount of current densities of the various origins, the authors made experiments to discover the influence of a direct current from an external transmitter on the signal of two electromagnetic sounding devices. The direct current density from external transmitter was essentially higher than the density of sounding currents of both devices.

The modern instrumentation for the measurement of specific electric conductivity have an improved spatial resolution with respect to older methods, featuring better local measurements, that are very important for the study of small objects. For instance, the multielectrode DC sounding systems are able to perform profiling down to h meters depth using $3h$ transmitting line length, where h can be as short as 1 m. Such a multielectrode system can be used to monitor the groundwater flow.

The near surface electromagnetic induction sounding device (NEMFIS) features a particularly good spatial resolution, with respect to other methods of nowadays [6]. The NEMFIS does not require galvanic contact with the media and is very immune to noise. The device is a TD-EM three-coils sensor controlled by a wireless remote control. The total length is 2.75 m (1.4 m when transported), weighting 8 kg. Data visualization is in the form of maps and cross-sections, that is now under development for a real time data representation together with GPS data binding.

7.2 Core Samples Electroosmosis

The widely known electrokinetic phenomena – electroosmosis is using for porous media dehydration, impregnation etc. The electrokinetic parameters of cemented sandstone samples were measured in the Institute. The permanent process of fluid flux was observed.

The aim of the work (assessment of the flux speed of a fluid volume in porous samples due to electroosmosis) was obtained by measuring:

- The transported fluid volume (ml)
- The elapsed process time (s)
- The electric field intensity (V/cm)
- The current in the sample (mA).

The fluid used was distilled water was used as the fluid. The temperature 22 °C, pressure 740 mm mercury, air moisture 30–40%. The sample cylinders were 38 ± 0.5 mm in diameter, 12 ± 0.5 mm in length.

For the averaged macroscopic section of capillary-porous media the basic electroosmosis equation [2] is:

$$\frac{V_{eo}}{I} = \frac{\varepsilon \cdot \beta}{4\pi \cdot \eta \cdot \sigma} \qquad (7.8)$$

where V_{eo} is the volumetric speed of electroosmosis flux under electric current I, ε is the dielectric permittivity of the fluid, β the electrokinetic potential including double layer potential ξ, η the viscosity and σ the specific electric conductivity of the electrolyte.

7 Electrokinetics for Groundwater Study

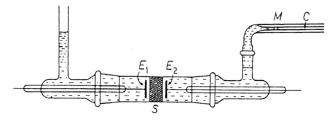

Fig. 7.1 Conventional device for the study of electroosmosis

Fig. 7.2 The device specifically built

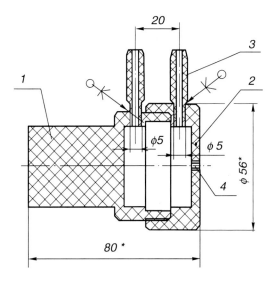

The V_{eo}/I ratio is the electroosmosis transportation. It can be experimentally measured and includes information about β [2].

The sketch of a conventional electroosmosis study device is shown in Fig. 7.1. Such device is not suitable for water based solutions, because of the influence of gases produced by the electrolysis, as a consequence, a specific device has been built by the authors (see Fig. 7.2).

The sample is inserted in the holder (pos. 1) with the screw cap (pos. 2). The fluid is fed through the hose connectors (pos. 3). The electrodes are located inside the connectors. Thus, the gases are not able to reach the measuring capillary, that is connected by a flexible hose to the hole in the center of the cap (pos. 4). Our device is equipped with a feed tank and measuring capillary installed higher than the holder, as well as the device shown in Fig. 7.1.

It must be noted that bases, i.e. KOH, are preferable than salts, such as NaCl, because they do not form the chlorine gels nearby the anode.

The experiments show a permanent fluid flux through the cemented sandstone. The ratios between the averaged water transportation (ml) due to electroosmosis

Table 7.1 Ratio between electroosmosis water transportation (ml) and electric charge (mA·s) for the presented experiments

Sample #	Electroosmosis	Potential, V
1660–93	0.169 · 10⁻⁴ ml/mA·s	260÷360
5138–92	0.250 · 10⁻⁴ ml/mA·s	200÷240
1812–93	0.100m · 10⁻³ ml/mA·s	80÷200
1796–93	0.074 · 10⁻³ ml/mA·s	160÷200

Table 7.2 Dried and water saturated weighing of the samples

Sample #	Water saturated weight, kg (m)	Dry weight, kg	Water weight in the sample, kg (m_g)	$n = m_g/m \cdot 100$, %
1660–93	0.033	0.0328	0.0002	0.6
5138–92	0.0316	0.0306	0.001	3.2
1812–93	0.0336	0.0314	0.0022	6.55
1796–93	0.0358	0.034	0.0018	5.03

and the electric charge (mA·s), obtained in the presented experiments, are shown in Table 7.1.

The porosity of samples was assessed by their dried and water-saturated weighing (see Table 7.2).

7.3 Electroosmosis and Permeability Index

Let us apply the Darcy law for our experiment. The fluid flux volumetric speed through a porous sample can be expressed as follows:

$$V_0 = k_{por} \frac{\Delta P S}{l \eta} \qquad (7.9)$$

where S is the section of the sample (m²), l is the sample length (m), ΔP is the difference of pressure at the sample edges (Pa), η is the fluid viscosity (Pa·s) and kpor is the porosity index.

The acting pressure of the fluid is:

$$P = \frac{F}{S} = \frac{mg}{S} = \frac{mg \cdot l}{S \cdot l} = \frac{W}{V_{sample}} \qquad (7.10)$$

where F is the acting force (N), m is the fluid mass (kg), g is the acceleration of gravity (m·s⁻²), W is the potential energy (J) and V_{sample} is the sample volume (m³). Expressing the difference of pressure in terms of energy, the Darcy law can be written as:

$$V_0 = k_{por} \frac{\frac{W}{Sl}}{l \eta} S = k_{por} \frac{W}{l^2 \eta} \qquad (7.11)$$

7 Electrokinetics for Groundwater Study

To express the potential energy by electric parameters we can use the mechanical power as a function of the energy $P = W/t$ (J/s) and electrical power $P = U \cdot I$ (V·A). Thus, the energy is $W = U \cdot I \cdot t$ (J), where t is the time of fluid flux. Considering the energy conservation law we can now write the Darcy law in terms of electric energy:

$$V_0 = k_{por} \frac{U \cdot I \cdot t}{l^2 \eta} \tag{7.12}$$

So, the basic equation of electroosmosis becomes:

$$\frac{V_0}{I} = k_{por} \frac{U \cdot t}{l^2 \cdot \eta} \tag{7.13}$$

Here, the volume of electroosmosis water flux is connected to the permeability index.

Since the energy is an additive physical parameter, the heating energy can be easily discriminated. The heat quantity that is needed to warm up the water is proportional to the water mass m and the temperature increment ΔT:

$$Q = c \cdot m \cdot \Delta T \tag{7.14}$$

where c is the specific heat, with $c = 4.19 \cdot J/(g \cdot K)$ for water. Considering the heating energy, the connection between electroosmosis and permeability is:

$$V_0 = k_{por} \frac{U \cdot I \cdot t - 4.19 \cdot m \cdot \Delta T}{l^2 \cdot \eta} \tag{7.15}$$

This relationship is essential for samples, while for in-situ works the heating energy can be neglected, being of insignificant value.

7.4 Electrometric Soundings and Electroosmosis

The influence of direct electric current on the results of electric and elec-tromagnetic soundings was studied. A direct current of 10 A was imposed between the casings of two wells. The distance between wells was 200 m and the water table was 2.5 m. All the measurements were made nearby the negative pole. The current flow time has been 80 min.

The apparent specific electric resistivity diagram measured by the NEMFIS sensor is shown in Fig. 7.3. The direct current makes the apparent resistivity decrease.

The apparent resistivity gradient is equal to the fluid flux speed. The speed can be assessed as an average resistivity decrease speed and media relaxation speed after the direct current cut off. The time of formation of a steady-flow process is the time of equalization of the electric energy with the potential energy and is equal to the fluid flux time.

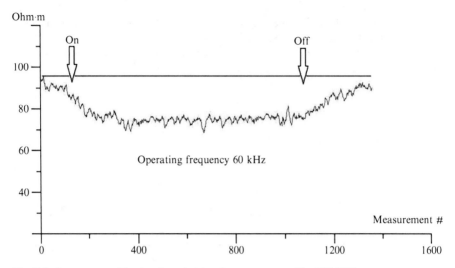

Fig. 7.3 Apparent specific electric resistivity diagram measured by NEMFIS

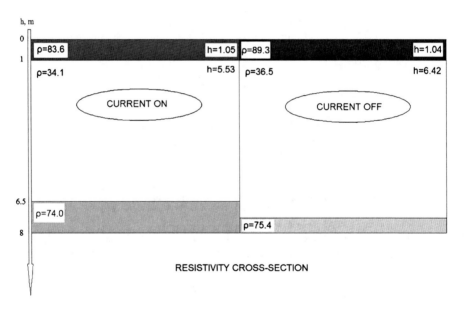

Fig. 7.4 Resistivity cross-section after 50 min of DC action

The DC electrical vertical sounding was performed before the direct cur-rent flow and during the steady-flow process (see Fig. 7.4). The data inversion in an horizontally layered model was performed. It is ascertained that after 50 min of direct current flow the soil becomes more conductive at the depth up to 6.5 m. The electric conductivity, as per S parameter which is the ratio of layer thickness to his specific resistivity, was increased up to 7%.

The increase of fluid volume in the system is the only one cause of the electric resistivity increase. Thus, both the experiments show the electric conductivity increase caused by electroosmosis fluid flux.

7.5 Results

The presented work shows the possibility of organization and control of the electroosmosis processes for large groundwater mass transportation. The electric conductivity change and the measurement of the volumetric speed of the fluid flux allow us to assess the water permeability of the media and permitted to introduce a new electric parameter: the electrokinetic potential.

References

1. Dobos D (1980) Electrochemical constants. Russian edition. Mir, Moscow, 365
2. Dukhin SS, Deryagin BM (1976) Electrophoresis. Nauka, Moscow, 332
3. Harned G, Owen B (1952) Physical chemistry off electrolytic solutions, Russian edition. Moscow, Izdatinlit, 628
4. Landau LD, Lifshiz EM (1982) Electrodynamics of continuum. Nauka, Moscow
5. Manstein AK, Epov MI et al (1999) Seismic velocities change under direct electric current. Geol I geofizika 40(3):465–473
6. Manstein AK, Epov MI et al (2000) Patent of Russian Federation No. 2152058 C1, G 01 V 3/10, dated 24.06.98. Method of induction frequency sounding. Russian patent agency, vol 18, p 1
7. Materials of 90-th convention of German Medical Doctors and Naturalists, Hamburg, September, 1928 Phys. Z. S. 29 751, 1928: I. M.Win (Jien). About digression of electrolytes from Ohm law; II. H. Ioos (Jien). Theoretic explanation of electrolytic conductivity from electric current voltage and frequency; III. E. Lange (Munich). New thermochemical and refractometry studies of strong electrolytes. http://www.ufn.ru/ru/articles/1929/3/e/

Chapter 8
Solid Particles Transport in Porous Media: Experimentation and Modelling

Hua-Qing Wang, Nasre-Dine Ahfir, Abdellah Alem, Anthony Beaudoin, Ahmed Benamar, Abdel Ghadir El Kawafi, Samira Oukfif, Samiara El Haddad, and Hui Wang

Abstract Solid particle detachment, transport and deposition in natural or artificial porous media have been the subject of an intense research effort in the last four decades. Particle-facilitated contaminants transport, accidents due to internal erosion in the hydraulic structures and permeability decreases of the oil wells, drinking water supply or artificial recharge of the aquifers, aroused a growing interest. In this study, results of two laboratory experimental systems for tracer tests in columns are presented. System 1 concerns step-input injection method where two studies were realized. The first study is devoted for studying deposition kinetics (K_{dep}) of the Suspended Particles (SP) and the second for evaluation the porous medium damage (clogging and release). However, system 2 concerns the pulse injection method whose aim was to study the SP deposition kinetics. The interpretation and analysis of the Break-Through Curves (BTCs) were obtained using the analytical and numerical solution of convection–dispersion equation (1D) including a source term (deposition and release term). Using system 2 results showed a decrease of the deposition kinetics coefficient with flow velocity until a critical velocity where K_{dep} decreases. For high injected volumes of the SP in system 1, the permeability decreases occurs throughout of the entrance of the porous medium.

Keywords Porous media • Solid particles transport • Deposition • Clogging • Modelling

---

H.-Q. Wang (✉) • N.-D. Ahfir • A. Alem • A. Beaudoin • A. Benamar • A.G. El Kawafi
• S. Oukfif • S. El Haddad • H. Wang
LOMC FRE CNRS 3102 – Université du Havre, 53 Rue de Prony, BP 540,
76058, Le Havre Cedex, France
e-mail: huaqing.wang@univ-lehavre.fr

8.1 Introduction

Solid particle detachment, transport and deposition in natural or artificial porous media have been the subject of an intense research effort in the last four decades. Particle-facilitated contaminants transport [15, 19], accidents due to internal erosion in the hydraulic structures [6] and permeability decreases (clogging due to particle deposition) of the oil wells [27], drinking water supply [11] or artificial recharge of the aquifers [7, 32, 33] aroused a growing interest.

Emerging pollution problems and water quality deterioration require an optimization of physical knowledge and a modelling of the phenomena of retention and transport of the pollutants in soil. Solid particles play a determining role in soils and aquifers contamination because of their adsorption capacity [25] and act as pollution carriers if they are transported easily in the flow, or in opposite, they remain a barrier to the migration of pollutants if their presence clogs the porous medium.

Clogging is an important process in both natural and engineered systems. It has been the subject of an intense research effort since the sixties of the last century [16, 18, 20]. Owing to the diversity and the complexity of the phenomena involved in porous media clogging, three mechanisms were retained: (i) physical clogging mechanisms restricting particles movement in soil and other porous media (straining and filtration) [26]; (ii) biological clogging as a result of the biological growth and the accumulation of by-products resulting from the decomposition of biological growth [8, 12, 17], and (iii) the chemical clogging, which controls the colloidal stability of the particles [9, 24].

Porous medium clogging is influenced by several factors: hydraulic conductivity [28], particles concentration of the influent [27], grain size distribution [27, 30] Grains geometry and surface are obviously the major parameters of clogging [5, 29, 31].

Experimental techniques are needed, in order to systematically investigate the influence of chemical and physical parameters on the transport and deposition of suspended particles in porous media. Such studies will ultimately contribute to a better understanding of its behaviour in subsurface environments. In literature, most of the studies were conducted in laboratory columns using step-input injections and the resulting breakthrough data were not evaluated in terms of deposition kinetics. Step-input injections have the disadvantage that relatively large amounts of particles are frequently introduced into the porous medium. As a result, blocking or filter ripening effects can occur and lead to severe changes in particle deposition rates with time [13, 22]. Recently, the pulse injection technique was conducted and showed a fast and accurate determination of particle deposition rates coefficients in porous media [21, 23, 34, 35]. Kretzschmar et al. [21] detailed the advantages of the pulse injection method on particle (colloid) deposition rates in laboratory columns.

In the present study, the transport and deposition of natural polydispersive suspended particles is explored. Two saturated granular porous media, under Darcy flow conditions, were used. Tracer tests were carried out by performing pulse and step-input injections in laboratory columns. The last two methods of injection let us study the deposition kinetics of the SP and the porous medium damage (clogging, release and permeability reduction). Analytical and numerical solutions of an advection-dispersion equation (1D), including a source term, describe the breakthrough of the SP. Hydrodynamic effects were considered, too.

8.2 Experimentation

Solid particles transport and deposition experiments were conducted using two different experimental systems (Fig. 8.1):
- System 1: column in vertical position packed with sand 1
- System 2: column in horizontal position packed with sand 2.

System 1 was used to study SP *deposition kinetics* and the *clogging* phenomena by the step input injection method. System 2 was used the pulse injection method for studying the *transport and deposition kinetics* of SP [2]. The injected suspended particles in the system 1 and system 2 consist of Kaolin P300 and silt, respectively. The experimental setup carried out used a Plexiglas column. The columns are fed by a reservoir containing water (pH of 6.80 ± 0.1) using a Master-Flex peristaltic pump (Cole-Parmer Instrument). A digital MacMillan numeric flow-meter is installed at the columns inlet to measure the flow rate. Pulse injections of the SP were performed with a syringe (system 2). For the system 1, the injected fluid consists of water with a fixed concentration of SP (Kaolin P300). A constant flow rate was applied from the upper (column inlet) to the down (column outlet). A series of tests were performed in order to observe the influence of the flow velocity on the transport and the clogging development in porous media. To monitor changes of the pressure in the porous medium during the tests, the column (system 1) was equipped with eight piezometers.

The detection system consists of a Fisher Instrument turbidimeter. The particle concentrations in the effluent were determined with the help of correlations made a priori between suspended particles concentrations in water and the turbidity. Table 8.1 summarises the material (porous media and SP) proprieties and the studied operating parameters.

In the conducted experiments the aim was the determination of the SP deposition kinetics; the injected volume of the suspension in the porous medium was limited to three pore volumes (Vp) for the step input injection method and 0.03Vp for the pulse injection method. However, in order to observe the clogging of the porous medium, the injected volume was $10 < Vp < 83$.

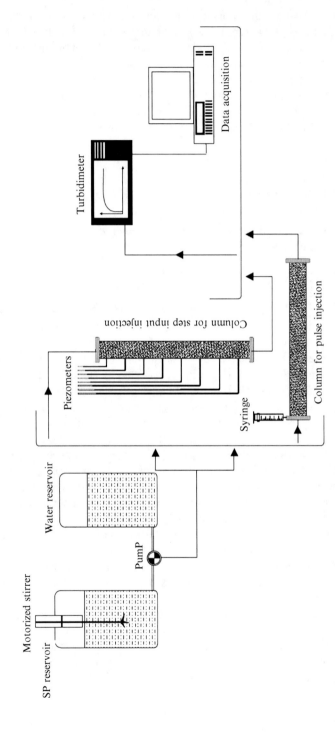

Fig. 8.1 Experimental set up (*arrows* indicate the direction of flow)

Table 8.1 Materials proprieties and the operating parameters

System 1			System 2		
Column	Length (mm)	400	Column	Length (mm)	330
	Diameter (mm)	45		Diameter (mm)	64
Sand 1	Range of the grains size (mm)	0.315–0.630	Sand 2	Range of the grains size (mm)	1.00–3.15
	d_{50} (mm)	0.410		d_{50} (mm)	2.35
	Porosity (%)	37		Porosity (%)	45
	Permeability (m/s)	2.1E-06		Permeability (m/s)	2.49E-02
SP 1: kaolin P300	C_0 (g/l)	1.00	SP 2: silt	C_0 (g/l)	0.10
	Size range (µm)	0.4–45		Size range (µm)	2–15
	Mean diameter (µm)	5.0		Mean diameter (µm)	6.0
Range of flow rate (ml/min)		40–200	**Range of flow rate (ml/min)**		40–930

8.3 Mathematical Modelling

Under steady-state and saturated flow conditions, the transport of particles through porous media can be described with a convection–dispersion equation (1D) including a source term including deposition and release kinetics [10]:

$$\frac{\partial\left[(\omega_0 - \omega_p)C\right]}{\partial t} = \frac{\partial}{\partial x}\left[D_L \frac{\partial(\omega_0 - \omega_p)C}{\partial x}\right] - \frac{\partial\left[u(\omega_0 - \omega_p)C\right]}{\partial x} - S(t) \quad (8.1)$$

$$S(t) = \frac{\partial(\rho_p \omega_p)}{\partial t} = q_{dep} + q_{rel} \quad (8.2)$$

where $S(t)$ is the source term, q_{dep} is the particle deposition term, q_{rel} is the particle release term, ρ_p is the bulk density of the particles [M·L^{-3}], ω_0 is the total porosity [L^3·L^{-3}], ω_p is the volume fraction of the deposited particles with respect to the total volume of the porous medium [L^3·L^{-3}], C is the particle concentration [M·L^{-3}], t is the time [T], x is the travel distance [L], D_L is the longitudinal dispersion coefficient [L^2·T^{-1}], and u is the pore flow velocity [L·T^{-1}].

Indeed, the difference $(\omega_0 - \omega_p)$ gives the effective porosity ω_c at a given time.

If the particle deposition was assumed to be irreversible, so their release is neglected. This assumption occurs under clean bed conditions (i.e., when the fraction of porous medium covered by suspended particles is small) and then the effective porosity is equal to the total porosity. This assumption is justified at sufficiently low particle concentration (i.e. no blocking or ripening) [13, 21]. Then, the source term becomes:

$$S(t) = q_{dep} = K_{dep}(\omega_0 - \omega_p)C \quad (8.3)$$

where K_{dep} is the deposition kinetics coefficient [T^{-1}].

- For experiments with *pulse injection method*, the initial and boundary conditions for semi-infinite medium are given by Eq. 8.4:

$$\left. \begin{array}{l} C(t=0, x) = 0 \\ C(t, x=0) = \dfrac{m}{Q} \delta(t) \\ C(t, x=\infty) = 0 \end{array} \right\} \quad (8.4)$$

where $\delta(t)$ is the Dirac function, m is the mass of the injected particles [M], and Q is the flow rate [L^3·T^{-1}]. The last boundary condition (C (t, $x=\infty$)$=0$) is not realistic for a limited column length, but necessary for providing an analytical solution.

- For experiments with *step-input injection method* the initial and boundary conditions for a semi-infinite medium are given by the Eq. 8.5:

$$\left. \begin{array}{l} C(t=0, x) = 0 \\ C(t, x=0) = C_0 \\ C(t, x=\infty) = 0 \end{array} \right\} \quad (8.5)$$

When the numbers of pore volumes injected in the porous medium are low (Vp<3), the porous medium is assumed to be homogeneous and the solution of Eqs. 8.1, 8.3 and 8.5 can be determined analytically. However, when the number of pore volumes injected in the porous medium is very high (Vp=83 in experiments dealing with clogging), the porous medium becomes heterogeneous, due to the deposition of particles. As a consequence, the porosity changes and release is not negligible. Then, terms q_{dep} and q_{rel} in Eq. 8.2 are written as follows:

$$\left\{ \begin{array}{l} q_{dep} = -K_{dep} \left(\omega_0 - \omega_p \right) C \\ q_{rel} = \left\{ \begin{array}{ll} \rho_p \cdot (\omega_0 - \omega_c) \cdot K_{rel} \cdot \left(\dfrac{\tau}{\tau_{cr}} - 1 \right) & \text{if } \tau > \tau_{cr} \\ 0 & \text{elsewhere} \end{array} \right. \end{array} \right. \quad (8.6)$$

where K_{rel} is the release kinetics coefficient [T^{-1}], τ is the shear stress in porous media [M·L^{-1}·T^{-2}], τ_{cr} is the critical shear stress above which the release of fine particles occurs [M·L^{-1}·T^{-2}].

Because of the nonlinearity of the different terms of the above equations, the use of numerical methods is required. In the present studies, the numerical method used is the "Particle Method".

The particle method is well adapted to solve correctly the problems dominated by the advection process. By using the dispersion velocity method, the transport equation (Eq. 8.1) is rewritten in the form of an advection equation [3], and will be written in a lagrangian framework yielding the discrete form:

$$\dfrac{dx_i}{dt} = u(t, x_i) + u_d(t, x_i) \quad (8.7)$$

8 Solid Particles Transport in Porous Media

$$\frac{d\Omega i}{dt} = \int_{|Pi|} S(t)ds \qquad (8.8)$$

where u_d is the dispersion velocity, Ωi is the numerical particle weight, and $|P_i|$ is the support of particles P_i. The Eq. 8.7 is an evaluation of characteristic lines, performed by using the explicit Euler scheme. The Eq. 8.8 describes the temporal evolution of numerical particle weight, estimated by means of a simple Euler scheme.

8.4 Results and Discussion

The mathematical model presented above was used to interpret the tracing experiments. The experimental Break-Through Curves (BTCs) obtained by the pulse injection of the suspended particles were fitted with the analytical solution of the Eqs. 8.1, 8.3 and 8.4. Fig. 8.2 shows comparisons between experimental data and those computed according to the model. Note that, the relative concentration C_R is expressed as:

$$C_R = \frac{CV_{Pm}}{V_{inj}C_0} \qquad (8.9)$$

where V_{pm} is the pore volume of the porous medium (L^3), and V_{inj} is the volume of each injection (L^3).

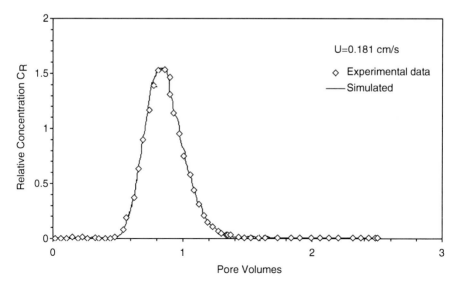

Fig. 8.2 Fitting of the experimental BTCs of the SP with the mathematical model for pulse injection method

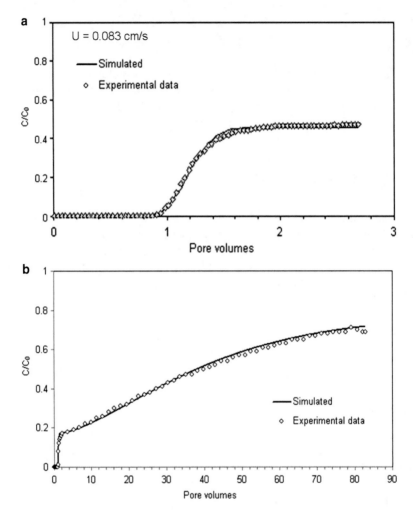

Fig. 8.3 Fitting of the experimental BTCs of the SP with the mathematical model for (**a**) step input injection method at low Vp injected (2.8 Vp), and (**b**) step input injection method at high Vp injected (83 Vp)

Figure 8.3 shows comparisons between experimental data from the step input injection method and those computed according to the model. As shown in Fig. 8.3 (a), the BTCs of the suspended particles when the number of pore volumes injected is low (Vp < 3) is well described by the analytical solution of the Eqs. 8.1, 8.3 and 8.5. When the number of pore volumes injected is high, the BTCs fitting is obtained by the numerical solution of the Eqs. 8.1, 8.5 and 8.6 as shown in Fig. 8.3b.

The investigations conducted within the framework of these studies show that, whatever the porous medium or the method of injection, a good fit of the experimental BTCs with those calculated by the model is obtained for all the flow velocities tested. Thus, the transport and the deposition parameters were determined.

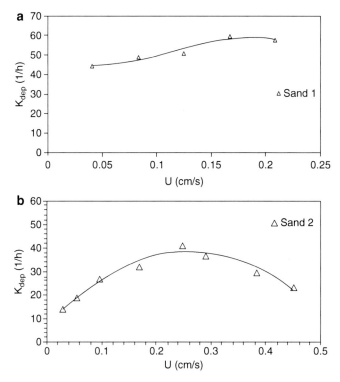

Fig. 8.4 Influence of flow velocity on deposition kinetics of the SP as a function of water flow velocity (U). (**a**) Step input injection method, and (**b**) pulse injection method

8.4.1 Influence of the Flow Velocity on the Deposition Kinetics

The SP deposition kinetics coefficient K_{dep} is determined by fitting the experimental BTCs with the analytical solution (Figs. 8.2 and 8.3a) of the convection–dispersion equation with first-order deposition kinetics (Eqs. 8.1 and 8.3).

In this study, the various tracer tests series carried out in the column packed with sand 1 (step input injection with Vp<3) and sand 2 (pulse injection) show that the deposition kinetics coefficient (K_{dep}) increases with Darcy's velocity (Fig. 8.4). This behaviour was already reported in literature [9, 21]. However, when the flow velocity is high (experiments with pulse injection), the obtained results showed that the deposition kinetics coefficient decreases beyond a critical velocity (Fig. 8.4b).

This behaviour is already observed by Benamar et al. [4] and Ahfir et al. [1] for silt particles (modal diameter of 14 μm) and the Rilsan particles (modal diameter of 25 μm). However, this behaviour is unusual, if compared to the results presented in literature concerning colloidal particles [14, 21, 22] and suspended particles [23, 34] where K_{dep} increases with the flow rate. The decrease of this parameter (K_{dep}), at high flow velocities, can be explained by the fact that the hydrodynamic forces dominate the gravitational forces applied on the suspended particles. The hydrodynamic forces play the double role of enhancing the transport rates (recovery rates) and reducing the particle deposition kinetics.

8.4.2 Porous Medium Damage

For the series of tests (system 1) where the number of the injected pore volumes were high (Vp > 10), five different flow rates were chosen for a fixed concentration (C_0 =1 g/l). The injection tests are carried out in the same porous medium and the volume of the injected SP equal to 83Vp. The porous medium (Sand1) clogging is analyzed on the basis of the recovery rate (retention) obtained and the changes in hydraulic characteristics of the porous medium during the injection. The time of tracer tests changed from 1 to 5 when the flow rate varies from 200 to 40 ml/min.

Therefore, it is preferable to present the evolution of the measured parameters (permeability) as a function of the number of the injected pore volumes (Vp) and not as a function of time. The two parameters (time and Vp) are related by the following equation:

$$Vp = \frac{Q.t}{\omega_0.V} \tag{8.10}$$

with V the total volume of the porous medium.

8.4.2.1 Permeability Reduction

Pressure measurements in the porous medium allow monitoring the spatial-temporal evolution of the porous medium permeability during the SP injection. To demonstrate the deposition consequence on the modification of the hydraulic characteristics of the porous medium, permeability reduction of the porous medium at the entrance of the column is plotted on Fig. 8.5.

For the flow rate 40 ml/min, the porous medium's permeability drop very quickly. After the injection of 83 pore volumes, permeability reduction was about 65% the initial permeability. Permeability reduces more slowly for the flow rate 200 ml/min. A reduction of 51% compared to the initial permeability is observed after the injection of 83 pore volumes. At high flow rates, SP deposition is low and more spread in the porous matrix. As a consequence, permeability decreases is less significant.

When the flow velocity is low, the deposition is important and focuses on a limited depth at the entrance of the porous medium. The permeability reduction is rapid and more pronounced. On the other hand, when the flow velocity is high, the deposition is smaller and more spread out in the porous medium. The clogging will be slower but thicker. The permeability reduction is less important.

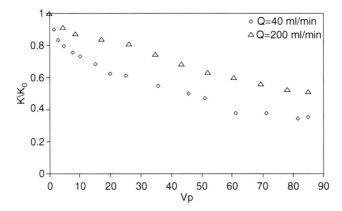

Fig. 8.5 Permeability reductions at the entrance of the column versus the number of pore volumes (Vp). K_0 and K are, respectively, the initial permeability and the porous medium permeability at a given time

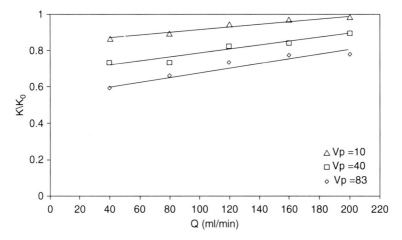

Fig. 8.6 Permeability reductions overall the porous medium for the different flow rates

Figure 8.6 indicates the evolution of the permeability as a function of the flow rate for different numbers of pore volumes injected. For a fixed volume injected, permeability increases linearly with the flow rate.

Also, it can be seen that the permeability decline due to particle deposition is not only a function of the injected pore volume, but also depends on the flow rate that controls the deposition kinetics. At higher flow rates, clogging is reduced and the overall permeability remains higher than at low flow rates. The decrease of permeability can be assessed with the following equation:

$$\frac{K}{K_0} = \alpha_1 Q + \alpha_2 \tag{8.11}$$

where α_1, α_2 are constants, whose value depends on the volume injected, and can be written as follows:

$$\alpha_1 = 6.10^{-6} \text{ Vp} + 0.0008$$
$$\alpha_2 = -0.00346 \text{ Vp} + 0.84$$

Knowing the concentration of particles and the number of pore volumes injected, the above relations can be used to evaluate the permeability decline of the porous medium.

8.5 Conclusions

Some issues of the transport and deposition of suspended particles (large size distribution) in saturated granular porous media were investigated. The breakthrough curves of the pulse and step input injections were competently described by the convection–dispersion equation with a source term (deposition or deposition-release kinetics).

The deposition kinetics increases with the flow velocity. At high flow velocities it decreases beyond a critical velocity. The decrease of this parameter, at high flow velocities, can be explained by the fact that the hydrodynamic forces dominate the gravitational forces applied on the suspended particles. The hydrodynamic forces play the double role of enhancing the transport rates (recovery rates) and reducing the particle deposition rates.

Permeability reduction takes place mostly in the inlet of the injected column. The permeability decline is affected by flow rate and the injected number of pore volumes (SP concentration).

The models presented were capable of simulating the transport, deposition and release of suspended particles in columns packed with porous media.

References

1. Ahfir N-D, Wang HQ, Benamar A, Alem A, Massei N, Dupont J-P (2007) Transport and deposition of suspended particles in saturated porous media: hydrodynamic effect. Hydrogeol J 15:659–668
2. Ahfir N-D, Benamar A, Alem A, Wang HQ (2009) Influence of internal structure and medium length on transport and deposition of suspended particles: a laboratory study. Transp Porous Med 76:289–307
3. Beaudoin A, Huberson S, Rivoalen E (2003) Simulation of anisotropic diffusion by means of a diffusion velocity method. J Comput Phys 186:122–135
4. Benamar A, Wang HQ, Ahfir N-D, Alem A, Massei N, Dupont JP (2005) Inertial effects on the transport and the rate deposition of fine particles in a soil. C R Geosci 337:497–504
5. Bhattacharjee S, Ko C-H, Elimelech M (1998) DLVO interaction between rough surfaces. Langmuir 14:3365–3375

6. Bonelli S, Brivois O, Borghi R, Benahmed N (2006) On the modelling of piping erosion. C R Mecanique 334:555–559
7. Bouwer H (2002) Artificial recharge of groundwater: hydrogeology and engineering. Hydrogeol J 10:121–142
8. Brovelli A, Malaguerra F, Barry DA (2009) Bioclogging in porous media: model development and sensitivity to initial conditions. Environ Modell Softw 24:611–626
9. Compere F, Porel G, Delay F (2001) Transport and retention of clay particles in saturated porous media: influence of ionic strength and pore velocity. J Contam Hydrol 49:1–21
10. Corapcioglu MY, Jiang S (1993) Colloid facilitated groundwater contaminant transport. Water Resour Res 29(7):2215–2226
11. Detay M (1993) Le Forage d'Eau: Réalisation, Entretien, Réhabilitation. Masson ed., Paris, 379 p
12. Dupin HJ, McCarty PL (1999) Mesoscale and microscale observations of biological growth in a silicon pore imaging element. Environ Sci Technol 33(8):1230–1236
13. Elimelech M, Gregory J, Jia X, Williams RA (1995) Particle deposition and aggregation: measurement, modeling, and simulation. Butterworth-Heinemann, Oxford
14. Grolimund D, Elimelich M, Borcovec M, Barmettler K, Kretzschmar R, Sticher H (1998) Transport of in situ mobilized colloidal particles in packed soil columns. Environ Sci Technol 32:3562–3569
15. Grolimund D, Borkovec M, Barmettler K, Sticher H (1996) Colloid facilitated transport of strongly sorbing contaminants in natural porous media: a laboratory column study. Environ Sci Technol 30:3118–3123
16. Guin JA (1972) Clogging of nonuniform filter media. Ind Eng Chem Fundam 11(3):335–349
17. Hand VL, Lloyd JR, Vaughan DJ, Wilkins MJ, Boult S (2008) Experimental studies of the influence of grain size, oxygen availability and organic carbon availability on bioclogging in porous media. Environ Sci Technol 42(5):1485–1491
18. Herzig JP, Leclerc DM, Le Goff P (1970) Flow of suspension through porous media: application to deep bed filtration. Ind Eng Chem 62:8–35
19. Kanti Sen T, Khilar KC (2006) Review on subsurface colloids and colloid-associated contaminant transport in saturated porous media. Adv Colloid Interface Sci 119:71–96
20. Kehat E, Lin A, Kaplan A (1967) Clogging of filter media. Ind Eng Chem Process Des Dev 6(1):48–55
21. Kretzschmar R, Barmettler K, Grolimund D, Yan YD, Borkovec M, Sticher H (1997) Experimental determination of colloid deposition rates and collision efficiencies in natural porous media. Water Resour Res 33:1129–1137
22. Kretzschmar R, Borkovec M, Grolimund D, Elimelech M (1999) Mobile subsurface colloids and their role in contaminant transport. Adv Agron 66:121–194
23. Massei N, Lacroix M, Wang HQ, Dupont JP (2002) Transport of particulate material and dissolved tracer in a highly permeable porous medium: comparison of the transfer parameters. J Contam Hydrol 57:21–39
24. Mays DC, Hunt JR (2007) Hydrodynamic and chemical factors in clogging by montmorillonite in porous media. Environ Sci Technol 41(16):5666–5671
25. McDowell-Boyer LM, Hunt JR, Sitar N (1986) Particle transport through porous media. Water Resour Res 22(13):1901–1921
26. McGechan MB, Lewis DR (2002) Transport of particulate and colloid-sorbed contaminants through soil, part 1: general principles. Biosyst Eng 83:255–273
27. Moghadasi J, Müller-Steinhagen H, Jamialahmadi M, Sharif A (2004) Theoretical and experimental study of particle movement and deposition in porous media during water injection. J Petrol Sci Eng 43:163–181
28. Reddi LN, Xiao M, Hajra MG, Lee IM (2005) Physical clogging of soil filters under constant flow rate versus constant head. Can Geotech J 42:804–811
29. Shellenberger K, Logan BE (2002) Effect of molecular scale roughness of glass beads on colloidal and bacterial deposition. Environ Sci Technol 36:184–189

30. Skolasinska K (2006) Clogging microstructures in the vadose zone – laboratory and field studies. Hydrogeol J 14:1005–1017
31. Tong M, Johnson WP (2006) Excess colloid retention in porous media as a function of colloid size, fluid velocity, and grain angularity. Environ Sci Technol 40:7725–7731
32. Vigneswaran S, Jeyaseelan S, DasGupta A (1985) Apilot-scale investigation of particle retention during artificial recharge. Water Air Soil Pollut 25:1–13
33. Vigneswaran S, Suazo Ronillo B (1987) A detailed investigation of physical and biological clogging during artificial recharge. Water Air Soil Pollut 35:119–140
34. Wang HQ, Lacroix M, Massei N, Dupont JP (2000) Particle transport in porous medium: determination of hydrodispersive characteristics and deposition rates. C R Acad Sci Paris Sci Terre Planèt 331:97–104
35. Weronski P, Walz JY, Elimelech M (2003) Effect of depletion interactions on transport of colloidal particles in porous media. J Colloid Interface Sci 262:372–383

Chapter 9
Challenges of Artificial Recharge of Aquifers: Reactive Transport Through Soils, Fate of Pollutants and Possibility of the Water Quality Improvement

Mohamed Azaroual, Marie Pettenati, Joel Casanova, Katia Besnard, and Nicolas Rampnoux

Abstract The unsaturated zone acts indeed as a natural reactive filter and can reduce or remove microbial and organic/inorganic contaminants through biogeochemical processes enhancing mass transfer between phases (soil – water – gases). The performance of the soil to purify the infiltrated water is based on both chemical, geo-biochemical and hydrodynamic coupled processes in a porous medium. The geochemical reactivity of soil minerals and the biodegradation of organic matter involving microbial mediated redox-reactions are the key reactions characterizing the water cleaning capacity of a soil. The reactive transport mechanisms induced by aquifer recharge using secondary or tertiary treated wastewaters still containing metals, metalloids and organic matter as pollutants is studied through laboratory and pilot experiments. This technology targets the geochemical reactivity and dynamics of soil to improve water quality while maintaining environment quality and protecting other resources (aquifers, agricultural production, soil, etc.). Obviously, the dilemma to meet these both constraints becomes a real challenge. This study aimed to develop a general concept based on the control of the physical, chemical and microbial keys processes easy to integrate in the numerical predictive and quantitative tools. The reactive transport modeling is carried out in order to identify the relevant processes controlling the filtration capability of the soil. Some results of ongoing projects based on the understanding of reactive transport processes will be presented. The technologic challenges emerged from the environmental safety issue and from the artificial recharge study will be discussed. Artificial groundwater recharge of aquifers by percolation through the unsaturated zone (UZ) is a technique

M. Azaroual (✉) • M. Pettenati • J. Casanova
BRGM, Water Division, 3 avenue Claude Guillemin BP 36009,
Orléans Cedex 2 45060, France
e-mail: m.azaroual@brgm.fr

K. Besnard • N. Rampnoux
Veolia Environnement, 10 rue Jacques Daguerre, Rueil-Malmaison, Paris 92500, France

to enhance the water quality for drinking water supplies. The performance of the UZ to purify the infiltrated water is based on chemical, geobiochemical and hydrodynamic coupled processes in a porous medium. The geochemical reactivity of soil minerals and the biodegradation of organic matter involving microbial mediated redox-reactions are the key reactions characterizing the epuration capacity of a soil. In order to improve our understanding of the physical and chemical phenomena controlling the efficiency of such process, a series of projects in a coastal aquifer in south-eastern France are built between Veolia and BRGM. The projects are based on the integration of numerical simulations with calibrated parameters on laboratory, pilot experiments and field aquifer characterization. The site characterizations and numerical simulations tend to show the development of "filtrating zones" by combination of various physico-chemical and thermokinetic processes. On the other hand, the mixing between infiltrating recharge waters and seawater can have important impact on the dissolution of carbonate minerals and precipitations of sulphate minerals. The results will be extrapolated to the real (industrial) system to elaborate exploitation scenarios and sensitivity analysis.

Keywords Unsaturated zone • Artificial recharge of aquifers • Reactive transport modeling • Microbial processes

9.1 Introduction

Climate change, population growth and economic development may be behind the acute tension over water resources, particularly in arid and semi-arid areas. In these areas, recourse to new sources of water has to be considered in order to safeguard the supply of water for drinking, irrigation and domestic using. Artificial groundwater recharge (AGR) using treated wastewater effluent accomplished by infiltration through the continuum unsaturated zone – saturated zone (UZ – SZ) appears as one of major solution to the recurrent issue of water scarcity. Artificial groundwater recharge is potentially a pragmatic way to restore water quality in the environment allowing to: (i) maintain necessary water supply levels, or store water, (ii) alleviate salt water intrusion into costal aquifers, (iii) preserve water from evaporation as opposed to storage in dams, and (iv) make it available at any time for all needs [26].

Reuse of treated wastewater through groundwater recharge has emerged as an integral part of water and wastewater management in many countries ([33], and references therein). During most of the twentieth century, wastewater treatment emphasized pollution abatement, public health protection, and prevention of environment degradation through removal of biodegradable material, nutrients, and pathogens [17]. Direct non-potable water reuse is currently the dominant mode for supplementing public water supplies for irrigation, industrial cooling systems, river flow augmentation, and other applications [4].

Management Aquifer Recharge (MAR) systems, such as Riverbank Filtration (RBF), Soil Aquifer Treatment (SAT), Sprinkling Infiltration (SI), and Vadose-Zone Injection (VZI) processes [6, 14, 18, 24, 28, 9] are widely used for projects of drinking water quality enhancement using source water that might be impaired by wastewater discharge. This study aimed to investigate the efficiency of the UZ treatment systems to improve the quality of treated wastewater through RBF, SAT, SI, and VZI concepts.

9.2 The Continuum Unsaturated – Saturated Zone Concept

Figure 9.1 shows the overall dynamic of the continuum UZ – ZS through the soil toward the deep aquifers (adapted from [21]). This is a comparison between a natural system of developed and mature soil, which provides the natural recharge of the aquifer and the severely disrupted soil by intense agricultural activity introducing new inputs, especially chemicals and solid particles. Agricultural practices introduce fertilizers and pesticides, herbicides, etc. Similarly, the artificial recharge introduces reactive chemicals (organic and inorganic) leading to thermodynamic disequilibria in the vadose zone and the underground aquifer. In the latter case, there is a strong disturbance producing gas exchange with the surrounding atmosphere and allowing the leaching or the fixation of pollutants. These new inputs can be nutrients for the soil microorganisms that contribute heavily to the kinetics of redox reactions and impact the physical chemistry of unsaturated porous media (water and gas phases), product chemicals (CO_2, N_2, NH_4, …) and pH.

Indeed, these microorganisms can have a fundamental role in the degradation of organic pollutants and the catalysis of a complex network of redox reactions that may be beneficial for the fixation of metal pollutants or in contrast inducing a negative effect in mobilizing these pollutants.

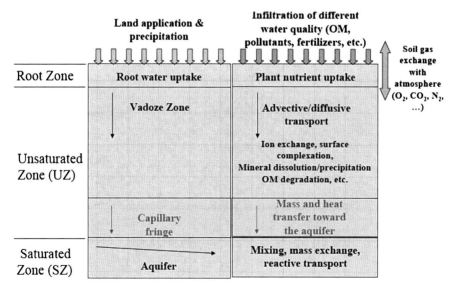

Fig. 9.1 Conceptual model of land application and unsaturated zone – saturated zone continuum ensuring the reactive transfer of mass and heat (modified from [21])

Generally, these microorganisms are concentrated in the capillary fringe by turning it into a true reactive barrier and an effective filter for pollutants. The extent of the capillary fringe depends on petrophysical properties of the soil and the relative humidity of the surrounding atmosphere or local climatic conditions. Finally, the contact of the artificial recharge water with aquifer native water, having contrasted physico-chemical properties, generates water mixtures that can interact strongly with the soil pore matrix.

Figure 9.2 shows the physical processes and physico-chemical key parameters that should be taken into account when the unsaturated zone is concerned by environmental issues, vulnerability of deep aquifers or even artificial recharge of underground aquifers. The figure on the right (from [27]) describes the main coupled phenomena that must be taken into account in an experimental study and numerical modelling of flow and reactive transfer through UZ. The following major phenomena are of concern:

- Dynamics of a gas phase (dissolution in the aqueous phase, degassing and boiling [32], transfer and mobility in the porous medium, …)
- Solid phases (dissolution/precipitation of mineral, ion exchange reactions, surface complexation reactions and adsorption/desorption of chemicals, …)
- Liquid phase (transfer of reactive and non reactive chemicals, …)
- Microorganisms and bacteria (geobiochemical catalysis reactions, biodegradation of organic matter, …)

Fresh et al. [11] have nicely schematized the dynamics of the unsaturated zone, which offers a very large exchange surface and brings into contact water and air which can interact and induce geochemical reactions or fluid flow strongly fractionated by mass exchanges. These phenomena generate heterogeneities and isolated aggregates communicating with the major fluid flowing paths only by diffusive phenomena and sometimes by evaporation processes.

9.3 Relevant Redox Couples and Geochemical Reactions in Soil and Aquifer Pore Waters

Figure 9.3 presents two families of half-reactions which can be triggered under recharge water percolating through the soil or the aquifer. Organic matter is often the main driver of direct or indirect reactions. The first family of reactions is the group of electron acceptors (A) mainly predominant in aerobic environments (oxic). The second family corresponds to the geochemical reactions involving electron donors (D). In function of concentrations of electron donors/acceptors in recharge water and vadose zone, the half-reactions in Fig. 9.3 will be associated in a global response to adenosine triphosphate (ATP) production from the adenosine diphosphate (ADP), by the reaction at the top of the figure, with the transfer of electrons between the electron donors (D) and electron acceptors (A). These reactions are catalyzed by bacteria enzymatic processes. The kinetics of these reactions network must be understood allowing assessing the performance of a recharge system using treated wastewater. Indeed, several redox

9 Challenges of Artificial Recharge of Aquifers

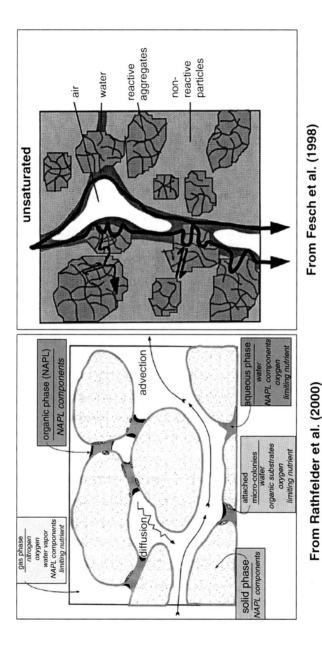

Fig. 9.2 Conceptual representation of the soil and fluid systems [27] and schematic illustration of an unsaturated zone [11]

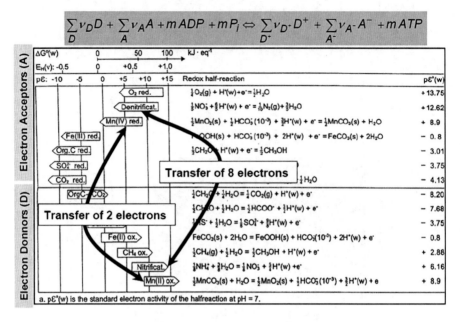

Fig. 9.3 Representative redox half reactions for soil and aquifer waters and their standard electron activity at pH = 7, T = 25°C and P = 1 atm

processes are relevant for artificial recharge of aquifers. Figure 9.3 shows the most relevant half reaction for reduction (electron-accepting) and for oxidation (electron-donating) and their corresponding standard electron activity at pH 7 [8].

A reduction half reaction of the upper part can, from a thermodynamical point of view, be combined with any oxidation reaction, if the pε of the reduction half reaction is higher than the pε of the oxidation half reaction (Fig. 9.3).

Figure 9.4 describes the order of consumption of electron acceptors in the infiltration of water rich in organic matter (electron donor) that can generate redox buffer zones (if the content in acceptor/electron donors is large enough and if the reaction kinetics is fast). In this case, the technology in MAR based on the principle of recharge through SAT, RBF, etc. technologies, becomes attractive to improve the quality of recharge water. From a thermodynamic point of view the reactive column is generally divided into three successive redox zones: Oxic (7 < pe < 12) suboxic (2 < pe < 7) and anoxic (−6 < pe < 2) for a pH value of 7.

9.4 The Soil Unsaturated Zone (UZ) Dynamics

The mobility of water and solutes in the unsaturated zone occurs through a complex continuum of pores ranging in size and shape [13]. An understanding of physical vadose zone processes is, therefore, an essential prerequisite for MAR concepts.

9 Challenges of Artificial Recharge of Aquifers

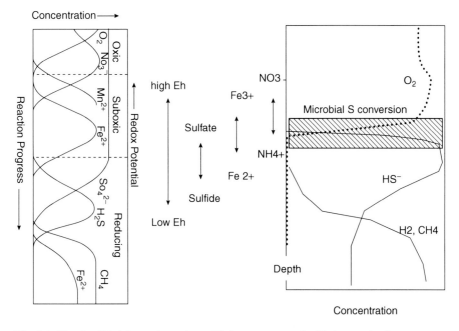

Fig. 9.4 The simplified thermodynamic equilibrium sequence of oxidation – reduction processes potentially operating during artificial recharge of aquifers (after [2])

The dynamics of this medium is controlled by physical, hydraulic, and geochemical disequilibrium conditions between large pores/fractures and the surrounding soil matrix (Fig. 9.1). Various coupled processes that exist between biogeochemical reactions, mass and heat transport processes, in particular gas phase transport, determines the structure and composition of soil minerals. The vadose zone is actively contributing in the dynamics of greenhouse gas emissions (agricultural activity, etc.). To explore quantitatively these feedback processes, many processes (multicomponent gas diffusion/advection, geobiochemical reactions, degassing/boiling, etc.) have been recently integrated in some reactive transport numerical modeling tools as STOMP [34] and MIN3P [22]. The integration of these processes for understanding the functioning of soil, which is the most reactive zone between the atmosphere and the subsurface geology, is a prerequisite to protect the environment, water, air and human life quality and biodiversity.

9.4.1 Reactive Transport Modeling in the Unsaturated Zone

Below the water table, in saturated zone (SZ), oxygen transport is supplied by purely advective flow whereas in unsaturated zone (UZ) gaseous diffusion may convey a much larger flux of oxygen. Pyrite oxidation depends strongly on moisture conditions ([5, 16], and references therein). In order to better understand the complex

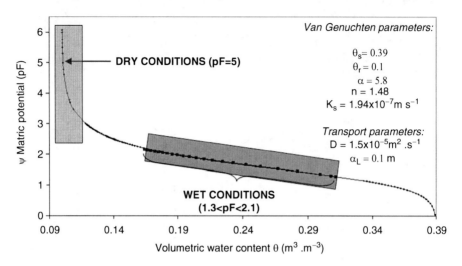

Fig. 9.5 Van Genuchten soil water retention curve for sandy clay loam (the saturation curve characteristics are taken from [7])

mechanisms of redox-reaction in UZ, we tried to determinate the impact of variably moisture conditions on pyrite oxidation in fictive sandy clay loam using MIN3P [20]. Solutions chemistry of the system is derived from the field data of Massmann et al. [19], see Tables 2 and 3. Figure 9.5 represents the typical van Genuchten [31] soil water retention curve of a fictive sample which is used to define case studies simulating reactive transport of chemicals through the unsaturated zone. The soil material considered is a sandy clay loam with two extreme water saturation conditions. The first case corresponds to the sandy clay loam which is completely unsaturated with very low water content simulating the dry conditions (Fig. 9.5). In contrast, the second case is based on the same soil material with high moisture content simulating the wet conditions (Fig. 9.5). For both cases, oxygen transport is essentially by diffusion mechanisms.

To highlight the reactivity of the UZ domain a simplified but very reactive mineralogical system composed of pyrite, ferrihydrites and calcite is considered. All mineral reactions are kinetically controlled. The MIN3P software [20] integrating the shrinking core model of Nicholson et al. [23] is used. This model is relevant to take into account the effect of the accumulation of alteration products on the mineral surface when the pyrite is oxidized by dissolved oxygen. The schematic representation of such complex reaction is often described as following:

$$FeS_2 + 3.5\ O_2\ (aq) + H_2O \rightarrow 2SO_4^{2-} + 2H^+ + Fe^{2+} \quad (9.1)$$

A rate expression based on the shrinking core model can be expressed as:

$$R_i^m = -10^3\ S_i\ \frac{r_i^p}{(r_i^p - r_i^r)r_i^r}\ \frac{D_{il}^m}{v_{il}^m}\ [O_2(aq)] \quad (9.2)$$

In this rate expression 10³ is a conversion factor [L.m⁻³], S_i is scaling factor including the tortuosity of the surface coating or altered rim on the mineral surface, r_i^p is the radius of the particle, r_i^r is the radius of the unreacted portion of the particle, D_{il}^m is the free phase diffusion coefficient of the primary reactant in water (in this case $O_{2(aq)}$) and υ_{il}^m is the stoichiometric coefficient of oxygen in the reaction Eq. (9.1). All parameters values are taken from the work of Massmann et al. [19].

Ferrihydrite and calcite are also considered as precipitating minerals in the UZ with another type of kinetic calculation:

$$CaCO_3 + H^+ \rightarrow HCO_3^- + Ca^{2+} \qquad (9.3)$$

$$Fe^{3+} + 3H_2O \rightarrow Fe(OH)_3 + 3H^+ \qquad (9.4)$$

$$R_i^m = -k_i^{m,eff}\left[1-\left(\frac{IAP_i^m}{K_i^m}\right)^m\right]^n \qquad (9.5)$$

Where $k_i^{m,eff}$ is the effective reaction rate constant (rate constant*reactive surface), IAP_i^m is the ion activity product and K_i^m is the equilibrium constant for the reaction [15].

9.4.2 Reactive Transport of Gases in the Soil

Results show a great difference between the two water saturation conditions (climate conditions). In dry case, O_2 is much more conveyed in soil profile than in wet case (Fig. 9.6). The consequence is the high concentration of sulphate and iron III due to oxic conditions at the surface of the profile and pH decrease due to much pronounced pyrite acidifying oxidation (Fig. 9.7) in dry context.

When water content is high, the mode of O_2 transport is essentially governed by advective transport of dissolved O_2 than diffusion. Protons are produced by pyrite oxidation and the enhancing of reaction by O_2 diffusion in dry conditions causes the strong acidification of pore solution. Transport of oxygen towards pyrite may become the limiting factor for pyrite oxidation and acidification depending strongly on moisture conditions. Because calcite is present in the system, pH is partially buffered by calcite dissolution (Eq. 9.3) and it leads to ferrihydrites stabilization in dry condition. This simulation shows that when the pyrite is subjected to wet conditions or dry conditions, the water chemistry is strongly affected. This study investigates the fate of O_2 diffusion in unsaturated media with different humidity conditions. The change of redox conditions with the low water content of dry media can have dramatic consequences of pore solution chemistry. This result is consistent with those of Greskowiak et al. [12], who find that the abrupt change from saturated to

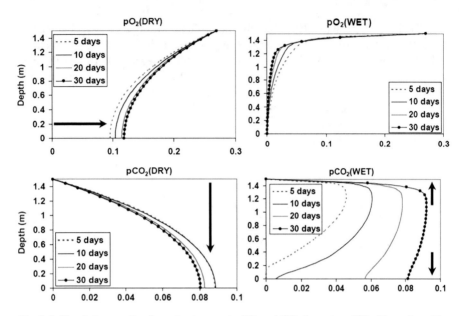

Fig. 9.6 Simulation results of reactive transport of O_2 and CO_2 through an UZ with pyrite oxidation and ferrihydrite precipitation (under dry and wet climate conditions)

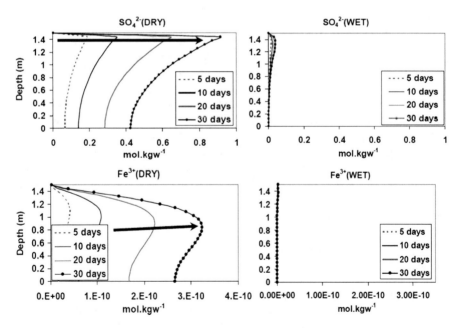

Fig. 9.7 Simulation results of reactive transport of SO_4^{2-} and Fe^{3+} through an UZ with pyrite oxidation and ferrihydrite precipitation (under dry and wet climate conditions)

unsaturated conditions causes the most significant geochemical changes, including the rapid oxidation of sulphides followed by the dissolution of calcite.

This study demonstrates the sensitive behaviour of the hydrochemical system, in particular the redox system, to the gas dynamics in the unsaturated zone. For example, the destabilization of ferrihydrites in dry conditions for pyrite-ferrihydrite system (no calcite for pH buffering), in contrast to the wet condition, is an important process in term of pollutants retention. Ferrihydrites are strong sorbents in nature because of their tendency to be finely dispersed and to coat other particles. In particular, they strongly adsorb arsenic and other inorganic compounds on their surface sites. In this case, ferrihydrites destabilization can conduct to the diminution of soil filter capacity towards inorganic pollutants.

9.5 Reactive Transport Through the Water Saturated Zone

Otherwise, the following objective of this study is to understand and describe the biogeochemical processes that taking place during artificial recharge with waste water near a coastal seawater intrusion without taking into account the complexity of hydraulic processes. In this way, calculations are performed using the geochemical computer program PHREEQC [25]. This code allows to easily implementing kinetics reaction rates by the user for both inorganic and organic processes. Values of equilibrium constants and sites densities for surface complexation reactions on Ferrihydrite are issued from literature [3, 10, 29]. The constants associated with cationic exchange are taken from PHREEQC database (file llnl.dat).

First, the methodology used to describe the waste water recharge system with vertical 1D PHREEQC column is to:

- Define recharge infiltration solution with chemical analysis of the measured infiltrating waste water composition
- Identify specific soil mineralogy using chemical and mineralogical (X-Ray diffraction) analyses of the experimental soil and potential mineralogical secondary phases from infiltration solution interacting with soil
- Determine composition of initial column pore water (i.e. before artificial recharge) assuming partial equilibrium with the soil mineralogy
- Assume a percolation velocity in order to performed the reactive coupled advective 1D transport.

  Two columns are performed:

- The first column does not integrate kinetically controlled redox reaction for organic matter degradation.
- The second column takes into account the succession of electron acceptation reactions based on Monod kinetics [30].

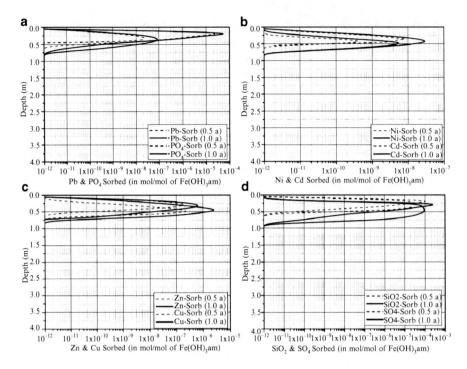

Fig. 9.8 Numerical simulation results of the accumulation on the first meter of Pb, PO$_4$, Ni, Cd, Zn, Cu, SiO$_2$ and SO$_4$ fixed on HFO surface sites along a 4 m reactive column of SAT system (under water saturation conditions)

9.5.1 Without Kinetic Degradation of Organic Matter: Equilibrium Approach (Column 1)

Figure 9.8 shows the evolution of the various physicochemical parameters and elements cited above. According to the hydraulic conditions applied to the system (see previous sections), the perturbation brought by the infiltration of waste water reaches a depth of 2.5 m after 1 year, as shown by pH (a) and pe (b) profiles, where "pe" is the electron activity. The pH of the solution progressively diminishes with time from the initial soil value to the waste water value. This evolution is proportional to the increase of pe, except at the redox front where pe decreases sharply until the top of the profile (b). This zone is the location of strongest perturbation by the waste water. Deeper, where pe increases softly from the bottom, the pe/pH ratio is kept constant, indicating a relative strong buffer potential of the soil given the high amounts of calcite.

At the top of the profile, the higher concentrations of DIC input by waste water are controlled by aragonite precipitation so that DIC infiltration is slightly retarded compared to pH and pe. The behaviour of sulphate results here in the return to equilibrium with barite, given the perturbation due to the input waste water and the involvement of Ba in exchange reactions within the clay fraction.

According to the rather reduced state of the system, Fe is essentially ferrous (amounts of dissolved ferric iron are lower than 10^{-8} mol/kgw in the whole profile during the whole simulation) so that ferric oxides cannot precipitate. This hinders provision of surface complexation sites for metallic pollutants retention. As dissolved metals concentrations are rather low, mineral precipitation of metal bearing minerals do not occur and only exchange reactions take place. When concerned, metals can be significantly retained by the clay fraction as is the case for Zn. But cations are adsorbed competitively and one may observe different behaviours depending on the affinity of the exchanger for each metallic cation. For instance, lead content in the infiltration water is much lower than Zn and its relative retention (i.e. compared to its input concentration) is lower. In the present work, Cr and Cu are extreme cases because they are not even concerned by cation exchange and they behave like tracers (not shown).

9.5.2 Kinetic Degradation of Organic Matter (Column 2)

This option is practically very computer time consuming and results presented here only concern the first 120 days of infiltration. But these results can give indications on the expected behaviour of the system at the scale of one year, and allow identification of differences compared to the "equilibrium" approach. They are shown in Fig. 9.9 and concern the evolution of the various physicochemical parameters and elements considered in the previous sub-section. After 30 days of infiltration, the perturbation from the waste water reaches a depth of 0.5 m as indicated by pH and pe. This depth is similar to that observed for the "equilibrium" approach after the same time period. But the amplitude of the perturbation seems smaller in the present case. This is also observed for DIC. The behaviour of sulphate is similar to that obtain in the "equilibrium" approach because here again it is not subjected to kinetic limitations. Iron is here again essentially in its reduced form according to the reducing conditions characterizing the system. The kinetically limited dissolution of green rust maintains a slightly higher concentration of dissolved iron at the top of the profile. At last, the behaviour of aqueous zinc is similar to that in the "equilibrium" approach since it is only controlled by cation exchange reactions which occur at equilibrium, i.e. with no kinetic limitations. In summary, differences between pH, pe and DIC responses to perturbation induced by the waste water have been pointed out only after 30 days of simulated infiltration. According to these observations, stronger discrepancies may be expected for simulations over longer time periods. Consequently, taking into account kinetic limitations for mineral dissolution/precipitation reactions is critical, particularly for artificial groundwater recharge modeling were hydrodynamics and chemical perturbations are much linked.

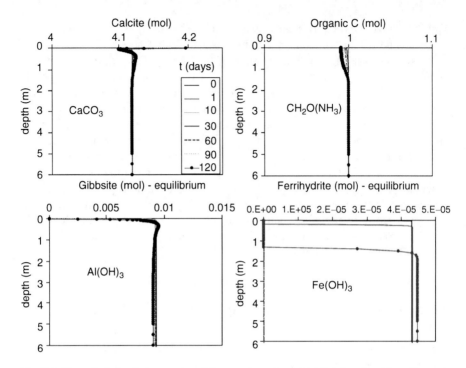

Fig. 9.9 Numerical simulation results of the reactive column highlighting the effect of organic matter oxidation on mineral profiles (calcite, Al(OH)$_3$ and Fe(OH)$_3$) along a 6 m reactive column of SAT system (under water saturation conditions)

9.6 Recharge of Coastal Aquifers

The next step concerns the horizontal 1D PHREEQC column used to describe geochemical processes occurring at seawater/freshwater interface. This column, allows describing de reactivity of zone between freshly infiltrated recharge water and aquifer and the "geochemical" capacity of mixing zone between aquifer freshwater and seawater (Fig. 9.10).

Chloride concentration shows the transport of the water ~40 m towards the coastline. Ca^{2+}, Mg^{2+} and Na$^+$ are high near the coastline showing the saline part of pore water in this area. A strong decrease of Ca^{2+} concentration is observed between 60 and 90 m. This decrease corresponds to the displacement of Mg^{2+}, K$^+$ and Na$^+$ derived from seawater, from the exchanger. This sequential cation displacement is in accordance with previous studies [1].

The depletion of Ca^{2+} in aqueous solution provokes calcite dissolution. A strong increase of alkalinity and pH is associated with this dissolution. This important modification of geochemical conditions at the seawater/freshwater interface such as pH and alkalinity has a strong influence on other minerals stability and consequently on porosity.

9 Challenges of Artificial Recharge of Aquifers

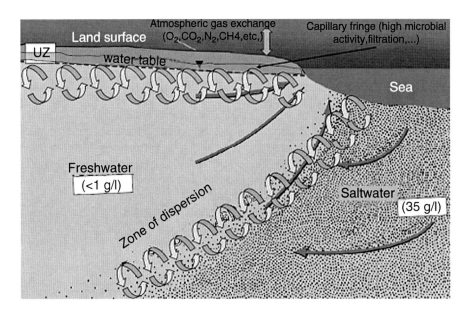

Fig. 9.10 Conceptual scheme of artificial recharge of coastal aquifers and seawater intrusion

Finally, in the context of coastal aquifers often overexploited can also target areas as potentially purifying mixture. In this case, the recharge water is characterized by a rather complex dynamics generated by the context of coastal aquifers. You can have a context with a capillary fringe well developed and rather reactive mixing zone between the recharge water still contains organic matter and other chemical reagents and water native to the aquifer. Meanwhile, water from the aquifer is constantly mixing with seawater cycle of active management based on the alternation period of overexploitation and artificial recharge resulting mobility of the mixing zone and the balance load can also generate mobile hotspots can create favourable reactions (purifying and fixing of pollutants) or unfavourable (mobilizer of pollutants). This type of aquifer vulnerability requires a fairly active management based on control of major geochemical processes of interaction at the borders of the aquifer.

9.7 Discussions

These preliminary results allow identification of key mechanisms which can possibly occur in a real system. From the aqueous solution point of view, the perturbation generated by the infiltration of waste water reaches a depth of about 3 m in a context of saturated porous medium. The cation exchanger shows different behaviours depending on the considered element, its concentration in the input waste water, and its implication within mineral reactions. The preliminary results of reactive transport simulations highlighted a strong reactivity of the ground receiving water infiltration. Indeed, depending on the availability of electron acceptors (O_2, NO_3, etc.) in the water infiltration as well as electron donors (($CH_2O(NH_3)$)), green-rust, hematite, pyrite, etc.) in the soil the system can develop very reactive redox zones, which could play the role of "real filters". This can also shape very favourable reaction paths to the fixing of metal pollutants and degradation of organic pollutants. In addition, the unsaturated zone water gives particularly beneficial dynamics for the oxygen allowing fast oxidation reactions. However, it is also important to study the impact of these reactions interaction and mass exchanges between phases (soil minerals – water – gases) which might develop permeability reduced to levels that could lead to a clogging of the soil. This clogging problem is the one of the critical need control to ensure the viability and sustainability of an industrial project of artificial recharge.

9.8 Conclusion

The understanding of key phenomena of reactive transport through the various subsystems of an artificial recharge site is a prerequisite for the industrial development of a viable and sustainable technology. Indeed, it is necessary to ensure the performance of the unsaturated zone to purify infiltrated water, generally of poor quality, while ensuring the adequacy of a continuum unsaturated zone – saturated zone performances. In addition, degradation of the quality of injected water has been postponed in some studies of artificial recharge when these phenomena were not controlled. In the sometime, the clogging problems are still the key of the success of an industrial exploitation of such technology.

We have initiated the development of a comprehensive methodology for treating the behaviour of a hydrogeochemical system simulating the artificial recharge conditions according to SAT, RBF, SI and VZI concepts. Field data compilation on mineralogical and chemical composition and the infiltrating water chemistry has helped to initiate a numerical simulation study of reactive transport phenomena. Obtained preliminary results highlighted several physicochemical phenomena potentially favourable to the creation of pollutant filtrating reactive redox zones. This study has confirmed the interest of the unsaturated zone in process water purification.

Consideration of flows in unsaturated porous medium is an essential aspect of understanding the artificial recharge system operation. This makes it possible to interpret, first on a simplified chemical system, the effect of water table rise up on the behaviour of redox couples. In this sense, the concept of cyclic alternating periods of intake of O_2 in the system and anoxic periods may also be approached to assess their effects on the dynamics of the system. More specifically, this dynamic may contribute to the preservation of iron hydroxides in the porous media and permits to create a complexation control (and hence filter) for trace and pollutant chemicals. Bacterial activity associated catalytic or inhibitor role of certain redox reactions are considered here with the Monod kinetics.

Acknowledgments This work is undertaken in the framework of the on-going multi-annual BRGM and "VEOLIA Environnement" partnership research and development projects (REGAL, Recharge).

References

1. Andersen MS, Jakobsen VNR, Postma D (2005) Geochemical processes and solute transport at the seawater/freshwater interface of a sandy aquifer. Geochim Cosmochim Acta 69: 3979–3994
2. Appelo CAJ, Postma D (2005) Geochemistry, groundwater and pollution, 2nd edn. A. A. Balkema Publishers, London, pp 649
3. Appelo CAJ, Weiden MJJVD, Tournassat C, Charlet L (2002) Surface complexation of ferrous iron and carbonate on ferrihydrite and the mobilization of arsenic. Environ Sci Technol 36:3096–3103
4. Asano T (1998) Wastewater reclamation and reuse, Water quality management library. CRC Press, Boca Raton, Pp 1570
5. Borek SL 1994 Effect of humidity on pyrite oxydation. In: Alpers CN and Blowes DW (ed) Environmental geochemistry of sulfide oxidation, ACS Symposium Series 550. American Chemical Society, Washington, DC, pp 31–44
6. Bouer H (2002) Artificial recharge of groundwater: hydrogeology and engineering. Hydrogeol J 10:121–142
7. Carsel RF, Parrish RS (1988) Developing joint probability distributions of soil water retention characteristics. Water Resour Res 24:755–769
8. Christensen TH, Bjerg PL, Banwart SA, Jakobson R, Heron G, Albrechtsen HJ (2000) Characterization of redox conditions in groundwater contaminant plumes. J Hydrol Contam 45:165–241
9. Dillon P (2009) Water recycling via managed aquifer recharge in Australia. Boletín Geológico y Minero 120(2):121–130
10. Dzombak DA, Morel FMM (1990) Surface complexation modeling – hydrous ferric oxide. Wiley, New York, pp 393
11. Fesch C, Lehmann P, Haderlein SB, Hinz C, Schwarzenbach RP, Flühler H (1998) Effect of water content on solute transport in porous medium containing reactive micro-aggregates. J Hydrol Contam 33:211–230
12. Greskowiak J, Prommer H, Massmann G, Johnston CD, Nützmann G, Pekdeger A (2005) The impact of variably saturated conditions on hydrogeochemical changes during artificial recharge of groundwater. App Geochem 20:1409–1426
13. Jardine GV, Wilson RJ, Luxmoore, J.P. Gwo, 2001, Conceptual models of flow and transport in the fractured vadose zone. U.S. National Committee for Rock Mechanics, National Research Council. National Academy Press, Washington, DC, pp 87–114
14. Johnson JS, Baker LA, Fox P (1999) Geochemical transformations during artificial groundwater recharge: soil-water interactions of inorganic constituents. Water Res 33:196–2006
15. Lasaga AC (1998) Kinetic theory in the earth sciences, Princeton series in geochemistry. Princeton University Press, Princeton, pp 811
16. Lassin A, Azaroual M, Mercury L (2005) Geochemistry of unsaturated soil systems: aqueous speciation and solubility of minerals and gases in capillary solutions. Geochim Cosmochim Acta 69:5187–5201
17. Levine AD, Asano T (2004) Recovering sustainable water from wastewater. Environ Sci Technol 208:201A–208A
18. Maeng SK, Ameda E, Sharma SK, Grützmacher G, Amy GL (2010) Organic micropollutant removal from waste water effluent-impacted drinking water during bank filtration and artificial recharge. Water Res XX:1–12

19. Massmann G, Greskowiak J, Dünnbier U, Zuehlke S, Knappe A, Pekdeger A (2006) The impact of variable temperatures on the redox conditions and the behaviour of pharmaceutical residues during artificial recharge. J Hydrol 328:141–156
20. Mayer KU, Frind EO, Blowes DW (2002) A numerical model for the investigation of reactive transport in variably saturated media using a generalized formulation for kinetically controlled reactions. Water Resour Res 38:1174–1194
21. Miller GR, Rubin Y, Mayer KU, Benito PH (2008) Modeling vadose zone processes during land application of food-processing waste water in California's Central Valley. J Environ Qual 37:43–57
22. Molins S, Mayer KU (2007) Coupling between geochemical reactions and multicomponent gas and solute transport in unsaturated media: a reactive transport modeling study. Water Resour Res 43:1–16
23. Nicholson RV, Gillham RW, Reardon EJ (1990) Pyrite oxidation in carbonate-buffered solution: 2. Rate control by oxide coatings. Geochim Cosmochim Acta 54:395–402
24. Nöjd P, Lindroos AJ, Smolander A, Derome J, Lumme I, Helmisaari HS (2009) Artificial recharge of groundwater through sprinkling infiltration: impacts on forest soil and the nutrient status and growth of Scots pine. Sci Total Environ 407:3365–3371
25. Parkhurst DL, Appelo CAJ (1999) User's guide to PHREEQC (version 2): a computer program for speciation, batch-reaction, one-dimensional transport, and inverse geochemical calculations. U.S. Geological Survey Water-Resources Investigations Report 99–4259, 312 p
26. Pyrne RDG (2005) Aquifer storage recovery – a guide to groundwater recharge through wells, 2nd edn. ASR systems, Gainsville, pp 608
27. Ratherfelder KM, Lang JR, Abriola LM (2000) A numerical model (MISER) fort the simulation of coupled physical, chemical and biological processes in soil vapor extraction and bioventing systems. J Hydrol Contam 43:239–270
28. Rauch-Williams T, Hoppe-Jones C, Drewes JE (2010) The role of organic matter in the removal of emerging trace organic chemicals during management aquifer recharge. Water Res 44:449–460
29. Swedlund PJ, Webster JG (1999) Adsorption and polymerisation of silicic acid on ferrihydrite, and its effect on arsenic adsorption. Water Res 33:3413–3422
30. Van Cappellen P, Wang Y (1996) Cycling of iron and manganese in surface sediments: a general theory for the coupled transport and reaction of carbon, oxygen, nitrogen, sulfur, iron, and manganese. Am J Sci 296:197–243
31. Van Genuchten M (1980) A closed form equation for predicting the hydraulic conductivity of unsaturated soils. Soil Sci Soc Am J 44:892–898
32. Visser A, Schaap JD, Broers HP, Bierkens MFP (2009) Degassing of ^{3}H/^{3}He, CFCs and SF$_6$ by denitrification: measurements and two-phase transport simulations. J Hydrol Contam 103:206–218
33. Westerhoff P, Pinney M (2000) Dissolved organic carbon transformations during laboratory-scale groundwater recharge using lagoon-treated wastewater. Waste Manag 20:75–83
34. White MD, Oostrom M (2000) STOMP subsurface transport over multiple phases: theory guide, PNNL-11216 (UC-2010). Pacific National Northwest Laboratory, Richland

Chapter 10
Geologic Storage of CO_2 to Mitigate Global Warming and Related Water Resources Issues

Yousif K. Kharaka and Dina M. Drennan

Abstract Global warming and resulting climate changes are arguably the most important environmental challenges facing the world in this century. Average global temperature is now approximately 0.8°C higher than during pre-industrial times, and is projected to increase by 2–6°C by 2100. Related climate changes with potential adverse environmental impacts may include: (i) sea-level rise from the alpine glaciers and ice sheets melting and from the ocean thermal expansion; (ii) increased frequency and intensity of floods, droughts, tropical storms and wildfires; and (iii) changes in the amount, timing, and distribution of rain, snow, and runoff. Results from global simulation modeling indicate that parts of both the southwestern USA and North Africa would experience increased drought frequency and water stress in the coming decades; model predictions indicate that precipitation in these two regions will decrease by about 20% by 2050. There is a broad scientific consensus that global warming and related climate changes are caused primarily by increased concentrations of atmospheric greenhouse gases (GHG), especially CO_2, emitted from the burning of fossil fuels. The amount of anthropogenic CO_2 currently added to the atmosphere is about 30 billion ton/year, and this could double by 2050. Capture and sequestration of CO_2 in depleted petroleum fields and saline aquifers in sedimentary basins is one plausible option to reduce GHG emissions and mitigate global warming. Currently there are five commercial projects that capture and inject about seven million tons of CO_2 annually; data from these, from enhanced oil recovery (EOR) operations and from pilot sites provide valuable experience for assessing the efficacy of carbon capture and sequestration. Detailed chemical and isotopic analyses of water, associated gases, and added tracers obtained from Frio field tests, Texas, proved powerful tools in: (i) tracking the successful injection and flow of CO_2 in the "C" sandstone; (ii) detecting some leakage of CO_2 into the overlying "B" sandstone; (iii) showing mobilization of metals and toxic organic compounds;

Y.K. Kharaka (✉) • D.M. Drennan
U.S. Geological Survey, Menlo Park, CA, USA
e-mail: ykharaka@usgs.gov

(iv) showing major changes in chemical and isotopic compositions of formation water, including a dramatic drop in calculated brine pH, from 6.3 to 3.0. Significant isotopic and chemical changes were also observed in shallow groundwater following CO_2 injection at the ZERT field site, Montana. These field tests indicate that highly sensitive chemical and isotopic tracers can effectively monitor injection performance and provide early detection of CO_2 and brine leakages into potable groundwater.

Keywords Geologic storage • GHG • Global warming • Water resources

10.1 Introduction

Fossil fuels–coal, oil, and natural gas–are essential sources of primary energy, currently supplying about 85% of the global energy [24]. Combustion of these fuels, however, releases large amounts of greenhouse gases (GHGs), primarily CO_2 to the atmosphere. Currently, close to 30 Gt/year of CO_2 are being added to the atmosphere from these sources, and the amount is projected to increase to 43 Gt/year by 2030 [24, 30]. Increased anthropogenic emissions of CO_2 have raised its atmospheric concentrations from about 280 ppmv during pre-industrial times to ~390 ppmv today, and based on several defined scenarios, CO_2 concentrations are projected to increase to up to 1,100 ppmv by 2100 [40, 74]. There is now a broad scientific consensus that current global warming (Fig. 10.1) and related climate changes are caused mainly by these atmospheric CO_2 increases. Surface temperature changes are highly variable, with the average global temperature being ~0.8°C higher now than during pre-industrial times (Fig. 10.1), and is projected to increase by 2–6°C by 2100 [40, 74]. Rising atmospheric CO_2 is also increasing the amount of CO_2 dissolved in ocean water, increasing its acidity (lowering its pH), with potentially disruptive effects on coral reefs, marine plankton and ecosystems [69].

Related climate changes with potential adverse impacts may include sea-level rise from the melting of alpine glaciers and polar ice sheets and from the ocean thermal expansion; increased frequency and intensity of wildfires, floods, droughts, and tropical storms; and changes in the amount, timing, and distribution of rain, snow, and runoff (Fig. 10.2). Morocco and North Africa have variable climates; from Mediterranean climate in the north near Spain to extreme aridity in the Sahara, to the semi-arid Atlas Mountains. Climate change, generally speaking, is a positive feedback system where wet climates get wetter and dry climates get drier [70]. Northern Africa is the most water-stressed region in the world with some transboundary basins between Morocco, Algeria, and Tunisia, and the large Nile basin. In the beginning of the twentieth century there was a drought event in North Africa about once every 10 years, whereas currently North Africa is experiencing five to six droughts every 10 years; the rain events are less frequent and more intense. A 1°C raise in temperature is predicted to decrease rainfall for the Saharan region by 10–20% (Fig. 10.2) [70].

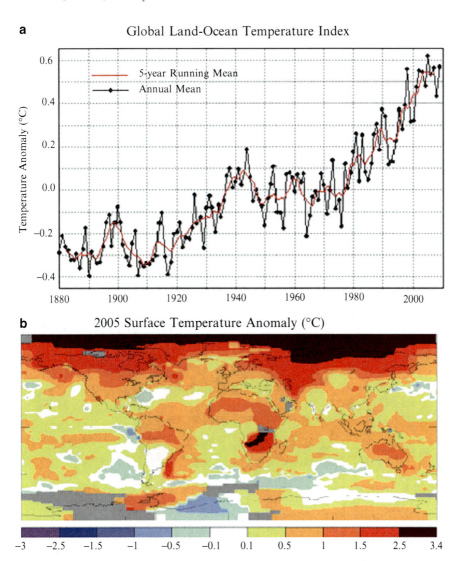

Fig. 10.1 (a) Annual mean and 5-year running mean global temperature index from 1880 to present, relative to 1951–1980 [28]; (b) surface temperature anomaly for 2005. Figures obtained on 7/23/2010 from http://data.giss.nasa.gov/gistemp/graphs/

Comparable precipitation decreases and other climate changes have also been observed in western North America, including California, and deep into Mexico (Fig. 10.2). The changes in the last decade include rising temperatures, declining late-season snowpack, northward-shifted winter storm tracks, increasing precipitation intensity, the worst drought since 1900, major declines in Colorado River reservoir storage, and sharp increases in the frequency of large wildfires. In the past decade, many locations, especially in the headwaters region of the Colorado River,

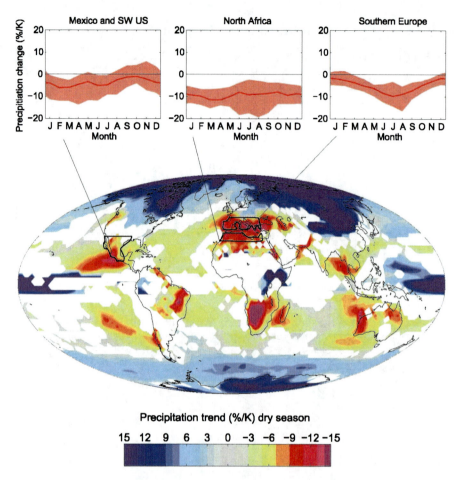

Fig. 10.2 Averaged percentage changes in global distribution in precipitation expected per degree of warming, relative to 1900–1950. Note the significant decreases in precipitation over North Africa and SW USA (Modified after Solomon et al. [67])

have been more than 1°C warmer than the twentieth century average [58]. This warming has been the primary driver in reducing late-season snowpack and the annual flow of the Colorado River [4]. These changes are consistent with projected anthropogenic climate change, but seem to be occurring faster than projected by the most recent [40, 41] climate change assessments.

Carbon dioxide sequestration, in addition to energy conservation and increased use of renewable and lower carbon intensity fuels, is now considered an important component of the portfolio of options for reducing greenhouse gas emissions to stabilize atmospheric levels of these gases and global temperatures at acceptable values [6, 40]. Sedimentary basins in general and deep saline aquifers in particular are being investigated as possible repositories for large volumes of anthropogenic CO_2 that must be sequestered to mitigate global warming [33, 44]. Currently there are five

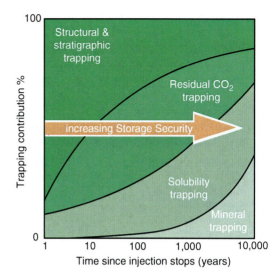

Fig. 10.3 Schematic diagram showing the proportions of CO_2 trapped over extended periods of time by various processes following injection. Note the general increases with time for solubility and mineral trapping (From Benson and Cook [6])

commercial projects operating worldwide that together capture and inject about seven million tons of CO_2 annually and that provide valuable experience for assessing the efficacy of carbon capture and sequestration (CCS). There are also more than 100 commercial 'enhanced oil recovery' (EOR) projects worldwide, primarily in the USA, where 5% of oil production is from EOR operations, and CO_2 sales for EOR reached ~85 million m^3/day in 2008 [21, 53]. In addition there are more than 30 geologic sequestration field demonstration projects worldwide at various stages of planning and deployment to investigate the storage of CO_2 in various rock formations using different injection schemes, and monitoring methods [48, 52].

Considerable uncertainties and scientific gaps, however, exist in understanding CO_2-brine-mineral interactions at reservoir conditions, because supercritical CO_2 is buoyant, displaces huge volumes of formation water and becomes reactive when dissolved in brine [30, 45]. Dissolved CO_2 is likely to react with the reservoir and cap rocks, causing dissolution, precipitation and transformation of minerals, and changing the porosity, permeability and injectivity of the reservoir, as well as impacting the extent of CO_2 and brine leakage that could contaminate underground sources of drinking water (USDW) [7, 44]. Reservoir capacity, performance and integrity are strongly affected by the four possible CO_2 trapping (Fig. 10.3) mechanisms [6]: (i) 'Structural and stratigraphic trapping', where the injected CO_2 is stored as a supercritical and buoyant fluid below a cap rock; (ii) 'residual trapping' of CO_2 by capillary forces in the pores of reservoir rocks away from the supercritical plume; (iii) 'solution trapping', where the CO_2 is dissolved in brine forming aqueous species such as $H_2CO_3^\circ$, HCO_3^-, and CO_3^{-2}; and (iv) 'mineral trapping', with the CO_2 precipitated as calcite, magnesite, siderite and dawsonite [5, 27, 51, 59, 64].

In this report, we discuss the geochemistry of CO_2 sequestration in sedimentary basins, emphasizing CO_2-brine-mineral interactions at reservoir conditions. We summarize geochemical results from commercial and pilot CO_2 injection sites, especially our reported results from the Frio Brine Pilot tests, a multi-laboratory field experiment located near Dayton, Texas [37, 44, 45].

We emphasize temporal changes in the chemical and isotopic composition of formation brine and gases that were used for 'deep monitoring' of fluid leakage from the injection sandstone "C" into the overlying "B" sandstone. Significant isotopic and chemical changes, including the lowering of pH, increases in alkalinity, and mobilization of metals, were also observed in samples obtained from shallow groundwater following CO_2 injection at the zero emission research and technology (ZERT) site, located in Bozeman, Montana [46, 68]. Results from both tests show that geochemical methods provide highly sensitive chemical and isotopic tracers that can be used to monitor injection performance, and for early detection of any CO_2 and brine leakages.

10.2 Sequestration of CO_2 in Sedimentary Basins

Sedimentary basins in general and deep saline aquifers in particular are being investigated as possible repositories for large volumes of anthropogenic CO_2 that must be sequestered to mitigate global warming and related climate changes [7, 74]. Currently there are a total of three commercial projects operating worldwide that capture and sequester close to three million tons of CO_2 annually and that provide valuable experience for assessing the efficacy of CCS. There are also two commercial projects (Weyburn-Midale, Canada, and Cranfield, MS) that capture and inject close to four million tons of CO_2 annually for EOR, but these differ from the more than 100 EOR projects worldwide, because they aim to integrate EOR with CO_2 sequestration [38, 53]. In addition there are approximately 25 geologic sequestration field demonstration projects in the US, and an equal number of projects in other countries that are investigating the storage of CO_2 in various clastic and carbonate rock formations using different injection schemes, monitoring methods, hazards assessment protocols, and mitigation strategies [17, 48]. Sedimentary basins are attractive for CO_2 storage because: (i) they have large estimated local and global (>10,000 Gt CO_2) storage capacities [11, 36]; (ii) they often have advantageous locations close to power plants and other major CO_2 sources [12, 33, 35]; and (iii) there is a great deal of geologic and other relevant information gained from petroleum exploration and production [3, 44].

The five commercial projects that currently capture and inject close to seven million tons of CO_2 annually are the Sleipner project, offshore Norway, the Weyburn-Midale EOR project in the Williston Basin of Saskatchewan, Canada, and the In Salah gas field project in Algeria; two projects started in 2008: The Snøhvit field in the Barents Sea offshore Norway and the Cranfield Oil field sequestration-EOR project in Mississippi. The Sleipner Project, operated by the StatoilHydro since 1996, is the world's first industrial-scale operation. Approximately one million tons of CO_2 is being extracted annually from the produced gas that has ~10% CO_2 to meet quality specifications, and stored ~1,000 m below sea level in the Utsira

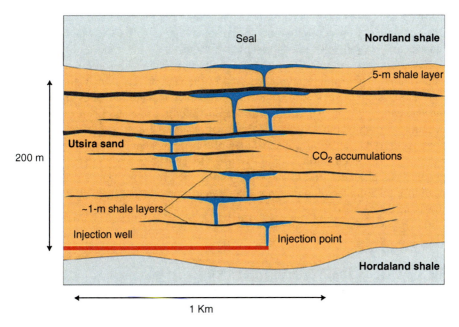

Fig. 10.4 Schematic diagram of carbon dioxide storage at the Sleipner Field, Norway based on seismic images [9]

Formation [32]. The Utsira Formation consists of thick and poorly consolidated sandstones of high porosity (35–40%) and permeability (1–3 darcies), and is overlain by the thick Nordland shale caprock (Fig. 10.4, [9]). The gas, though injected into the bottom of the Utsira sandstone, has migrated through intervening shale beds into nine sandstone layers (Fig. 10.4). However, results from the seismic investigations show that no CO_2 is leaking out of the formation [9, 16]. Reactive transport modeling of CO_2 injected at the Sleipner site indicates that after 10,000 years, only 5% of the injected gas will be trapped into minerals while 95% will be dissolved in the brine [1].

The Weyburn-Midale project (Williston Basin, Canada) started injecting CO_2 into the Weyburn field in 2000, combining CO_2 sequestration with EOR by injecting the gas and brine to increase fluid pressure and oil production [23]. The CO_2 used is obtained from the Dakota gasification plant near Beulah, North Dakota, It is the first anthropogenic source of CO_2 being used for EOR. The oil field consists of shallow marine carbonate (depth of ~1,500 m) and the injection rate is now >3 million tons of CO_2/year. The chemical and C-isotope composition of both brine and H_2O exhibit significant changes resulting from CO_2 injection; geochemical modeling and monitoring of the aquifer indicate that the injected CO_2 reacts with the brine and the host rock, dissolving carbonate minerals, increasing alkalinity, and lowering the pH of produced water; slow dawsonite precipitation is also predicted [13].

The In Salah Gas Joint Venture, located in the Algerian Sahara, injects ~1 million tons/year of CO_2 into the water leg of the Carboniferous Krechba sandstone gas-reservoir (20 m thick) via three horizontal wells at a depth of ~1,900 m [39]. The CO_2 is obtained from several gas fields that contain up to 10% CO_2, which has to be reduced to 0.3% [65]. CO_2 injection started in August 2004 into sandstones with relatively low porosity (11–20%) and permeability (~10 md) with some fractures and small faults in both the reservoir unit and in the Carboniferous mudstones of the caprock. Despite the evidence of fractures, the thick caprock provides an effective mechanical seal for the CO_2. Time-lapse satellite images (using PSInSAR Technology), which measure ground deformation, show a surface uplift ~5 mm/year above active CO_2 injection wells and the uplift pattern extends several km from the wells. The observed surface uplift is used to constrain the coupled reservoir-geomechanical model, and to show that surface deformations from InSAR can be useful for tracking the fluid pressure and for detection of a leakage path through the overlying caprock [65].

10.3 CO_2-Brine-Mineral Interactions in the Frio Formation, Texas

The Frio Brine pilot was the first field study to investigate the potential for geologic storage of CO_2 in saline aquifers and to develop geological, geochemical and geophysical tools, and multi-phase simulation programs to track the injected CO_2 and predict its interactions with reservoir brine and minerals [22, 37]. For Frio-I, ~1,600 tons of CO_2, were injected during October 2004 at a depth of 1,500 m into a 24-m thick "C" sandstone (Fig. 10.5) of the Oligocene Frio Formation [44]. Using a variety of tools, fluid samples were obtained before CO_2 injection for baseline geochemical characterization, during the CO_2 injection to track its breakthrough into the observation well, and after injection to investigate changes in fluid composition and leakage into the overlying "B" sandstone. New geophysical [18, 19] and geochemical tools were deployed and additional detailed tests were carried out during the Frio-II Brine test (September, 2006 to October 2008), where ~300 tons of CO_2 were injected into a 17-m thick Frio "Blue" sandstone, located ~120 m below the Frio "C". Geochemical tools and methods deployed for Frio-II, included online pH, EC and temperature probes, field determinations of Fe^{2+} and Fe^{3+} and collection and analyses for a large number of metals using ICP-MS and organic compounds [45].

10.3.1 Results and Discussion

Results of chemical analyses for Frio-I samples prior to CO_2 injection show that the brine is a Na-Ca-Cl type water, with a salinity of ~93,000 mg/l TDS (Fig. 10.6), with relatively high concentrations of Mg and Ba, but low values for SO_4, HCO_3, DOC and organic acid anions. The high salinity and the low Br/Cl

10 Geologic Storage of CO_2 and Water Issues   137

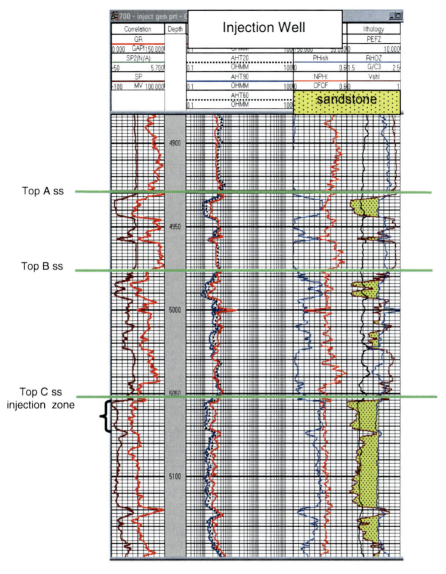

Fig. 10.5 Open-hole logs of the injection well. Note the relatively thick beds of shale and siltstone between the injection zone, Frio "C", and the overlying monitoring sandstone, Frio "B" (Modified from Kharaka et al. [45])

ratio (0.0013) relative to sea water indicate dissolution of halite from the nearby salt dome (e.g., [42]). The brine has 40–45 mM dissolved CH_4, which is close to saturation at reservoir conditions (60 °C and 150 bar), and CH_4 comprises ~95% of total gas, but the CO_2 content is low at ~0.3% (Table 10.1).

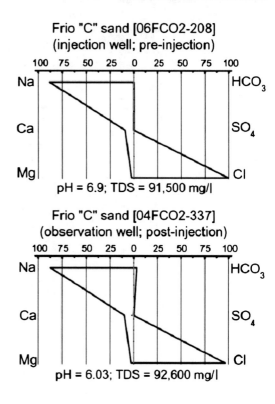

Fig. 10.6 Modified Stiff diagrams showing concentrations (equivalent units normalized to 100%) of major cations and anions, together with salinity and pH of Frio I brine from the "C" sandstone before and after CO_2 injection

Table 10.1 Composition (vol-%) of gas obtained from frio "C" and "B" sandstones. Note the low CO_2 content in background gas, and higher value from the "B" sandstone (From Kharaka et al. [45])

Gas	"C"[a]	"C"[b]	"B"[c]
He	0.0077	0	0.01
H_2	0.040	0.19	0.92
Ar	0.041	0	0.13
CO_2	0.31	96.8	2.86
N_2	3.87	0.037	151
CH_4	93.7	2.94	94.3
C_2H_6+	1.95	0.0052	0.12

[a]From C of the injection well before CO_2 injection
[b]From C of the observation well after CO_2 breakthrough
[c]From B of the observation well 6 months after CO_2 injection

During CO_2 injection, October 4–14, 2004, on-site measurements of electrical conductance (EC) exhibited only a small increase from a pre-injection value of ~120 mS/cm at ~22 °C. However, when the CO_2 reached the observation well, there was a sharp drop in pH, from 6.5 to 5.7, and large increases in alkalinity, from 100 to 3,000 mg/l as HCO_3^- [44]. Additionally, laboratory determinations showed major increases in dissolved Fe (from 30 to 1,100 mg/l) and Mn, (Fig. 10.7) and

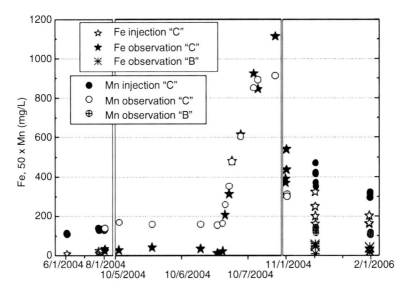

Fig. 10.7 The concentrations of Fe and Mn in selected brine samples from Frio I. Note the high values for Fe and Mn following CO_2 injection

significant increases in the concentration of Ca. The most dramatic changes in chemistry occurred at CO_2 breakthrough 51 h after injection, as evidenced also by PFT tracer analysis [62] and on-site analysis of gas samples from the U-tube system [25] that showed CO_2 concentrations increasing from 0.3% to 3.6% and then quickly to 97% of total gas (Table 10.1). The variations in the field measured pH proved the most sensitive and rapid parameter for tracking the arrival of CO_2 at the observation well, especially when online pH probe was successfully deployed during Frio II (Fig. 10.8).

Results of geochemical modeling, using updated SOLMINEQ [43] indicate that the Frio brine in contact with the supercritical CO_2 would have a pH of ~3 at subsurface conditions, and this low pH causes the brine to become highly undersaturated with respect to carbonate, aluminosilicate and most other minerals present in the Frio Formation (Fig. 10.9). Because mineral dissolution rates are generally higher by orders of magnitude at such low pH values, the observed increases in concentrations of HCO_3^- and Ca^{2+} likely result from the rapid dissolution of calcite via the reaction:

$$H_2CO_3^\circ + CaCO_{3(s)} \leftrightarrow Ca^{2+} + 2HCO_3^- \tag{10.1}$$

The large increases in concentrations of dissolved Fe and equivalent bicarbonate alkalinity likely were by dissolution of the observed iron oxyhydroxides, depicted in redox-sensitive reaction:

$$2Fe(OH)_{3(s)} + 4H_2CO_3^\circ + H_2^\circ \leftrightarrow 2Fe^{2+} + 4HCO_3^- + 6H_2O. \tag{10.2}$$

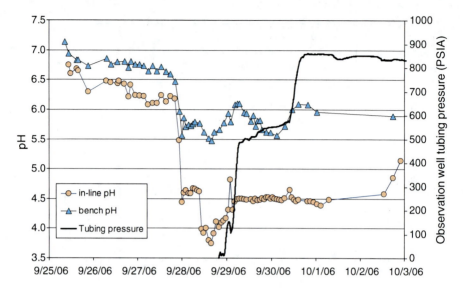

Fig. 10.8 Bench and in-line pH values obtained from Frio II brines before and following CO_2 breakthrough at the observation well. Note the sharp drops of pH, especially values from in-line probe following the breakthrough of CO_2

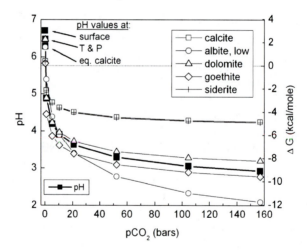

Fig. 10.9 Computed pH values and saturation states of selected minerals in Frio I brine as a function of CO_2 partial pressure at subsurface conditions. Note the sharp initial drop of pH from ~6.4, the average value computed at T & P, and calcite saturation before CO_2 injection (Modified from Kharaka et al. [44])

However, some of the increase in dissolved Fe and bicarbonate could also result from corrosion of pipe and well casing that contact low pH brine [34, 45], as indicated by the redox-sensitive reaction:

$$Fe_{(s)} + 2H_2CO_3^0 \leftrightarrow Fe^{2+} + 2HCO_3^- + H_{2(g)} \qquad (10.3)$$

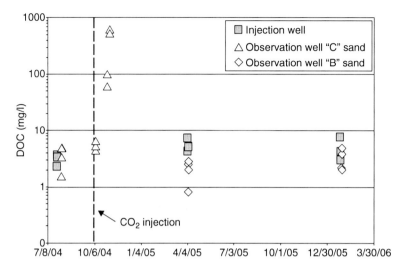

Fig. 10.10 Concentration of dissolved organic carbon (DOC) in Frio I brine. Note the extremely high values obtained on November, 2004, ~20 days after the end of CO_2 injection

There were also increases in the concentration of other metals, including Mn, Zn, Pb and Mo, which are generally associated (sorbed and coprecipitated) with iron oxyhydroxides, but could also be obtained from the low-carbon steel pipe used in petroleum wells [46].

10.3.2 Dissolved Organics

Dissolved organic carbon (DOC) values obtained in Frio-I samples before CO_2 injection are expectedly low (1–5 mg/l); the values obtained during the CO_2 injection increased to 5–6 mg/l. DOC values, however, increased unexpectedly by a factor of 100 on samples collected ~20 days after injection stopped (Fig. 10.10). The concentrations of organic acid anions and BTEX in these samples were low (<1 mg/l), but values of formate, acetate, and toluene were generally higher in the enriched DOC samples. Results from a more detailed sampling protocol for samples collected 6 and 15 months after injection show oil and grease values below detection limit, low levels (up to 30 ppb phenol) of volatile organic compounds (VOCs) and semi-VOCs (30 ppb naphthalene), and slightly elevated DOC values (5–8 mg/l) values (Fig. 10.10).

The high DOC values observed in both the Frio I and II likely represent a 'slug' of organic matter mobilized by the injected CO_2, as reported in laboratory experiments simulating CO_2 storage in coal beds [47] and as happens in EOR operations [66]. Additional investigations are required in this area, but these results suggest that mobilization of organics from oil reservoirs and non oil-bearing aquifers could have major implications for the environmental aspects of CO_2 storage. The concern here is warranted as high concentrations of toxic organic compounds, including benzene, toluene (up to 60 mg/l for BTEX), phenols (20 mg/l), and polycyclic aromatic hydrocarbons (10 mg/l for PAHs), have been reported in oil-field waters [42].

10.3.3 Isotopic Composition of Water and Gases

Significant shifts were observed in the isotopic compositions of H_2O and DIC following CO_2 injection, but only subtle changes in the δD and $\delta^{13}C$ values of CH_4. The $\delta^{13}C$ values of DIC became profoundly lighter, shifting from −3‰ to −33‰, reflecting the fact that the injected CO_2 is the dominant C source and is depleted, with $\delta^{13}C = -44$‰ to −34‰, depending on the mixing proportions of the two gas sources. The $\delta^{18}O$ values of brine became isotopically lighter with time, shifting from 0.80‰ to −11.1‰, and there was a corresponding increase in the $\delta^{18}O$ values of CO_2, from 9‰ to 43‰. Because water and CO_2 rapidly exchange oxygen isotopes even at low temperature, it is possible to use their $\delta^{18}O$ values in mass balance equations to estimate the brine to CO_2 mass and volume ratios in the reservoir [44, 45].

10.4 Potential Impacts and Risks of Geologic Storage of CO_2

The major environmental risks associated with a CO_2 storage site, potentially may include induced seismicity in vulnerable locations, and leakage of CO_2 and brine that may damage USDW or cause harm to humans, animals and ecosystems [71]. Fluid leakage could occur along unmapped fracture systems and faults, improperly sealed abandoned and orphaned wells [15], corroded well casings and cements [14] or even via pathways created in the rock seals as a result of CO_2-brine-rock interactions [44, 45]. Maintaining reservoir integrity that limits CO_2 leakage to very low levels (less than 0.01%) is essential to the long term success of injection operations [31]. Preventing brine migration into USDW is equally important, because it will have toxic metals, inorganic and organic components [45].

To evaluate these risks requires a detailed geologic site characterization and an improved understanding of formation properties and how the injected CO_2 spreads and interacts with the rock matrix and reservoir fluids [2, 26]. Geologic formations typically consist of layers of rock with different porosities, thicknesses, and brine and mineral compositions. All of these factors, together with the presence of faults and fracture systems, affect the suitability of the formation as a site for CO_2 sequestration. Porosity and thickness determine the storage capacity of the formation, and chemical composition determines the interaction of CO_2 with the minerals in place. Also, an impervious cap rock and absence of high-permeability faults are necessary to prevent the sequestered buoyant CO_2 from migrating to the surface or to overlying fresh water aquifers.

10.4.1 Environmental Impacts

Leakage of CO_2 is the most serious potential environmental problem. Leakage to the atmosphere negates the original environmental benefit and economic effort expended in sequestering the CO_2 [26]. Another potential problem is accumulation of CO_2 in pockets in soil and on the surface of the earth, where it could present a health hazard to humans, animals and ecosystems. Furthermore, CO_2 could migrate into other strata, with the potential for contaminating USDW or causing other problems. Results from the Frio Brine tests indicate that the injected supercritical CO_2, which is a very effective solvent for organic compounds [47], could mobilize and transport organics, including BTEX, phenols, PAHs and other toxic organic compounds that have been reported in relatively high concentrations (10–60 mg/l) in oil field brines [42, 45]. Preventing brine migration into overlying USDW is equally important, because dissolution of minerals following CO_2 injection would mobilize Fe, Mn and other metals, in addition to the high concentration of metals and other chemicals (salinity 5,000 to more than 200,000 mg/l TDS) present in the original brine [42]. Finally, if a project does not operate within prescribed injection rates and pressures, there is potential for initiating seismic activity [26].

10.4.2 Health and Safety Impacts

The National Institute of Occupational Safety and Health (NIOSH/OSHA) has determined that human time weighted average (8-h TWA) exposure limit to CO_2 of 1% (10,000 ppm), and short term exposure limit (15 min STEL) of 3% are appropriate [55, 56]. Human exposure to elevated levels of CO_2 can be hazardous in two ways – by a reduction in the oxygen content of the ambient air causing hypoxia or through direct CO_2 toxicity [20, 71]. In most cases of hazardous CO_2 exposure, the gas is presumed to act as a simple asphyxiant. Concentrations higher than 10% have caused difficulty in breathing, impaired hearing, nausea, vomiting, a strangling sensation, sweating, and loss of consciousness within 15 min. As O_2 concentration drops below 17%, increasingly severe physiological effects occur until below 6% O_2, loss of consciousness and death take place within minutes. If H_2S is sequestered along with CO_2, health risks are significantly increased, as H_2S is highly toxic. Also, the health impacts would increase substantially if significant amounts of BTEX, PAHs and other mobilized organics are carried with the leaking CO_2 [45].

In order to minimize the environmental hazards and risks associated with CO_2 storage, a rigorous program should be implemented for the measurement, monitoring, and validation (MMV) of the injected CO_2 and associated brine [6, 22]. MMV is concerned with the capability to measure the amount of CO_2 stored at a specific site, map its spatial disposition through time, develop techniques for surface and subsurface monitoring for the early detection of leakage, and verify that the CO_2 is stored or isolated as intended and will not adversely impact the host ecosystem, including USDW. MMV for geologic sequestration consists of three areas: (i) Modeling and analysis of the geology and hydrology of the total injection system before injection occurs; (ii) tracking and monitoring the movement of the CO_2 and brine plumes; and (iii) measurements that verify that the CO_2 remains sequestered [71]. Results from pilot studies highlight the importance of using the more sensitive geochemical markers [44, 45, 72] and the importance of subsurface monitoring for detecting any early leakage of injected CO_2 and brine to minimize damage to groundwater and the local environment.

10.5 Geochemical Monitoring

10.5.1 Subsurface Monitoring

Monitoring at and close to the ground level for CO_2 and brine leakage into soil gas and shallow (~30 m depth) groundwater was not effective at the Frio site, primarily because of perturbation caused by injection operations [45, 54]. Because of such difficulties, and the long time that could be required for a potential CO_2 and/or brine leakage to reach near the surface, a rigorous program for deep subsurface monitoring was carried out. Results of brine and gas analyses of fluid samples obtained from the "B" sandstone [45], first perforated and sampled 6 months after CO_2 injection, showed: (i) Slightly elevated concentrations of bicarbonate, Fe, and Mn; (ii) significantly depleted $\delta^{13}C$ values (−17.5‰ to −5.9‰ vs. ~−4‰) of DIC; and (iii) somewhat depleted $\delta^{18}O$ values of brine, relative to pre-injection values obtained for the "C" samples. A more definitive proof of the migration of injected CO_2 into the "B" sandstone was obtained from the presence of the two PFT tracers (PMCH and PTCH) that were added to the injected CO_2, and migrated with the initial CO_2 breakthrough [62]. Additional proof of the migration of injected CO_2 into the "B" sandstone is obtained from the high concentration (2.9 vs. ~0.3%) of CO_2 in dissolved gas obtained from one of the two downhole Kuster samples (Table 10.1). Results of samples collected from Frio "B" in January 23–27, 2006 (approximately 15 months after CO_2 injection) gave brine and gas compositions that are approximately similar to those obtained from the "C" sandstone before CO_2 injection. The overall results indicate the absence of significant additional amounts of injected CO_2 in the "B" fluids sampled.

Results from the "B" sandstone clearly show some CO_2 migration from the "C" sandstone after about 6 months following injection. These results are comparable to those observed at Sleipner, where CO_2 is injected into the bottom of the Utsira sandstone, but it has migrated through intervening shale beds (one extensive and ~5 m thick) into nine different sandstone layers (Fig. 10.4). Results from the seismic investigations show that no CO_2 is leaking out of the cap rocks at Sleipner [9]. Leakage of CO_2 injected into the Shatuck sandstone (Permian) in a pilot study similar to Frio was measured at surface at the West Pearl Queen field, NM, where 2,090 tons of CO_2, tagged with PFT tracers, were injected at a depth of 1,400 m. Using an array of PFTs capillary absorption tubes placed 2 m below ground, a leakage rate of 0.009% of injected CO_2 per year was measured by Wells et al. [72].

10.5.2 Near-Surface Monitoring at the Zert Site, Montana

The Zero Emission Research and Technology (ZERT) facility has been developed at a field site in Bozeman, Montana, USA, to allow controlled studies of near surface CO_2 transport and detection technologies. A slotted horizontal well divided into six zones was installed horizontally 2–2.3 m deep. Controlled releases of CO_2 tagged with PFTs and other tracers were performed in the summers of 2007–2010. A wide variety of detection techniques, including soil gas flux, composition and isotopes, eddy covariance measurements, hyperspectral and multispectral imaging of plants, and differential absorption measurements using laser based instruments were deployed by collaborators from many institutions. Even at relatively low CO_2 fluxes, most techniques were able to detect elevated levels of CO_2 in the soil or atmosphere [68]. As part of this ongoing research, 80 samples of water were collected during 2008 season from 10 shallow monitoring wells (1.5 or 3.0 m deep) installed 1–6 m from the injection pipe, and from two distant monitoring wells [46]. Approximately 300 kg/day of food-grade CO_2 was injected through the perforated pipe during July 9-August 7, 2008 at the field test. Samples were collected before, during and following CO_2 injection. The main objective of study was to investigate changes in the concentrations of major, minor and trace inorganic and organic compounds during and following CO_2 injection.

Field determinations showed rapid and systematic changes in pH (7.0 to 5.6), alkalinity (400 to 1,330 mg/l as HCO_3) and electrical conductance (600 to 1,800 μS/cm) following CO_2 injection in samples collected from the 1.5 m-deep wells (Fig. 10.11). Laboratory results show major increases in the concentrations of Ca (90 to 240 mg/l), Mg (25 to 70 mg/l), Fe (5 to 1,200 ppb) and Mn (5 to 1,400 ppb) following CO_2 injection. These chemical changes, especially the easily measured field parameters, could provide early detection of CO_2 leakage into shallow groundwater from deep storage operations.

Dissolution of observed carbonate minerals and desorption-ion exchange resulting from lowered pH values following CO_2 injection are the likely processes responsible for the observed increases in the concentrations of solutes; concentrations generally decreased temporarily following four significant precipitation events. The DOC values obtained are 5 ± 2 mg/l, and the variations do not correlate with CO_2 injection. CO_2 injection, however, is responsible for detection of BTEX (e.g. benzene, 0 to 0.8 ppb), mobilization of metals, the lowered pH values, and increases in the concentrations of other solutes in groundwater (Fig. 10.11). Sequential leaching of core samples is being carried out to investigate the source of metals and other solutes.

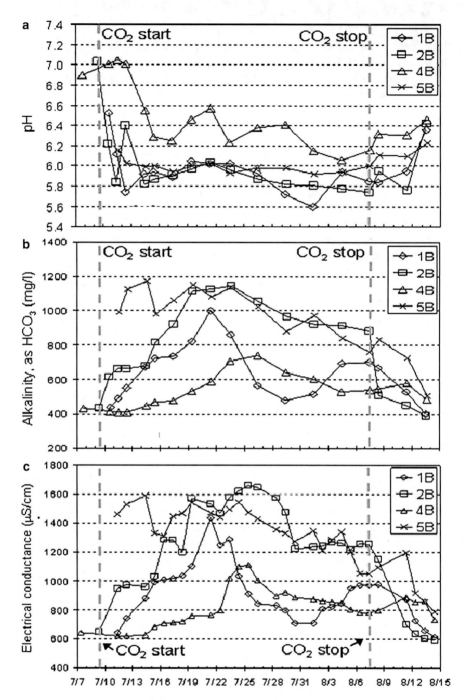

Fig. 10.11 Field measured groundwater pH values (**a**), alkalinities (**b**), and electrical conductance (**c**), obtained from selected ZERT wells as a function of time. Note the systematic decrease in pH values from ~7.0 before CO_2 injection to values as low as 5.6 during injection, and subsequent pH increases after CO_2 injection was terminated. Alkalinities increased from ~400 mg/l to ~1,200 mg/l as HCO_3, and EC increased from ~600–1,600 μS/cm (From Kharaka et al. [46])

10.6 Geochemical and Multi Phase Reactive Transport Modeling

Mathematical models and numerical simulation tools of various complexity play an important role in evaluating the feasibility of geologic storage of CO_2. Geochemical codes such as SOLMINEQ [43] or PHREEQC [60] use brine analyses, together with formation mineralogy to assess their equilibrium states. Kinetic codes, such as PATH [61] or EQ3/6 [75] are then used to assess the reactivity and geochemical trapping potential over time when CO_2 is injected into a depleted oil reservoir or deep saline aquifer. Transport codes, such as GEOCHEMIST WORKBENCH [8] and TOUGHREACT [76] can be used to assess the distribution of CO_2 in time and space. Finally, hydrogeological codes such as MODFLOW [29], or STOMP [73] can model the regional implications due to displacement of large volumes of brine. These codes can be used sequentially to assess implications for dynamic CO_2 storage, followed by code improvement and integration.

Numerous studies of water-rock interaction have shown that without the close connection between experimental results and well-constrained natural systems, geochemical models can be highly inaccurate because of the complex interactions between mineral dissolution, mineral precipitation, aqueous chemistry (e.g., organics) and transport processes [50]. There are several challenges in modeling systems with high ionic strength and mixtures of CO_2 and brine [49]. The credibility of theoretical models for injecting CO_2-rich solutions into the subsurface with the intent of successful hydrodynamic, solubility, or mineralogical sequestration [57] is dependent on the adopted thermodynamic and transport parameters, the theoretical algorithms, initial and boundary conditions, and related assumptions and approximations [10].

10.7 Concluding Remarks

Results from global simulation modeling indicate that parts of both the southwestern USA and North Africa would experience increased drought frequency and water stress in the coming decades as a result of global warming; models predict that precipitation in these two regions will decrease by about 20% by 2050. Capture and sequestration of CO_2 in depleted petroleum fields and saline aquifers is one plausible option to reduce GHG emissions and mitigate global warming and related global climate changes, including water resources issues. Since supercritical CO_2 is buoyant and reactive in water, it will have a tendency to flow upward in the geologic section towards the earth's surface. Therefore, despite the fact that many deep geological formations may be suitable for geological carbon sequestration, supercritical and gaseous CO_2 carries the possibility of leakage, which would negate the benefits of sequestration and introduce elements of risk to groundwater and biota. Contamination of groundwater that provides approximately 21% of all water and 50% of US drinking water by the leakage of injected CO_2 and displaced brine should be prevented because once contaminated, groundwater remediation is very difficult and expensive [63].

CO_2 leakage risk will not be uniform across all sites, thus CO_2 storage sites will have to demonstrate minimal risk potential in their site characterization plans [11]. A small percentage of sites might end up having significant leakage rates during injection, which will require subsurface monitoring [45] and/or near surface monitoring using the variety of geophysical and geochemical tools such as those deployed at the ZERT site [46, 68]; storage sites would require validation as well as mitigation plans. Based on analogous experience in CO_2 injection such as acid gas disposal and EOR, these risks are relatively minor [6]. Moreover, if storage projects are not sited near seismically active faults, and operate within prescribed injection rates and pressures, then the potential for initiating significant seismic activity likely would be small. Finally, the likelihood and extent of any potential CO_2 leakage should slowly decrease with time after injection stops, because the formation pressure will begin to drop to pre-injection levels, as more of the injected CO_2 dissolves into the pore brine and begins the long-term process of forming chemically stable carbonate minerals.

References

1. Audigane P, Gaus I, Czernichowski-Lauriol I, Pruess K, Xu TF (2007) Two-dimensional reactive transport modeling of CO_2 injection in a saline Aquifer at the Sleipner site, North Sea. Am J Sci 307:974–1008
2. Bachu S (2003) Screening and ranking of sedimentary basins for sequestration of CO_2 in geological media in response to climate change. Environ Geol 44:277–289
3. Bachu S, Bonijoly D, Bradshaw J, Burruss R, Holloway S, Christensen NP, Mathiassen OM (2007) CO_2 storage capacity estimation: methodology and gaps. Int J Greenhouse Gas Control 1:430–443
4. Barnett TP, Pierce DW, Hidalgo H, Bonfils C, Santer B, Das T, Bala G, Wood A, Nozawa T, Mirin A, Cayan D, Dettinger M (2008) Human-induced changes in hydrology of the western United States. Science 319:1080–1083
5. Bénézeth P, Méncz B, Noiricl C (2009) CO_2 geological storage: Integrating geochemical, hydrodynamical, mechanical and biological processes from the pore to the reservoir scale. Chem Geol 265:1–2
6. Benson SM, Cook P (2005) Underground geological storage. In: Carbon dioxide capture and storage: special report of the Intergovernmental Panel on Climate Change (IPCC). Cambridge University Press, Interlachen, Switzerland, pp 5–1 to 5–134
7. Benson SM, Cole DR (2008) CO_2 sequestration in deep sedimentary formations. Elements 4 (5), In: Cole DR, Oelkers EH (eds) Carbon dioxide sequestration, pp 305–310
8. Bethke CM (2007) The geochemist's workbench ® (Version 7.0). Hydrogeology program, University of Illinois, Urbana
9. Bickle MJ (2009) Geological carbon storage. Nat Geosci 2(12):815–818
10. Birkholzer JT, Zhou Q (2009) Basin–scale hydrogeologic impacts of CO_2 storage: capacity and regulatory implications. Int J Greenhouse Gas Control 3:745–756
11. Bradshaw J, Bachu S, Bonijoly D, Burruss R, Holloway S, Christensen NP, Mathiassen OM (2007) CO_2 storage capacity estimation: issues and development of standards. Int J Greenhouse Gas Control 1:62–68
12. Burruss R, Brennan ST, Freeman PA, Merrill MD, Ruppert LF, Becker MF, Herkelrath WN, Kharaka YK, Neuzil CE, Swanson SM, Cook TA, Klett TR, Nelson PH, Schenk CJ (2009) Development of a probabilistic assessment methodology for evaluation of carbon dioxide storage. USGS Open-file Report No 09–1035

13. Cantucci B, Montegrossi G, Vaselli O, Tassi F, Quattrocchi F, Perkins EH (2009) Geochemical modeling of CO_2 storage in deep reservoirs: the Weyburn Project (Canada) case study. Chem Geol 265:181–197
14. Carey JW, Wigand M, Chipera SJ, Wolde G, Pawar R, Lichtner PC, Wehner SC, Raines MA, Guthrie GD (2007) Analysis and performance of oil well cement with 30 years of CO_2 exposure from the SACROC Unit, West Texas, USA. Int J Greenhouse Gas Control 1:75–85
15. Celia MA, Kavetski D, Nordbotten JM, Bachu S, Gasda SE (2006) Implications of abandoned wells for site selection. In: Proceedings of the CO2SC 2006 international symposium on site characterization for CO_2 geological storage, Lawrence Berkeley National Laboratory, Berkeley, CA, 20–22 Mar 2006, pp 157–159
16. Chadwick RA, Zweigel P, Gregersen U, Kirby GA, Holloway S, Johannessen PN (2004) Geological reservoir characterization of a CO_2 storage site: the Utsira Sand, Sleipner, northern North Sea. Energy 29:1371–1381
17. Cook PJ (2009) Demonstration and deployment of carbon dioxide capture and storage in Australia. Energy Procedia 1:3859–3866
18. Daley TM, Solbau RD, Ajo-Franklin JB, Benson SM (2007) Continuous active-source monitoring of CO_2 injection in a brine aquifer. Geophysics 72(5):A57–A61
19. Daley TM, Myer LR, Peterson JE, Majer EL, Hoversten GM (2008) Time-lapse crosswell seismic and VSP monitoring of injected CO_2 in a brine aquifer. Environ Geol 54:1657–1665
20. Deel D, Mahajan K, Mahoney CR, McIlvried HG, Srivastava RD (2006) Risk assessment and management for long-term storage of CO_2 in geologic formations. United States Department of Energy R&D. J Systemics Cybern Inform 5:79–85
21. DOE/NETL (2008) Storing CO_2 with enhanced oil recovery. DOE/NETL-402/1312/01-070-08, Advanced Resources International Inc, 7 Feb 2008
22. Doughty C, Freifeld BM, Trautz RC (2008) Site characterization for CO_2 geologic storage and vice versa: the Frio brine pilot, Texas, USA as a case study. Environ Geol 54:1635–1656
23. Emberley S, Hutcheon I, Shevalier M, Durocher K, Mayer B, Gunter WD, Perkins EH (2005) Monitoring of fluid-rock interaction and CO_2 storage through produced fluid sampling at the Weyburn CO_2-injection enhanced oil recovery site, Saskatchewan, Canada. Appl Geochem 20:1131–1157
24. Energy Information Administration (EIA) (2009) Annual energy outlook 2010 early release overview. Report #:DOE/EIA-0383. Washington DC. http://www.eia.doe.gov/oiaf/ieo/pdf/emissions_tables.pdf
25. Freifeld BM, Trautz RC, Kharaka YK, Phelps TJ, Myer LR, Hovorka SD, Collins DJ (2005) The U-tube: a novel system for acquiring borehole fluid samples from a deep geologic CO_2 sequestration experiment. J Geophys Res 110:B10203
26. Friedmann SJ (2007) Geological carbon dioxide sequestration. Elements 3:179–184
27. Gunter WD, Perkins EH, McCann TJ (1993) Aquifer disposal of CO_2-rich gases: reaction design for added capacity. Energy Convers Manage 34:941–948
28. Hansen J, Sato M, Ruedy R, Lo K, Lea DW, Medina-Elizade M (2006) Global temperature change. Proc Natl Acad Sci 103:14288–14293
29. Harbaugh AW (2005) MODFLOW-2005, the U.S. geological survey modular ground-water model: the ground-water flow process. U.S. Geological Survey techniques and methods 6-A16, variously p
30. Haszeldine SR (2009) Carbon capture and storage: how green can black be? Science 325:1647–1652
31. Hepple RP, Benson SM (2005) Geologic storage of carbon dioxide as a climate change mitigation strategy: performance requirements and the implications of surface seepage. Environ Geol 47:576–585
32. Hermanrud C, Andresen T, Eiken O, Hansen H, Janbu A, Lippard J, Bolas HN, Simmenes TH, Teige GMG, Ostmo S (2009) Storage of CO_2 in saline aquifers: lessons learned from 10 years of injection into the Utsira Formation in the Sleipner area. Energy Procedia 1:1997–2004
33. Hitchon B (ed) (1996) Aquifer disposal of carbon dioxide. Geoscience Publishing Ltd, Sherwood Park, p 165

34. Hitchon B (2000) "Rust" contamination of formation waters from producing wells. Appl Geochem 15:1527–1533
35. Hitchon B (ed) (2009) Pembina cardium CO_2 monitoring pilot: a CO_2-EOR project, Alberta, Canada. Geoscience Publishing, Sherwood Park, pp 360
36. Holloway S (1997) An overview of the underground disposal of carbon dioxide. Energy Convers Manage 38:193–198
37. Hovorka SD, Benson SM, Doughty CK, Freifeld BM, Sakurai S, Daley TM, Kharaka YK, Holtz MH, Trautz RC, Nance HS, Myer LR, Knauss KG (2006) Measuring permanence of CO_2 storage in saline formations: the Frio experiment. Environ Geosci 13:105–121
38. Hovorka SD, Choi JW, Menckel TA, Trevino RH, Zeng H, Kordi M, Wang FP, Nicot J-P (2010) Measured and modeled CO_2 flow through heterogeneous reservoir – early results of SECARB test at Cranfield Mississippi. In: Abstract, ninth annual conference on carbon capture and sequestration DOE/NETL, May 10-13, 2010, Pittsburgh, PA, p 214
39. Iding M, Ringrose P (2009) Evaluating the impact of fractures on the long-term performance of the In-Salah CO_2 storage site. Energy Procedia 1:2021–2028
40. Intergovernmental Panel on Climate Change (IPCC) (2007) The physical science basis. Contribution of working group I to the fourth assessment report of the Intergovernmental Panel on Climate Change. Cambridge University Press, Cambridge/New York, 996 p
41. Karl TR, Melillo JM, Peterson TC (eds) (2009) Global climate change impacts in the United States. Cambridge University Press, New York
42. Kharaka YK, Hanor JS, Kharaka YK, Hanor JS (2007) Deep fluids in the continents I: sedimentary basins. In: Drever JI (ed) Surface and ground water, weathering and soils, vol 5. Treatise on Geochemistry. Elsevier, San Diego, pp 1–48
43. Kharaka YK, Gunter WD, Aggarwal PK, Perkins EH, DeBrall JD (1988) SOLMINEQ.88: a computer program for geochemical modeling of water-rock interactions. U.S. Geological Survey Water Resources Investigations Report, pp 88–4227
44. Kharaka YK, Cole DR, Hovorka SD, Gunter WD, Knauss KG, Freifeld BM (2006) Gas-water-rock interactions in Frio Formation following CO_2 injection: implications to the storage of greenhouse gases in sedimentary basins. Geology 34:577–580
45. Kharaka YK, Thordsen JJ, Hovorka SD, Nance HS, Cole DR, Phelps TJ, Knauss KG (2009) Potential environmental issues of CO_2 storage in deep saline aquifers: geochemical results from the Frio-I Brine Pilot test, Texas, USA. Appl Geochem 24:1106–1112
46. Kharaka YK, Thordsen JJ, Kakouros E, Ambats G, Herkelrath WN, Birkolzer JT, Apps JA, Spycher NF, Zheng L, Trautz RC, Rauch HW, Gullickson KS (2010) Changes in the chemistry of shallow groundwater related to the 2008 injection of CO_2 at the ZERT field site, Bozeman, Montana. Environ Earth Sciences 60:273–284
47. Kolak JJ, Burruss RC (2006) Geochemical investigation of the potential for mobilizing non-methane hydrocarbons during carbon dioxide storage in deep coal beds. Energy Fuels 20(2):566–574
48. Litynski JT, Plasynski S, McIlvried HG, Mahoney C, Srivastava RD (2008) The United States Department of Energy's Regional Carbon Sequestration Partnerships Program Validation Phase. Environ Int 34:127–138
49. Lu C, Han WS, Lee SY, McPherson BJ, Lichtner PC (2009) Effects of density and mutual solubility of a CO_2-brine system on CO_2 storage in geological formations: "warm" vs. "cold" formations. Adv Water Resour 32(12):1685–1702
50. Maher K, Steefel CI, Depaolo DJ, Viani BE (2006) The mineral dissolution rate conundrum: insights from reactive transport modeling of U isotopes and pore fluid chemistry in marine sediments. Geochim Cosmochim Acta 70(2):337–363
51. Marini L (2007) Geological sequestration of carbon dioxide: thermodynamics, kinetics and reaction path modeling. Elsevier, Amsterdam, p 470
52. Michael K, Arnot M, Cook P, Ennis-King J, Funnell R, Kaldi J, Kirste D, Paterson L (2009) CO_2 storage in saline aquifers I: current state of scientific knowledge. Energy Procedia 1:3197–3204
53. Moritis G (Dec 2009) Special report: more CO_2-EOR projects likely as new CO_2 supply sources become available. Oil Gas J 107(45):70–76

54. Nance HS, Rauch H, Strazisar B, Bromhal G, Wells A, Diehl R, Klusman R, Lewicki JL, Oldenberg CM, Kharaka YK, Kakouros E (2005) Surface environmental monitoring at the Frio Test site. In: Proceedings of the fourth annual conference on carbon capture and sequestration. Alexandria, 2–5 May, 16 p, CD-ROM
55. NIOSH/OSHA (1981) Occupational health guidelines for chemical hazards. Department of Health and Human Services (National Institute for Occupational Safety and Health) Publication No. 81-123, United States Government Printing Office, Washington, DC
56. NIOSH/OSHA (1989) Toxicologic review of selected chemicals: carbon dioxide. http://www.cdc.gov/niosh/pel88/124-38.html
57. Oelkers EH, Gislason SR, Matter J (2008) Mineral Carbonation of CO_2. Elements 4(5):333–337
58. Overpack J, Udall B (2010) Dry times ahead. Science 328:1642–1643
59. Palandri JL, Rosenbauer RJ, Kharaka YK (2005) Ferric iron in sediments as a novel CO_2 mineral trap: CO_2-SO_2 reaction with hematite. Appl Geochem 20:2038–2048
60. Parkhurst DL, Appelo CAJ (1999) User's guide to PHREEQC (Version 2): a computer program for speciation, batch-reaction, one-dimensional transport, and inverse geochemical calculations. USGS Water-Resources Investigations Report 99-4259, Denver
61. Perkins EH, Gunter WD (1995) A user's manual for β PATHARCH.94: a reaction path-mass transfer program. Alberta Research Council Report, ENVTR pp 95–11, 179
62. Phelps TJ, McCallum SD, Cole DR, Kharaka YK, Hovorka SD (2006) Monitoring geological CO_2 sequestration using perfluorocarbon gas tracers and isotopes. In: Proceedings of fifth annual conference on carbon capture and sequestration, Alexandria, 8–11 May, p 8, CD-ROM
63. Reilly TE, Dennehy KF, Alley WM, Cunningham WL (2008) Ground-water availability in the United States. U.S. Geological Survey Circular 1323, 70. Available at http://pubs.usgs.gov/circ/1323/
64. Rosenbauer RJ, Koksalan T, Palandri JL (2005) Experimental investigation of CO_2-brine-rock interactions at elevated temperature and pressure: implications for CO_2 sequestration in deep-saline aquifers. Fuel Process Technol 86:1581–1597
65. Rutqvist J, Vasco DW, Myer L (2009) Coupled reservoir-geomechanical analysis of CO_2 injection at In Salah, Algeria. Energy Procedia 1:1847–1854
66. Shiraki R, Dunn TL (2000) Experimental study on water-rock interactions during CO_2 flooding in the Tensleep Formation, Wyoming, USA. Appl Geochem 15:265–279
67. Solomon S, Gian-Kasper P, Reto K, Pierre F (2010) Irreversible climate change due to carbon dioxide emissions. Proc Natl Acad Sci USA 106(6):1704–1709
68. Spangler LH, Dobeck LM, Repasky K, Nehrir A, Humphries S, Barr J, Keith C, Shaw J, Rouse J, Cunningham A, Benson S, Oldenburg CM, Lewicki JL, Wells A, Diehl R, Strazisar B, Fessenden J, Rahn T, Amonette J, Barr J, Pickles W, Jacobson J, Silver E, Male E, Rauch H, Gullickson K, Trautz R, Kharaka YK, Birkholzer J, Wielopolski L (2009) A controlled field pilot for testing near surface CO_2 detection techniques and transport models. Energy Procedia 1:2143–2150
69. Sundquist ET, Ackerman KV, Parker L, Huntzinger DN (2009) An introduction to global carbon cycle. In: McPherson BJ, Sundquist ET (eds) Carbon sequestration and its role in the global carbon cycle, vol 183, Geophysical Monograph. American Geophysical Union, Washington, DC, pp 1–23
70. UNEP (2009) Assessment of transboundary freshwater vulnerability in Africa to climate change. UNEP Water Research Commission report, 104 p
71. U. S. Environmental Protection Agency (2008) Vulnerability evaluation framework for geologic sequestration of carbon dioxide. EPA430-R-08-009, p 85
72. Wells AW, Diehl JR, Bromhal G, Strazisar BR, Wilson TH, White CM (2007) The use of tracers to assess leakage from the sequestration of CO_2 in a depleted oil reservoir, New Mexico, USA. Appl Geochem 22:996–1016
73. White MD, Oostrom M (2006) STOMP subsurface transport over multiple phases, Version 4.0: user's guide, PNNL-15782. Pacific Northwest National Laboratory, Richland, Washington

74. White CM, Strazisar BR, Granite EJ (2003) Separation and capture of CO_2 from large stationary sources and sequestration in geological formations: coalbeds and deep saline aquifers. J Air Waste Manage Assoc 53:645–715
75. Wolery TJ, Daveler SA (1992) A computer program for reaction path modeling of aqueous geochemical solutions: theoretical manual, user's guide, and related documentation (version 7.0). Report UCRL-MA-110662 PT IV Lawrence Livermore National Laboratory, Livermore
76. Xu T (2008) TOUGHREACT user's guide: a simulation program for non-isothermal multiphase reactive geochemical transport in variably saturated geologic media: V1.2.1. Lawrence Berkeley National Laboratory Paper LBNL-55460-2008. Accessed 12/31/08 at http://repositories.cdlib.org/lbnl/LBNL-55460-2008

Chapter 11
Environmental Security Issues Related to Impacts of Anthropogenic Activities on Groundwater: Examples from the Real World

Brunella Raco, Andrea Cerrina Feroni, Simone Da Prato, Marco Doveri, Alessandro Ellero, Matteo Lelli, Giulio Masetti, Barbara Nisi, and Luigi Marini

Abstract Correct management of groundwater resources must be based on the knowledge of their geological-hydrogeological-geochemical framework. As an example from the real world of environmental security issues determined by anthropogenic activities we will consider the strong impact on the quality of the groundwater hosted into the shallow coastal aquifers of the plain of the rivers Cecina and Fine in Tuscany, Italy. This area was recently investigated through a multidisciplinary geological-hydrogeological-geochemical approach in the framework of the Significant Subterranean Water Bodies (SSWBs) Project funded by the Tuscany Region. This study presents the results obtained for the first three considered SSWBs, namely the 32CT010 SSWB – "Coastal acquifer between the Cecina river and San Vincenzo", the 32CT030 SSWB – "Coastal acquifer between the Fine river and Cecina river" and 32CT050 SSWB – "Aquifer of the Cecina valley".

Keywords Central-southern Tuscany • Geology • Hydrogeology • Hydrogeochemistry • Water management

---

B. Raco (✉) • A.C. Feroni • S. Da Prato • M. Doveri • A. Ellero
• M. Lelli • G. Masetti • B. Nisi
CNR Institute of Geoscience and Earth Resources (CNR-IGG), Via Moruzzi 1,
Pisa 56124, Italy
e-mail: b.raco@igg.cnr.it

L. Marini
CNR Institute of Geoscience and Earth Resources (CNR-IGG), Via Moruzzi 1,
Pisa 56124, Italy

Dip.Te.Ris., Università di Genova, Corso Europa 26, Genova 16132, Italy

11.1 Introduction

According to the Deliberation of the Regional Committee no. 225 on 10 March 2003 (DGRT 225/2003), the Tuscany Region was formally divided into 45 Significant Subterranean Water Bodies (SSWBs) to protect the water resources hosted in these aquifers and to undertake remediation/reclamation actions if needed.

A first geological-stratigraphic study was carried out by the Institute of Geosciences and Georesources of CNR, technical support of the LaMMa Consortium for geology and related disciplines, for the Tuscany Region, with the contribution of other regional agencies (ARPAT, CGT of Siena University, Arno River Basin Authority). Reconstruction of the boundaries, top and bottom surfaces of each SSWBs were reconstructed through the interpretation of available stratigraphic data [5, 15]. The present work was conducted to characterize the SSWBs from the geological, hydrogeological, and hydrogeochemical point of view, through the collection, elaboration and interpretation of available data for the first three considered SSWBs, namely the 32CT010 SSWB – "Coastal acquifer between the Cecina River and San Vincenzo", the 32CT030 SSWB – "Coastal acquifer between the Fine River and Cecina River" and 32CT050 SSWB – "Aquifer of the Cecina valley".

Several geological, hydrogeological, and geochemical studies were devoted to these three SSWBs, to understand different aspects of high scientific relevance and for practical implications, as the water from these aquifers is used as potable water by a very large number of people.

In addition to the collection, processing and interpretation of the published geological, hydrogeological, and hydrogeochemical data [1–3, 6–8, 10, 11, 14], and some not published data produced up to now for these SSWBs, a conceptual hydrogeological model was reconstructed, through the comparison and synthesis of the indications obtained from the different disciplines.

11.2 Geological and Hydrogeological Background

The investigated area regarding the three Significant Subterranean Water Bodies (SSWBs) 32CT050, 32CT010, and 32CT030, is situated in the coastal area between Rosignano and San Vincenzo (central-western Tuscany, Italy), which includes the terminal paths of the Cecina and Fine rivers (Fig. 11.1).

The SSWB of the Cecina valley is hosted into the permeable layers of the alluvial sequence, which are prevailingly made up of gravels and pebbly sands. The underlying bedrock is represented by impermeable Mio-Pliocene clays, in most of the drainage basin, except in its terminal part, where they are substituted by chiefly sandy lithotypes [4, 9]. Consequently, in the terminal part of the valley, important water exchanges can occur between the aquifer and its permeable bedrock (Fig. 11.2).

The total volume of permeable deposits is approximately 280×10^6 m^3. The potentiometric surface of this water body is controlled by the main rivers and its slope and experiences a change of ca. 0.6 m between the end of the dry season and

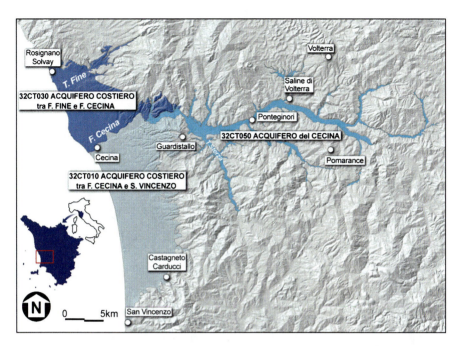

Fig. 11.1 Significant subterranean water bodies (SSWB) in the Cecina Valley alluvial deposits: coastal aquifer between Fine and Cecina rivers (32CT030), coastal aquifer between Cecina river and S. Vincenzo (32CT010) and Cecina river aquifer (32CT050)

the end of the rainy season. The stored water volume varies between 28.9×10^6 and 31.3×10^6 m^3. Owing to the presence of poorly permeable rocks in most of the drainage basin, the recharge of the Cecina valley SSWB is mainly provided by runoff water, which infiltrates in the alluvial sequence situated at the bottom of the valley. However, lateral contributions, from the permeable lithotypes cropping out on the flanks of the valley, become important in its terminal part.

The coastal SSWBs are constituted by a sequence of permeable, gravel and sand layers separated by impermeable silty-clayey deposits. The bedrock of this sequence is represented by the sands and clays with Arctic islandica in the northern sector, by the low-permeability Ligurian Units in the southern sector, and by clayey deposits of uncertain stratigraphic position in the sector between the Cecina river and Bolgheri. The total volume of permeable deposits is approximately $4,630 \times 10^6$ m^3 This multi-layer system behaves similar to a single-layer aquifer, as suggested by the potentiometric surface of these two coastal water bodies, owing to both the discontinuous nature of the impermeable deposits and the presence of several boreholes connecting the permeable layers situated at different depths. The stored water volume is in the order of 394×10^6–398×10^6 m^3. The coastal SSWBs are recharged by both local precipitation, mainly in the plains, and meteoric water infiltrating in the nearby hills, especially where permeable rocks (e.g., Pleistocene sands) crop out. This is in line with both the shape of the potentiometric surface and available $\delta^{18}O$ values [6, 12, 13].

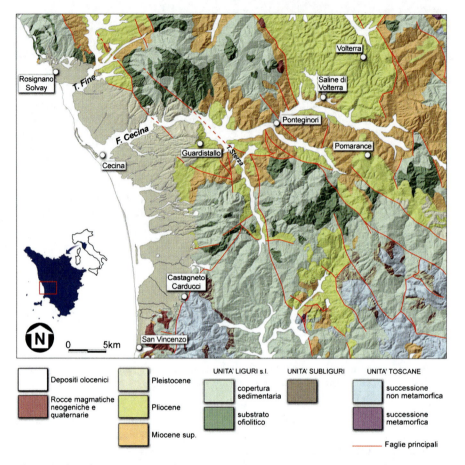

Fig. 11.2 Geological sketch map of the study area (from: Geological map of Tuscany, 1:250.000, modified – In Italian)

All the three SSWBs have to be considered hydraulically connected, as no clear hydrogeological boundary can be recognized among them.

11.3 Hydrogeochemical Characterization

Available geochemical data for the three water bodies of interest were processed together. Since data are frequently incomplete, it was necessary to estimate some parameters, such as pH (assuming saturation with respect to calcite) and silica concentration (through multiple regression analysis of existing data). Most groundwater samples collected in the considered SSWBs have chemical composition from Ca-HCO_3 to Ca(Mg)-HCO_3 and originate through interaction with calcite and

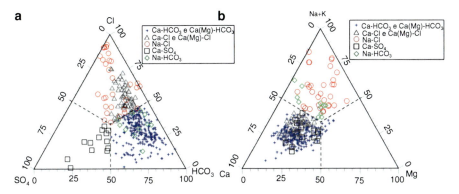

Fig. 11.3 Triangular diagrams of (**a**) SO$_4$, Cl and HCO$_3$ and (**b**) (Na+K), Mg and Ca for the waters relative to the 32CT010+32CT030 and 32CT050 SSWBs (significant subterranean water bodies)

dolomite, whose dissolution rates are much higher than those of silicates and Al-silicates (Fig. 11.3). Virtually all the groundwaters are saturated with respect to these carbonate minerals, for CO$_2$ fugacities between 10^{-3} and 10^{-1} bar. This condition acts as an effective geochemical barrier, preventing the attainment of equilibrium with the primary minerals bearing Ca (e.g., plagioclases), Mg (e.g., serpentine and chlorites) and both Ca and Mg (e.g., pyroxenes), as suggested by the activity diagrams for the systems CaO-SiO$_2$-Al$_2$O$_3$-CO$_2$-H$_2$O and MgO-SiO$_2$-Al$_2$O$_3$-CO$_2$-H$_2$O (Fig. 11.4).

Calcium-sulfate waters are locally present. They are produced through dissolution of gypsum and/or anhydrite, contained either in evaporite rocks cropping out in the nearby hills or as clastic constituents into alluvial deposits.

The interpretation of geochemical data highlighted some environmental problems, such as those linked to the elevated concentrations of: (i) dissolved nitrate, up to ca. 300 mg/l (Fig. 11.5), which is due mainly to the intense animal farms and use of fertilizers and secondarily to domestic and civil wastewaters [10]; (ii) boron, chiefly coming from the drainage basin of the Possera Creek, originally of geothermal origin and more recently dispersed in the environment from chemical plants (Fig. 11.6); (iii) hexavalent chromium, whose origin is still the subject of detailed studies.

The main problem is represented by seawater intrusion in coastal aquifers, which locally generates Na-Cl waters, with high concentrations of dissolved chloride, up to a maximum value of 13,500 mg/l. As a consequence of this process, Na-Ca and Na-Mg ion exchanges take place and water chemistry changes from Na-Cl to either Ca-Cl or Ca(Mg)-Cl. The reverse process (freshening) occurs when low-salinity Ca-HCO3 groundwater flushes saltwater from the aquifer. Adsorption of Ca^{2+} ion and concurrent release of Na$^+$ ion lead to generation of Na-HCO$_3$-type water, which are also present in the study area (Fig. 11.7).

Intrusion of seawater, due to either direct inland displacement of the saline wedge or inflow of saltwater along the stream channels and subsequent entrance in the nearby shallow aquifers, is largely controlled by the depressions in the potentiometric

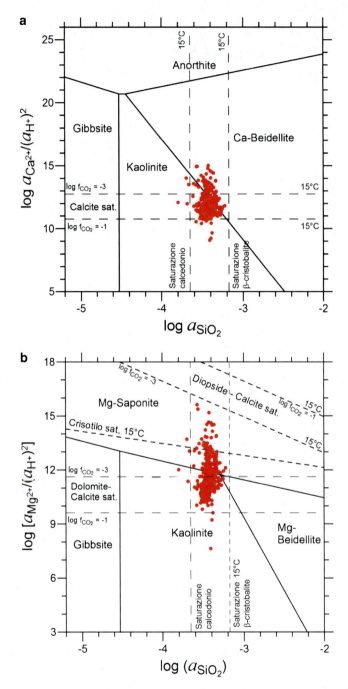

Fig. 11.4 (a) Activity diagram for the system CaO-SiO_2-Al_2O_3-CO_2-H_2O (b) activity diagram for the system MgO-SiO_2-Al_2O_3-CO_2-H_2O

11 Environmental Security and Anthropogenic Activities 159

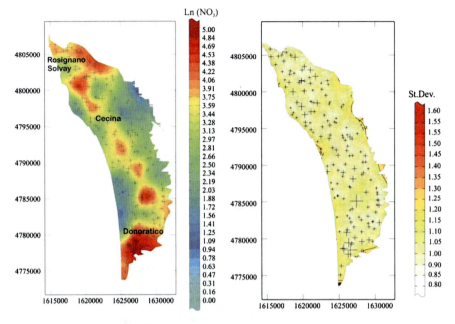

Fig. 11.5 Maps of ln(NO3) (mg/l) and standard deviation (St.Dev.) for the waters relative to the 32CT010+32CT030 and 32CT050 SSWBs (significant subterranean water bodies)

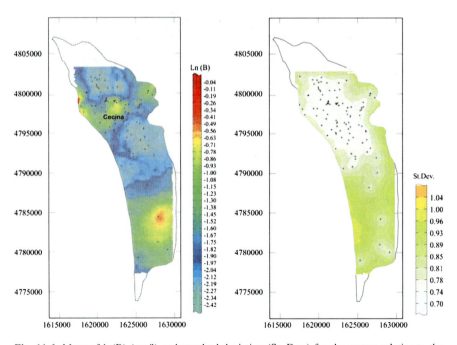

Fig. 11.6 Maps of ln(B) (mg/l) and standard deviation (St. Dev.) for the waters relative to the 32CT010+32CT030 and 32CT050 SSWBs (significant subterranean water bodies). The white dotted areas refer to those zones for which the concentrations of Boron were not available

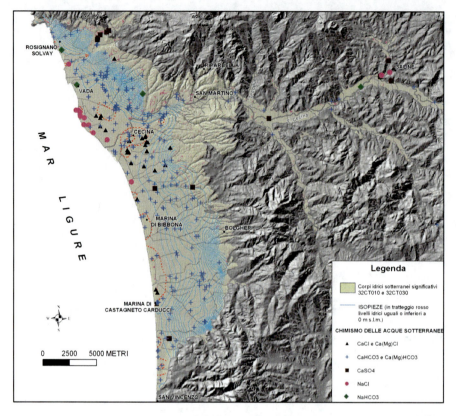

Fig. 11.7 Chemical composition and piezometric map of the study area (SSWBs 32CT010+ 32CT030 and 32CT050 – In Italian)

surface, which are present near Cecina and along the coast, between Marina di Bibbona and Castagneto Carducci as well as near San Vincenzo. There is no doubt that these wide cones of depression are caused by excessive groundwater pumping. Consequently, there is an urgent need to change the management of the coastal aquifers situated between the Fine river and San Vincenzo to avoid further deterioration of groundwater quality.

11.4 Conclusion

This case study indicated that the elevated anthropogenic pressure locally has caused a strong deterioration of water quality, mainly due to: (i) intrusion of marine water; (ii) use of N-bearing fertilizers in agricultural practices; (iii) synthesis and refinement of borate salts. A further complication derives from the presence of hexavalent chromium in some areas, possibly due to oxidative dissolution of Cr-bearing minerals, probably driven by trivalent and tetravalent Mn oxides.

The results obtained from this multidisciplinary approach represent the necessary information to define a conceptual hydrogeological model to be used as input datum for computer modelling of water flow and reactive transport. Moreover these results provide the technical support for any political decision related to the optimal exploitation of water resources.

This work highlights the importance of water chemistry database that should be designed and updated in order to provide a basic knowledge of the territory. Continuous monitoring of critical areas should be planned and carried out, in order to control local problems causing the deterioration of water quality and to set up possible solutions. In particular, detailed knowledge of both water origin (by interpretation of isotope and chemical data) and impact of anthropogenic activities on water resources can be extremely useful for the optimization of potable-water production, and for reducing costs and energy consumption.

Acknowledgements The authors wish to warmly thank Dr. Scozzari Andrea for his technical support. We also gratefully acknowledge Prof.Yousif Kharaka for his comments and suggestions, which greatly improved an early version of the manuscript.

References

1. AF-Geoscience and Technology Consulting (1997) Bilancio idrogeologico del Bacino del Fiume Cecina. Relazione finale, 35pp. 8 all
2. Arpat (2003) Studio idrogeologico con utilizzo di modelli numerici di simulazione per la definizione dei meccanismi d'arricchimento in nitrati delle acque sotterranee nell'area compresa tra gli stradoni del Lupo, del Tripesce, la SS. n° 206, Vada e S. Pietro in Palazzi. Relazione Finale, 40pp., 24 Figg. f.t
3. Capri E, Civita M, Corniello A, Cusimano G, De Maio M, Ducci D, Fait G, Fiorucci A, Hauser S, Pisciotta A, Pranzino G, Trevisan M, Delgado Huertas A, Ferrari F, Frullini R, Nisi B, Offi M, Vaselli O, Vassallo M (2009) Assessment of nitrate contamination risk: the Italian experience. J Geochem Explor. doi:10.1016/j.gexplo.2009.02.006
4. Cerrina Feroni A, Patacca E, Plesi G (1973) La Zona di Lanciaia fra il Cretaceo inferiore e l'Eocene inferiore. Atti Soc Tosc Sc Nat Mem 73:412–468
5. Cerrina Feroni A, Ellero A, Masetti G, Otaria G, Pardini E, Doveri M, Lelli M, Raco B, Romanelli S, Menichetti S, Viti ML (2008) I Corpi Idrici Sotterranei Significativi della Regione Toscana DGRT 225/2003. Monografia della Regione Toscana
6. Frullini R, Gardin L, Morini D, Nevini R, Pranzini G (2007) Carta della vulnerabilità all'inquinamento della pianura costiere fra Rosignano e San Vincenzo (LI): Confronto fra diversi metodi di valutazione del parametro suolo. IGEA Ingegneria e Geologia degli Acquiferi 22:113–126, 7
7. Grassi S, Squarci P (2004) La contaminazione da boro lungo il Fiume Cecina. Atti Soc Tosc Sci Nat Mem 109:21–28
8. Grassi S, Cortecci G, Squarci P (2007) Groundwater resource degradation in coastal plains: the example of the Cecina area (Tuscany – Central Italy). Appl Geochem 22:2273–2289
9. Mozzanti R (1984) Il punto sul Quaternario della fascia costiera e dell'arcipelago di Toscana. Boll Soc Geol It 102:419–556

10. Nisi B, Capecchiacci F, Frullini R, Huertas Delgado A, Vaselli O, Pranzini G (2007) Inquinamento naturale ed antropico delle acque di falda della pianura costiera livornese fra Rosignano e San Vincenzo (Toscana Centro-occidentale): evidenze geochimiche ed isotopiche. Acque Sotterranee 110:11–20
11. Pennisi M, Gonfiantini R, Grassi S, Squarci P (2006) The utilization of boron and strontium isotopes for the assessment of boron contamination of the Cecina River alluvial aquifer (central-western Tuscany, Italy). Appl Geochem 21:643–655
12. Pranzino G (2004a) Studio Idrogeologico della Pianura Costiera fra Rosignano e San Vincenzo. Relazione inedita per Autorità di Bacino Toscana Costa, 174pp
13. Pranzino G (2004b) Studio idrogeologico del Bacino del Fiume Cecina. Relazione Gennaio 2004, Regione Toscana, Dipartimento delle politiche territoriali e ambientali – Autorità di Bacino Toscana Costa, 55pp
14. S.I.R.A. Sistema Informativo Regionale Ambientale della Toscana. http://sira.arpat.toscana.it/sira/acqua.htm
15. Website of: "Servizio Geologico Della Regione Toscana". http://www.regione.toscana.it/regione/export/RT/sito-RT/Contenuti/sezioni/ambiente_territorio/geologia/rubriche/piani_progetti/visualizza_asset.html_1113854852.html

Chapter 12
Overview on the Occurrence and Seasonal Variability of Trace Elements in Different Aqueous Fractions in River and Stream Waters

Rosa Cidu

Abstract The distribution of trace elements, including toxic and harmful elements, in different aqueous fractions can be investigated via their determination in non filtered and filtered water samples. Concentrations determined in non filtered samples can be roughly regarded as *total* amounts. Filtration through 0.45 μm pore-size filters is conventionally used to remove the matter in suspension, thus concentrations determined in the water fraction <0.45 μm should be regarded as *dissolved* amounts. However, in this fraction inorganic contaminants may occur either as *truly dissolved* species or hosted in fine particles of <0.45 μm size, such as clay and colloidal materials. Results of hydrogeochemical surveys carried out on river and stream waters from Sardinia (Italy) showed significant differences in concentrations of specific elements under different seasonal conditions. Variations in the dissolved amount of trace elements appear to depend on the composition of rocks drained, and the occurrence of mineral deposits, as well as on hydrological conditions, such as runoff, flow and turbulence. The elements B, Li, Rb, Sr, Ba, As, Sb, Mo, and U in the studied waters showed small differences in concentrations determined either in non filtered or filtered water samples. *Dissolved* concentrations were higher in summer, when the contribution of rainwater to the rivers was minimum; concentrations were often positively correlated with Total Dissolved Solids (TDS) and/or major ions. These elements occurred as *truly dissolved* species, which concentrations appeared related to the intensity of water-rock interaction processes. The elements Al, Fe, Pb, Zn, Cd, Co, Ni, Cs, Y, REE and Th were not related to TDS and/or major ions; they showed higher concentrations under high flow conditions; marked differences occurred between *total* and *dissolved* amounts; concentrations in the water filtered through 0.015 μm were much lower than in the water filtered through 0.4 μm, especially

R. Cidu (✉)
Dipartimento di Scienze della Terra, via Trentino 51, Cagliari I-09127, Italy
e-mail: cidur@unica.it

when sampling was carried out after storm events that enhanced the load of solid matter in the water, hence indicating an aqueous transport via sorption processes on very fine particles. Considering that seasonal variability in concentrations of inorganic contaminants may affect the water quality, results from this study can be useful to understand the human- and aquatic-life risk exposure to toxic and harmful elements.

Keywords Trace elements • Aqueous fractions • Filtration • Surface water • Sardinia

12.1 Introduction

The determination of aqueous inorganic contaminants is fundamental in pollution and toxicological studies because some elements might degrade the quality of water even if present at very low (trace) concentrations, notably As, Cd, Hg and Pb [20]. The study of trace elements and their physicochemical properties in river waters is also important for understanding their fate in the ocean. Therefore, significant research efforts have been dedicated to understanding the transport of trace contaminants in the water system, and to use this knowledge in the development of environmental policy and regulation.

Physical and chemical factors may affect the mobility and dispersion of trace elements at given conditions. Spatial variability in the concentration of elements in aquatic systems is mainly related to lithological-morphological heterogeneity and weathering degree. Temporal variability due to climatic factors and hydrodynamic processes may cause relevant seasonal changes on the aqueous amount of inorganic contaminants. There is convincing evidence that the worldwide climate is changing, and that these changes are not part of natural processes or cycles [13]. Climate change is likely to have wide ranging impacts on water resources in many regions of the world [11, 14], but it is difficult to predict impacts on water quality. This has increased the need for robust information on how climate change could affect the fate and transport of inorganic contaminants. The transport of contaminants from sediments and soils into aquifers is sensitive to the intensity and seasonality of rainfall events. More frequent and intense storm events can promote the erosion of solid particles, thus allowing an increase of the suspended load in the water. Discharge peaks in running waters usually result in a drop of concentrations of dissolved major ions due to dilution by rainwater, but the abundant fine materials may act as carrier phases for several trace elements. These processes can be particularly relevant in river water [12, 15], but the sorbed contaminants can also migrate to the unsaturated and saturated zone and may pose a hazard in water bodies used to supply drinking water.

The aim of this study was to investigate the distribution of trace elements in different aqueous fractions of surface waters. River and stream water samples collected in Sardinia under different seasonal conditions have been considered as a case study. Implications in the water quality and management will be briefly discussed.

12.2 The Study Area

Sardinia, an island with surface of 24,090 km^2, is located in the Mediterranean Sea between 38°40′N and 41°40′N, and between 7°50′E and 9°40′E. The highest elevations (up to 1,800 m a.s.l.) occur in the eastern part of the island. Sardinia has variable climatic conditions. Mean rainfall in the 1922–1992 period ranged from <500 mm/a in the plains and coastal areas to about 1,000 mm/a in the mountains, with a mean of <50 and 90 rainy days/a, respectively. Climatic conditions in the region are characterized by long periods of heat and drought interrupted by relatively short rainy periods, with occasional heavy rain events that have been more frequent in the past decade [6]. Accordingly, the flow of rivers and streams is strongly dependent on rainfall. In Sardinia, surface waters constitute an important resource. The river water collected in artificial basins represents about 70% of the water supply for agricultural, industrial and domestic uses for a population of approximately 1.6 million people.

A large variety of rocks outcrops in Sardinia. The main geologic features can be summarized as follows [2]. A Palaeozoic basement, consisting of granitic and metamorphic rocks, extends nearly continuously from north to south in the eastern part, and it is interrupted by Tertiary sediments and volcanic rocks in the western part of the island. Remains of Mesozoic carbonate rocks lay on the basement. A prominent structure, marked by important regional faults and known as the Sardinian Rift, extends from north (the Gulf of Asinara) to south (the Gulf of Cagliari). The Sardinian Rift has been filled by volcanic rocks (Oligo-Miocene andesite, and Plio-Pleistocene basalt), marine and lacustrine sediments (Miocene conglomerate, sandstone, carbonate and marl sequences) and recent continental sediments.

The occurrence of metal (mainly Pb, Zn, Cu) deposits is particularly relevant in Sardinia. Sulphide ores and gossans were exploited intensively since 1880. Most mines were closed in 1980–1990. The residues of mining exploitation and processing have been left on site, frequently without an adequate disposal; they constitute potential hazards and may cause contamination in the aquatic system [5]. Water samples that drain abandoned mining sites were not considered in this study.

12.3 Materials and Methods

A total of 183 river and stream water samples, collected from 1997 till 2008 in Sardinia, were considered in this study. Their flow shows a large range (<10–>10,000 l/s). Detailed sampling was carried out in the Flumendosa [8] and Cixerri [6] basins. At selected sites, sampling was repeated under different seasonal conditions. Among the studied waters, 55 samples refer to waters collected under high rain conditions. The location of water samples is shown in Fig. 12.1.

The physical-chemical parameters and major components, that may influence aqueous speciation, need to be known for understanding the geochemical behaviour

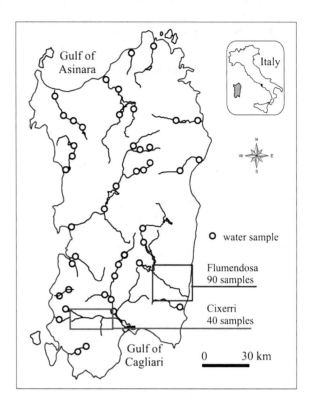

Fig. 12.1 Sardinia map showing the location of river and stream waters considered in this study

of trace elements. Therefore, comprehensive chemical analyses of the studied waters were carried out.

Water samples were collected at sites away from stagnant water or abundant vegetation. At the sampling sites, the physical–chemical parameters and alkalinity were measured. Immediately after collection, the water was filtered through 0.4 μm pore-size Nuclepore polycarbonate filters into pre-cleaned high-density polyethylene bottles. Different aliquots were collected for determining specific components. An aliquot of filtered sample was used for the determination of Cl^-, NO_3^- and SO_4^{-2} by ion chromatography. An aliquot filtered and acidified to 1% (v/v) with HNO_3 (Suprapure grade, Carlo Erba) was used for trace metal analyses by quadrupole ICP-MS (Perkin–ElmerSCIEX ELAN 5000 and ELAN DRC-e) using Rh as an internal standard. In this aliquot, concentrations of Ca, Mg, Na, K, Si, B, Mn, Fe, Zn, Sr and Ba were also determined by ICP-OES (ARL Fisons 3520). An aliquot of filtered sample was acidified to 0.5% (v/v) with HCl (Suprapure grade, Carlo Erba) for the determination of As by hydride generation ICP-MS.

In this study, *dissolved* components refer to those concentrations measured in the filtered (F) sample, that is in the fraction <0.4 μm. At selected sites, two additional aliquots were collected: one non filtered and acidified to 1% (v/v) with HNO_3, for the determination of *total* amounts; and another one filtered first through 0.4 μm then through 0.015 μm pore-size Nuclepore polycarbonate filters (FF) and acidified to 1% (v/v) with HNO_3, for the determination of elements *truly dissolved*.

To check potential contamination during sampling and analyses, blank solutions were prepared in the field with ultra-pure water (MILLI-Q, specific conductance: 0.06 μS/cm) and processed using the same procedure used for the water samples. Certified standard reference solutions were used to estimate the analytical errors [4, 8].

Calculation of aqueous speciation and chemical equilibrium with respect to solid phases was performed using the computer program PHREEQC [16] using the included thermodynamic database *lllnl*.

12.4 Results and Discussion

Table 12.1 reports the minimum, maximum, median, mean and standard deviation (SD) values for temperature, pH, conductivity, total dissolved solids (TDS), nitrate and some *dissolved* trace components in the studied waters; the mean values for world rivers are reported for comparison. A large range of values can be observed for many parameters.

The water temperature was dependent on the air temperature at the time of sampling: the lowest value was recorded in January and the highest in July. Slightly alkaline pH was observed in the majority of samples, as indicated by the small difference between the median and the mean pH values (Table 12.1). The redox potential (not reported) was in the range of 0.40–0.52 V indicating oxidizing conditions.

Table 12.1 Minimum, maximum, median, mean and standard deviation (SD) values for physical-chemical parameters, total dissolved solids (TDS), and selected trace components in river and stream waters collected in Sardinia. Mean values for world rivers are reported for comparison

	Unit	Sardian waters (183 samples; <0.4 μm)					World rivers	
		Min	Max	Median	Mean	SD	Mean	
							<0.4 μm[a]	<0.2 μm[b]
T	°C	8	29	14.0	14.6	5.0		
pH		6.5	9.0	7.8	7.7	0.4		
Cond	mS/cm	0.15	3.20	0.57	0.65	0.48		
TDS	mg/l	82	150	318	369	290		
NO_3^-	mg/l	0.1	29.2	1.1	5.4	8		
Al	mg/l	2	1,400	24	**98**	265	**50**	32
Fe	mg/l	4	1,440	45	**118**	253	**40**	66
Mn	mg/l	1	460	7.0	**23**	48	**8.2**	34
Zn	mg/l	2.7	365	13	**24**	38	**30**	0.6
Cd	mg/l	0.02	2.50	0.10	**0.21**	0.30	**0.02**	0.08
Pb	mg/l	0.09	45	0.60	**2.4**	5.1	**0.1**	0.08
Cu	mg/l	0.1	10.7	1.6	**2.1**	1.6	**1.5**	1.5
Ni	mg/l	0.1	17	0.8	**1.3**	1.5	**0.5**	0.8
Co	mg/l	0.04	13.5	0.15	**0.31**	1.01	**0.2**	0.15

[a]Chester [3]
[b]Gaillardet et al. [10]

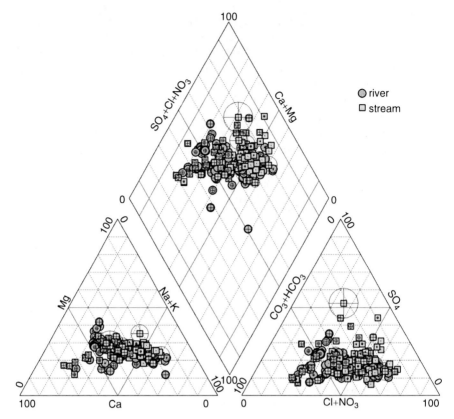

Fig. 12.2 Piper diagram showing the chemical composition of river (>100 l/s) and stream (<100 l/s) waters collected in Sardinia (total samples 183; the largest circle corresponds to 1.5 g/l TDS)

Conductivity varied from 0.2 to 1.8 mS/cm. The lower values were observed in the upper courses of rivers and in catchments dominated by silicate rock outcrops. Conductivity generally increased from the source to the mouth of rivers. At sites where sampling has been repeated, conductivity under low flow condition was higher than values observed under high flow condition.

The Piper diagram shown in Fig. 12.2 summarizes the water chemistry in the Sardinian river and stream waters. Many waters show a dominant sodium-chloride composition and low salinity (TDS: 0.1–0.4 g/l), most of them drain granite and volcanic rocks. Some waters have a dominant calcium-bicarbonate composition and TDS 0.6–0.8 g/l, they drain carbonate formations. The other samples show an intermediate composition between sodium-chloride and calcium-bicarbonate and generally have higher salinity (TDS: 0.6–1.2 g/l).

Significant sulphate contribution and relatively high TDS values correspond to waters flowing on marl-carbonate sediments of Tertiary age.

The relative proportions of major ions at each site did not significantly change when the water was sampled under different climatic conditions, but the concentrations of

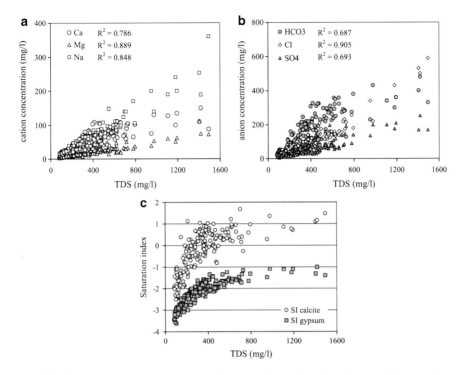

Fig. 12.3 Plots showing concentrations of (**a**) major cations, (**b**) major anions, and (**c**) saturation index (SI) values with respect to calcite and gypsum versus total dissolved solids (TDS) in river and stream waters of Sardinia

major ions generally increased in summer under low-flow condition. This increase may be due to the low contribution of rainwater in summer, to evaporation processes, and/or to the input of more saline groundwater to the rivers [7].

Figures 12.3a, b show the dissolved concentrations of major cations and anions, respectively, versus TDS in the Sardinian river and stream waters. Positive correlations can be observed. Calcium and bicarbonate showed lower correlations versus TDS; this can be explained taking into account that as TDS values increase the waters reached equilibrium or supersaturation with respect to calcite, while all waters were undersaturated with respect to gypsum (Fig. 12.3c).

Concentrations of nitrate species varied from <1 to 29.2 mg/l (Table 12.1), that is below the maximum value (50 mg/l) established by Italian regulations for drinking water. Figure 12.4a shows the box-plots for nitrate concentrations in waters collected during the rainy (high rain) and dry (low rain) seasons. It can be observed that the median NO_3^- concentration at high rain is higher than the median value at low-rain condition. Nitrate flushing mainly occurred in waters draining cultivated and farmed areas, and was often associated to relatively high K contents (Fig. 12.4b). Peak winter values exceeding 100 mg/l NO_3^- concentrations have been observed in other rivers draining intensively farmed areas, such as East Anglia in the UK [1].

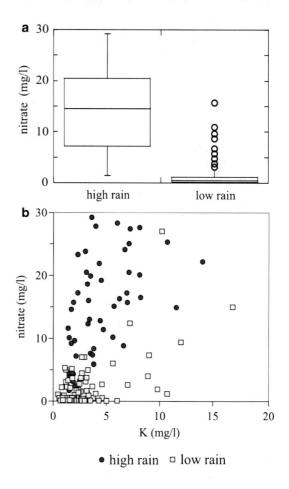

Fig. 12.4 (a) Box-plots showing nitrate concentrations in river and stream waters collected under high and low rain conditions in Sardinia; the line in each box indicates the median value. (b) Nitrate versus potassium concentrations for the same waters

The wide range in concentrations and high standard deviation values reported in Table 12.1 highlight marked spatial variations of trace elements in the studied waters. Mean concentrations of trace elements in the Sardinian waters are similar to those estimated for the world rivers. Exceptions are Cd and Pb being significantly higher in the Sardinian waters (Table 12.1). This might be due to the diffused occurrence of base-metal deposits in the region, although waters known to drain abandoned mining sites were not considered in this study. At some sites, abandoned electric appliances, car batteries and car frames discharged close to the rivers were observed [4]; these wastes might have contributed to enhance the Cd and Pb mean values.

Two major groups of elements can be distinguished for their different behaviour in the studied waters. The first group includes the elements B, Li, Rb, Sr, Ba, As, Sb, Mo and U. Concentrations of these elements generally increased with increasing TDS. This group is characterized by small differences between *total* and *dissolved* concentrations. Also, when comparing different pore-size filters, differences between

Fig. 12.5 Plots showing elements determined in two aliquots of the same water sample: one (FF) filtered through 0.015 μm pore-size filters versus another (F) filtered through 0.4 μm pore-size filters

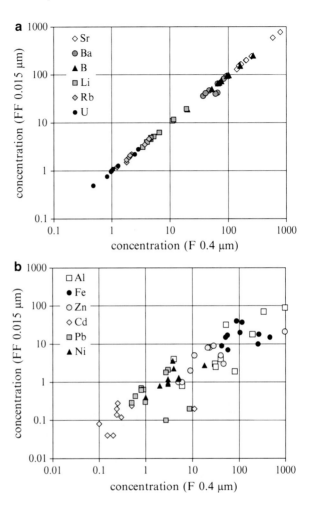

concentrations determined in the <0.4 μm (F) and <0.015 μm (FF) fractions were found within analytical errors (e.g. Fig. 12.5a).

At sites where time series are available, it has been observed that concentrations of these elements increase under low flow conditions. Sampling soon after storm events resulted in lower concentrations (Fig. 12.6a), likely due to the dilution effect of rain water. The calculated and most abundant aqueous species were: free cations (Li^+, Rb^+, Sr^{+2}, Ba^{+2}), like as the major alkaline (Na^+, K^+) and alkaline-earth (Ca^{+2}, Mg^{+2}) cations; oxy-anions (e.g. $HAsO_4^{-2}$, MoO_4^{-2}), like as the major SO_4^{-2} anion, and negatively charged complexes, e.g. $UO_2(CO_3)_3^{-4}$. The above observations indicate that the first group of elements is transported as *truly dissolved*, which is consistent with predictions based on the ionic potential values of these elements. Concentrations of the first group of elements appear to depend on the composition of rocks drained and on the intensity of water-rock interaction processes.

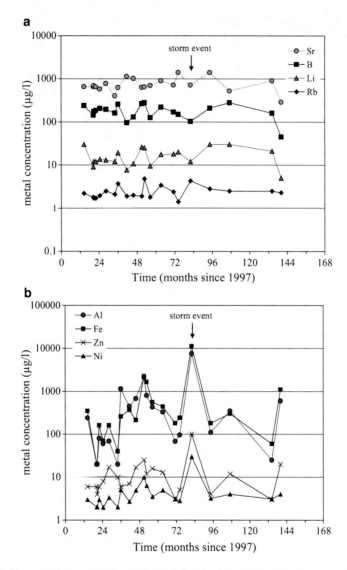

Fig. 12.6 Plots showing temporal variations of *dissolved* metal concentrations at one sampling site in a Sardinian river. All concentrations were determined in aliquots filtered through 0.4 μm pore-size filters. The arrow indicates water sampling carried out soon after a storm event

The second group of elements Al, Fe, Pb, Zn, Cd, Co, and Ni showed large differences between *total* and *dissolved* concentrations. When comparing different pore-size filters, concentrations determined in the <0.015 μm (FF) fraction were often much lower than those determined in the <0.4 μm (F), as it can be seen in Fig. 12.5b. These elements were related neither to TDS (e.g. Fig. 12.7a) nor to major ions.

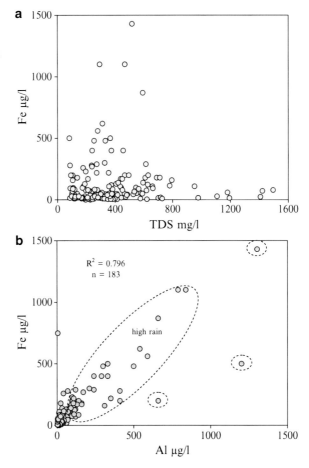

Fig. 12.7 Plots showing (**a**) concentrations of Fe versus total dissolved solids, and (**b**) versus Al, in river and stream waters of Sardinia. Both Fe and Al concentrations were determined in aliquots filtered through 0.4 μm pore-size filters. Dashed lines in plot (**b**) indicate water samples collected under high rain conditions

On the basis of speciation computation, the most abundant aqueous species were: $Al(OH)_3^0$, $Al(OH)_4^-$, $Fe(OH)_3^0$, $Fe(OH)_2^+$, $PbCO_3^0$, $PbHCO_3^+$, Pb^{+2}, Zn^{+2}, Cd^{+2}, Co^{+2}, and Ni^{+2}.

At sites where time series were available, it has been observed that variations in concentrations occurred seasonally, mainly depending on the contribution of runoff to the river. Much higher concentrations were observed in waters collected during rainy periods (Fig. 12.7b), and dramatic increases occurred when sampling was carried out soon after storm events (Fig. 12.6b). Under such conditions, the water showed high turbidity. The above observations indicate that the second group of elements is mainly transported in the studied waters via fine particles and colloids. Investigations on trace metals in other river waters have already documented a strong peak of colloidal concentrations of trace metals during the spring flood [9].

When detected, Cs, Y, all rare earth elements (REE) and Th behaved similarly to the second group of elements [4, 6].

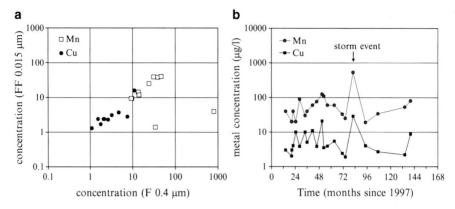

Fig. 12.8 Plots showing: (**a**) Mn and Cu determined in two aliquots of the same water sample (FF filtered through 0.015 μm versus F filtered through 0.4 μm pore-size filters); (**b**) temporal variations in *dissolved* concentrations of Mn and Cu in a Sardinian river water (the arrow indicates water sampling carried out soon after a storm event)

The elements Mn and Cu in the studied waters showed an intermediate behaviour between the two groups of elements distinguished above, probably reflecting their variability in speciation. The most abundant aqueous species were Mn^{+2}, $Mn(CO_3)^0$, $Cu(CO_3)^0$, $Cu(OH)^+$, and Cu^{+2}. Figure 12.8a compares Mn and Cu determined in the F and FF aliquots; either small or marked differences may occur between concentrations determined in 0.4 and 0.015 μm pore-size filtered aliquots. Figure 12.8b shows that both Mn and Cu increase under high rain condition, like the second group of elements.

On the whole, results of this study are in good agreement with results obtained using ultrafiltration methods for distinguishing elements associated with colloidal materials, either in Arctic river waters [18] or in soil column percolated solutions [17].

12.5 Conclusions

It is known that both size and composition of colloids may influence the aqueous concentration of elements depending on the adsorptive affinity and aqueous speciation of individual elements [19]. Colloids do not easily settle down out of suspensions because of their small size (1 nm to 1 μm). Thus, the solid–water interface established by these micro-particles plays a key role in regulating the concentrations of most elements and many pollutants in soils and natural waters [19].

In this study, variations in the *dissolved* amounts of trace elements appear to depend on the composition of rocks and soils drained, as well as on hydrological conditions, such as runoff, flow and turbulence. All trace elements do not behave and interact in the same way with the surrounding medium. Seasonal variations at specific sites resulted in contrasting trends for different elements: concentrations in a first group of elements (B, Li, Rb, Sr, Ba, As, Sb, Mo and U) showed small variations with respect to a second group (Al, Fe, Pb, Zn, Cd, Co, and Ni). Then, concentrations of the latter elements increased more than one order of magnitude when sampling was carried out after heavy rain, whereas the first group showed a reverse trend.

Summarizing, the first group of elements does not appear prone to be mobilized via the colloidal pool in the studied waters; these elements are transported as *truly dissolved*; peak concentrations are expected to occur under low flow conditions. In contrast, the second group seems much more involved in an aqueous transport controlled by colloids, either inorganic or organic; peak concentrations are expected to occur under heavy rain conditions when high runoff usually brings abundant solid particles to the river water.

The results of the present study may be useful for understanding the seasonal variability of trace element concentrations, and for evaluating the relative proportions of suspended, colloidal and *truly dissolved* (potentially bioavailable) trace elements. In turn, information derived from these results could allow estimation of the risk of exposure to the toxic and harmful elements, both to humans and aquatic life. Also, the potential peak concentrations of some contaminants under heavy rain events should be taken into account in the management of river and stream waters, especially those destined to domestic purposes.

Acknowledgments The dataset used in this study derives from several research programs coordinated by the Author, and funded by MIUR (PRIN 2004 and PRIN 2007 Cidu), the University of Cagliari and the Fondazione Banco di Sardegna. Analyses by ion chromatography were carried out by G. Contis. Thanks to reviewers for useful suggestions.

References

1. Alloway BJ, Ayres DC (1997) Chemical principles of environmental pollution, 2nd edn. Blackie Academic and Professional, London, 395 pp
2. Carmignani L, Oggiano G, Barca S, Conti P, Eltrudis A, Funedda A, Pasci S (2001) Geologia della Sardegna – Memorie descrittive della carta geologica d'Italia, vol. LX. Servizio Geologico d'Italia, Roma
3. Chester R (1990) Marine geochemistry. Unwin Hyman Ltd, London
4. Cidu R, Biddau R (2007) Transport of trace elements at different seasonal conditions: effects on the quality of river water in a Mediterranean area. Appl Geochem 22:2777–2794
5. Cidu R, Fanfani L (2002) Overview of the environmental geochemistry of mining districts in southwestern Sardinia, Italy. Geochem Explor Environ Anal 2:243–251
6. Cidu R, Frau F (2009) Distribution of trace elements in filtered and non filtered aqueous fractions: Insights from rivers and streams of Sardinia (Italy). Appl Geochem 24:611–623
7. Cidu R, Biddau R, Manca F, Piras M (2007) Hydrogeochemical features of the Sardinian rivers. Periodico Mineral 76:41–57
8. Cidu R, Caboi R, Biddau R, Petrini R, Slejko F, Flora O, Stenni B, Aiuppa A, Parello F, Valenza M (2008) Caratterizzazione idrogeochimica ed isotopica e valutazione della qualità delle acque superficiali e sotterranee campionate nel Foglio 549 Muravera. In: Ottonello G (ed.) GEOBASI. Pacini Editore, Pisa, pp 149–183. ISBN 978-88-7781-9260

9. Dahlqvist R, Andersson K, Ingri J, Larsson T, Stolpe B, Turner D (2007) Temporal variations of colloidal carrier phases and associated trace elements in a boreal river. Geochim Cosmochim Acta 71:5339–5354
10. Gaillardet J, Viers J, Dupré B (2003) Trace elements in river waters. In: Drever JI (ed) Surface and Ground Water, Weathering and Soils. In: Holland HD, Turekian KK (eds) Treatise on Geochemistry, vol 5, Elsevier–Pergamon, Oxford, pp 225–272
11. IPCC (Intergovernmental Panel on Climate Change) (2008) Climate change and water: technical paper of the intergovernmental panel on climate change. IPCC Secretariat, Geneva, p. 210
12. Kimball BA, Bianchi F, Walton-Day K, Runkel RL, Nannucci M, Salvadori A (2007) Quantification of changes in metal loading from storm runoff, Merse River (Tuscany, Italy). Mine Water Environ 26:209–216
13. King DA (2004) Climate change science: Adapt, mitigate, or ignore. Science 303:176–177
14. Le TPQ, Garnier J, Gilles B, Sylvain T, Van Minh C (2007) The changing flow regime and sediment load of the Red River, Viet Nam. J Hydrol 334:199–214
15. Nordstrom DK (2009) Acid rock drainage and climate change. J Geochem Explor 100:97–104
16. Parkhurst DL, Appelo CAJ (1999) User's guide to PHREEQC (version 2) – A computer program for speciation, batch-reaction, one-dimensional transport, and inverse geochemical calculations. US Geological Survey Water-Resources Investigations Report 99-4259
17. Pédrot M, Dia A, Davranche M, Bouhnik-Le Coz M, Henin O, Gruau G (2008) Insights into colloid-mediated trace element release at the soil/water interface. J Colloid Interface Sci 325:187–197
18. Pokrovsky OS, Viers J, Shirokova LS, Shevchenko VP, Filipov AS, Dupré B (2010) Dissolved, suspended, and colloidal fluxes of organic carbon, major and trace elements in the Severnaya Dvina River and its tributary. Chem Geol 273:136–149
19. Stumm W, Morgan JJ (1996) Aquatic chemistry. Wiley–Interscience, New York
20. WHO (2006) Guidelines for drinking-water quality, 4th edn, Annex 4. World Health Organization, Geneva. ISBN 9241546964

Chapter 13
A Worldwide Emergency: Arsenic Risk in Water. Case Study of an Abandoned Mine in Italy

Luca Fanfani and Carla Ardau

Abstract Water is an essential resource for the development of human societies. Inadequate availability of good quality water especially in the future will probably cause transboundary and social conflicts. Presently, As contamination in water represents the most relevant risk for many populations in the world, because the diffusion of the contaminant has been facilitated by inappropriate water management, and the effects on the health are serious, though delayed in time. Risk mitigation is expensive and sometimes not completely successful. The case study concerns a small hilly area around an abandoned mine in Sardinia (Italy), and the downstream coastal plain where a subsistence agriculture is developed. Uncontrolled dispersion of wastes from the exploited area and the processing plant, transported downstream, is still affecting (after 40–50 years) the quality of the water in the catchment with concentration of As up to 0.9 mg/l. A study of the As dissolved content in the excavated wells of the plain reveals an irregular spatial distribution with higher values (up to 1 mg/l) in low-lying zones covered by contaminated sediments overflooded from the nearby river. The remediation plan, limited to the hilly area, intends to reduce the supply of contaminant downstream and includes the building of a save dump, where most of the waste-rocks and tailings will be collected, treating the acid waters from the adits on-site, and reclamating in situ small old dumps in the lower part of the catchment.

Keywords Water resources • Arsenic contamination • Mining • Remediation

L. Fanfani (✉) • C. Ardau
Dipartimento di Scienze della Terra, Università di Cagliari, Via Trentino 51,
Cagliari I-09127, Italy
e-mail: lfanfani@unica.it

13.1 Introduction

Environmental security requires actions devoted to avoid social and transboundary conflicts. The target to guarantee enough water of good quality for domestic, agricultural, industrial and recreational use is undoubtedly one of the primary aims to pursue inside any community and any international scenario.

In this respect we must take in mind that scarce and unsafe water already causes transboundary contrasts in the Middle East and North Africa, and the difficult access to safe water is cause of extreme poverty in several Asian, African and South-American regions.

The supply of arsenic free water represents a definitive challenge for many countries such as Bangladesh, where the last Government elected on January 6, 2009 committed in its election manifesto to the fact that the arsenic problem will be taken to ensure the supply of safe drinking water for all the population by 2011 [11].

In the preface of the book "Arsenic Pollution. A global synthesis", K. Richards [20] affirms that "the most severe effect of human impact on environmental system is not climate change but arsenic crisis." He continues: "One similarity between global warming and the arsenic crisis is that in both, human actions accentuate risks associated with otherwise natural phenomena. Another is that the consequences affect the global poor most severely. Indeed, this dimension is already obvious from the history of the arsenic crisis. Nearly 50 million people in south and east Asia have, for some decades, drunk water contaminated with arsenic at the levels above the old WHO standard of 50 ppb. Many already have clinical symptoms of arsenicosis, leading to this being referred to as history's largest mass poisoning. By contrast, the USA has diverse sources and types of arsenic contamination in water supplies, but little evidence that this has a significant effect, because of better water treatment and better general health in the population."

Adverse health effects have been documented in China, India (West Bengala), Taiwan and United States of America, besides Bangladesh. In many other countries like Argentina, Australia, Chile, Hungary, Mexico, Peru and Thailand arsenic in drinking water has been detected at concentrations higher than 0.010 mg/l, the guideline value recommended by WHO.

Arsenic is widespread in natural environments but, as mentioned above, its mobility depends on many factors related to human activity. Arsenic is documented in fossil fuels, but contamination derives from burning them; arsenic is documented in ores, but important risks derive only when ores are exploited; in most cases (often the most dangerous ones) arsenic contamination has a geologic origin, but the lack of appropriate groundwater exploitation and of a wise management of the risk are at the origin of the health emergency. Modeling arsenic mobility in different aquatic environments becomes an important task to assess the arsenic-derived health risk and may help to individuate technological mitigation options.

The case study we are going to present here examines the effect of an unwise management of a mining activity, which, combined with the lack of any mitigation plan for 40 years after the mine closure, poses serious difficulties to maintain a poor agricultural economy, but of great importance for some farmers. The case may be considered of scarce importance when its dimension is compared with several situations at world scale, but useful to comprehend the development of the risk, if not adequately monitored and contrasted.

13.2 Arsenic in Mine Areas

Arsenic is documented in ores (mainly in sulphide and sulphosalt minerals), but important contamination is only observed when they are mined. Modern mining can be seen as a process that begins with the exploration and discovery of a mineral deposit, and continues, through the ore extraction and the treatment processing up to the closure and the remediation of the worked-out site [22]. This last aspect, which is crucial for the consequences on environment and public health, has been often disregarded in the past.

Exploitation results in the production of immense quantities of waste rocks, mill tailings and wastes from refining processes, representing potential sources of As contamination. Differently from natural sources of contamination, mining is responsible for As diffusion in several environmental matrixes: surface water, groundwater, soil, vegetation, and air ([9] and references therein).

Health risk in mine areas affects not only workers, but the whole population living around the areas, in particular children, so that millions of people in the world are estimated to be exposed to metals in mine areas. Differently from other kind of land degradation, chemical contamination may be not adequately perceived because frequently invisible. In some countries, population around chemically compromised areas still now has no conscience or knowledge of a possible danger to human health, so that does not care to undertake any precaution against As exposure. Not many studies were conducted about the risk of arsenic exposure in mine areas. Yáñez et al. [22] conducted a study on the risk of As and Pb exposure of children living in the urban area of Villa de la Paz (Mexico), near the homonymous mine; in that case, contaminated household dust is the main pathway of exposure (through ingestion of particles adhering to food, toys, and surfaces in the houses). They found that urinary arsenic in children in Villa de la Paz was significantly higher than that found in children from a less exposed community 15 km away from Villa de la Paz. Asante et al. [4] conducted a study to assess the contamination from trace elements, especially As, in mine workers and inhabitants of Tarkwa, an Au mining town in Ghana, where people usually drink river water. Some samples from the river showed As concentrations above the WHO guideline values for drinkable use (As ranges: 0.5–73 µg/l), while urinary As concentrations in population were comparable with those in some well-known arsenic-affected countries of the world. The presence of other sources of arsenic, as the ingestion of contaminated food, was hypothesised.

A study conducted on the abandoned tungsten mine in Shantou (southern China) [17] showed a severe As contamination in the agriculture soil of the surrounding area. Data monitoring health impact indicated high As concentrations in hair and urine as a potential health risk for the local residents.

Even when drinking water is not directly involved in arsenic poisoning of population, water can represent a preferential pathway for As contamination as in the case of the paddy rice which represents large part of Asian people diet.

13.3 Case Study

In the past century, up to 1980s, mining was the driving sector of the economy in south-eastern Sardinia (Italy). In the name of the immediate profit and the need of facing unemployment, and due to the scarce care of the environment and human health, mining was allowed by license to cause land degradation, which affected traditional economic sectors as agriculture or sheep farming, at the moment considered less profitable. When mining activity stopped, as a consequence of the crisis of the sector, tons of sulphide-bearing wastes from mining and mineral processing, including waste rocks or mill tailings, were abandoned in piles or impoundments near the mine works, and/or used for mine tunnels refilling. Waste piles pose serious threats to the environment; sulphide minerals exposed to air and water originate oxidation products, which generate acid effluents and release Fe, SO_4 and potentially toxic elements into surface water and groundwater. Such materials, often maintain the capability to supply contaminants to the environment for many decades to centuries, if effective remedial programs are not undertaken. Toxic element dispersion occurs both from dissolution and solid transport. In Sardinia poor management has led to intense heavy metal contamination of water and soil over large areas (e.g., [6, 7]). An emblematic case is represented by the deposit of Baccu Locci (south-eastern Sardinia), exploited for about a century (1873–1965) for Pb and As. In the years of more intense activity the mine employed up to 600 workers, many from the near villages, silencing any possible criticism from local government about the unwise water management in the area (over all the practice of discharging tailings from the flotation plant directly into the Baccu Locci stream). Environmental studies (e.g., [2, 10, 16, 19] and references therein) aimed to establish the level of contamination were conducted only in the last decade. These studies pointed out a diffuse, severe arsenic contamination affecting surface water and groundwater as far as the alluvial plain of the river Quirra, which receives the inflow of Baccu Locci stream (BLS) not far from the Tyrrhenian coast.

13.4 Description of Study Area

13.4.1 Geography, Hydrography and Ore

The Baccu Locci mine is located in the Sarrabus-Gerrei mining district, in the Quirra region, (south-eastern Sardinia), in a hilly territory with an altitude from 170 to 430 m, cut deeply by a well-developed network of streamlets, all tributaries of BLS.

Upstream, there is the highland of Monte Cardiga (early eocenic conglomerates), downstream, the alluvial coastal plain of Quirra. BLS crosses the mine area from north-west to south-east, and continues its flow in east direction toward the Tyrrhenian sea, changing its name in Corr'e Cerbus stream; then it flows into the Quirra river, at the coastal plain (Fig. 13.1). BLS is characterised by a torrential regime (discharge: <0.01–100 m^3/s), with long periods of minimum flow and sporadic, sometimes catastrophic, floods. During the prolonged dry periods, the stream water flows underground for long stretches, and the bed sediments remain exposed to the air. BLS joins the Quirra river (discharge: 0.1–1,000 m^3/s) underground, except in the periods of high flow.

Within the mine area there is a reservoir (capacity 55,000 cubic meters) built in '40s on a BLS tributary, to supply water for the mining operations.

The sulphide deposit occurs in a Cambrian to Devonian volcano-sedimentary sequence. It mainly consists of black phyllitic shales, grey shales, phyllites, metasandstones, metarhyolites and metarhyodacites, associated with late Hercynian magmatic products such as dioritic porphyrites and lamprophyres in dikes [8]. Ore bodies occur either as concordant lenses or as discordant veins within the Palaeozoic sedimentary rocks, and also as stockworks within the igneous rocks. The mineralization is composed of prevalent arsenopyrite and galena in about equal proportions, and minor amounts of sphalerite, pyrite, chalcopyrite and pyrrhotite; gangue minerals are quartz, minor calcite and siderite, and rare fluorite [5, 23].

After mine closure about 40 dumps, for a total of 42,000 m^3 of waste material were abandoned on the land and exposed to the weathering action; at least 180 tailings hardpans (for a total of 6,000 m^3) are still visible along the middle-lower BLS course [1, 12]. At present, it is impossible to estimate the amount of tailings intrinsically mixed, as loose material, with stream-bed sediments, as well as wastes and tailings transported as far as the alluvial plain (in particular during flooding events) and afterwards covered by sediments and soil.

13.4.2 After Mine Closure

Since the mine closure in the 1960s the area remained uninhabited, and the nearby military base of Perdasdefogu preserved the landscape around the abandoned mine from considerable anthropogenic modifications. Later on, the Baccu Locci mine was included in one of the eight districts of the Geomining Park of Sardinia, UNESCO World heritage [21]. Minerals of uncommon beauty or rarity such some selenium minerals (calcomenite [$CuSeO_3 \cdot 2H_2O$], francisite [$Cu_3Bi(SeO_3)2O_2Cl$] and orlandiite [$Pb_3Cl_4(SeO_3) \cdot H_2O$]), attracted many collectors. On the other side, poor management of the abandoned mining works led to a severe heavy metal (in particular As) contamination over a large area around the mine, which was unconsidered until the last decade. The first diffusion of scientific data on As contamination in the Baccu Locci area goes back to 2001 [3, 14]. Only in 2002 warning signs were positioned to notify population for the risk of "heavy metals contamination" in the area.

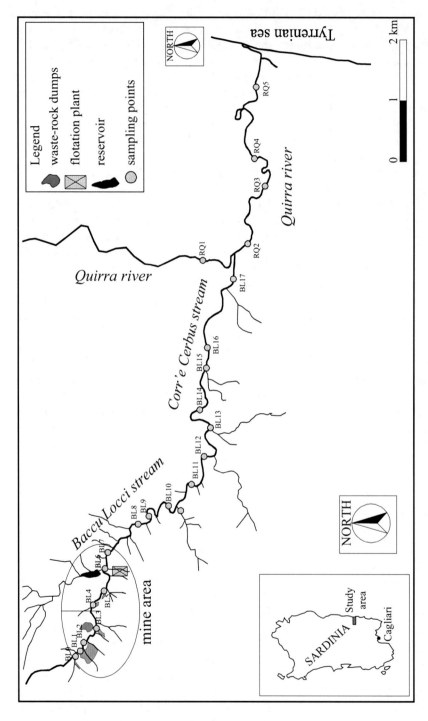

Fig. 13.1 Schematic map of the study area with location of sampling points. BL=Baccu Locci stream, RQ=Quirra river (Modified from [12])

13 Arsenic Risk in Water   183

Thirty-eight years after mine closure, every activity concerning land and water use was forbidden for a distance of 1 km from both side of BLS banks (including pasture, harvest but also the normal maintenance of the roads). A big tourist project aimed to revalue the area was suspended. However, the coastal plain of Quirra (approximately 10 km^2), which represents the area of major concern (because inhabited) is so far excluded from any interdiction. Here, a little rural community (about 150 inhabitants), cares moderate agricultural activity (mainly olive and orange trees), in small properties which extract water, in the dry season, from excavated shallow wells.

The geochemical-mineralogical investigations aimed to individuate and characterize As contamination sources and to model the geochemical processes that affect the chemistry of the waters extended for a distance of about 10 km from the mine and an area of about 100 km^2. The diffuse presence of waste rock dumps (As up to 100 g/kg in the <2 mm fraction) and tailings hardpans (As up to 24 g/kg) represent the main contamination sources in the area [13].

A description of main results of environmental studies [2, 10, 12] conducted in the area will be given in the following paragraphs. A short description of a rehabilitation plan over a large area – including a portion of territory external to the mine site- aimed to reduce the risk of As and heavy metals dispersion [1] will be given at the end.

13.5 Legacies from the Abandoned Mine

13.5.1 *Surface Water Quality*

The chemistry of BLS water changes over a distance of about 10 km, evolving from a low-metal Ca-Mg-HCO$_3$ type to a high-metal Ca-Mg-SO$_4$ type, as soon as the BLS drains the mine area (Fig. 13.2). Indeed, the chemistry of the BLS water is affected by the mixing with some tributaries highly contaminated by the direct interaction with the mine works and the waste-rock dumps, and the inflow of acid drainage from adits. This effect is only mitigated by mixing with low-metal Ca-Mg-HCO$_3$ or Ca-Mg-HCO$_3$-SO$_4$ water from few seasonal tributaries and the effluent of the reservoir built to supply water to the plant.

Concentration of most of the metals (e.g. Al, Cd, Cu, Fe, Mn, Ni, Pb, and Zn) shows a fast increase in the upper BLS course due to inputs from the contaminated tributaries or point sources of contamination (e.g. run-off water and seepages from the waste-rock dumps). This increase is followed by a strong attenuation downstream the flotation plant. In fact, the diffuse presence of tailings along the middle-lower BLS course does not affect the metal concentration significantly.

The main factors attenuating the metal concentration along the BLS course are: (i) a rapid (co-)precipitation of prevalently amorphous metal-bearing phases owing to sudden conditions of oversaturation or sorption onto Fe-Mn-Al hydroxides (several metals are not stable at the near neutral conditions of BLS water and tend to be rapidly removed from the solution); (ii) a dilution process from low-metal

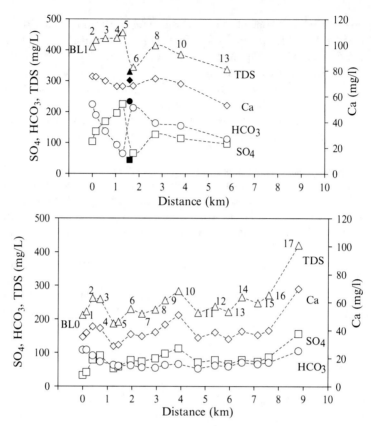

Fig. 13.2 Chemical evolution of the Baccu Locci stream water over 9 km distance. Graphics refer respectively to sampling campaigns in a minimum flow period (*top*) and in a rainy period (*bottom*). In the minimum flow period the stream water flows underground downstream BL6 sampling point. *Solid symbols* refer to the reservoir

surface water (from the reservoir or some seasonal tributaries) and groundwater during the dry season.

The distribution of As dissolved content along BLS highlights a behaviour that is substantially different from that of metals, showing an increase in concentration along the course of the stream (Fig. 13.3). Only in the mine area significant local variations are observed. This irregular trend can be related to local As inputs from contamination sources followed by precipitation of arsenic-bearing sulphate phases and/or arsenic sorption onto Fe-Mn-Al hydroxides. Downstream the mine area the concentration of As increases constantly up to a maximum of about 0.9 mg/l (March 2000) due to the interaction between BLS water and tailings accumulated along the banks and mixed with stream sediments. The release of As from the tailings can be mainly attributed to desorption processes from the surface of Fe(III)-hydroxides occurring as coatings around silicate grains [12]. Such processes are favoured by circum-neutral pH values and the competition of HCO_3^{-1} for sorption sites [15].

13 Arsenic Risk in Water

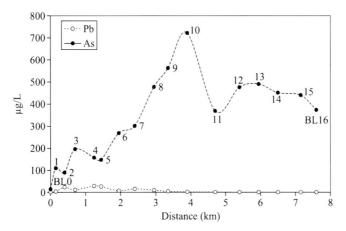

Fig. 13.3 Pb and As distribution along the Baccu Locci stream course. Data refer to a rainy period (Modified from [12])

More downstream, mixing with uncontaminated water from a tributary reduces the concentration of arsenic, though still at very high levels (on average 0.45 mg/l).

Speciation studies point out that arsenic occurs in solution exclusively as the less toxic form As(V). The As(III) form is not stable at the pH and Eh conditions of the BLS water and is rapidly oxidised to As(V).

13.5.2 Groundwater Quality

Sediments in the plain at depth more than 2 m exhibit As contents lower than 50 mg/kg; very high values of As up to 1,300 mg/kg are observed in the upper part of soils and shallow sediments lying in the north-western part of the plain and/or relatively close to the banks of the Quirra river. Generally, anomalous As values are located in low-lying zones covered by contaminated sediments overflooded from the river [18].

Groundwater sampling campaigns were carried in both wet and dry season. Samples were collected from excavated wells relatively homogenously distributed over the whole plain, but with a denser sampling net in the central part where higher As contents were detected.

No large differences in chemical data arise from wet and dry season samplings. Only a slight increase in salinity with uniformly higher dissolved amounts of major constituents is observed after the dry season. All samples show approximately neutral pH. Bicarbonate-chloride terms prevail in the western and central part of the plain while water has an alkaline-chloride composition and a reducing Eh (around 50 mV) in the eastern area close to the coast; this seems to be related to the presence of clay mud sediments with low permeability originated by reclamation of the coastal salt marsh.

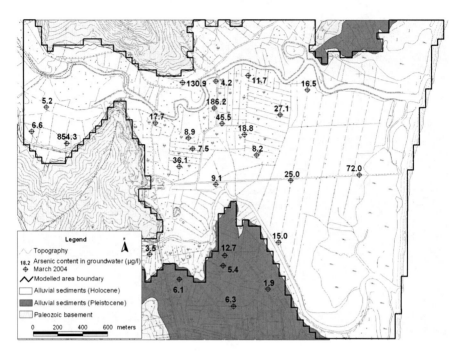

Fig. 13.4 Dissolved content of arsenic in groundwater (March, 2004) – (From [10])

More than twenty minor and trace elements were analyzed in the water samples of the two campaigns with detection limits between 1 and 0.1 µg/l. From the data it is observed: (i) the dissolved content of As is high in several wells (between 0.1 and 1 mg/l) (Fig. 13.4) with an irregular space distribution, similar in the two campaigns (higher values are observed in the dry season); (ii) the dissolved content of As is low in the Pleistocene aquifer in the southern part of the plain; (iii) though lead is the second contaminant (in amount) originating from Baccu Locci mine, its dissolved content in the wells of the coastal plain of Quirra is generally low, ranging from less than the detection limit (0.1 µg/l) up to 12 µg/l.

13.5.3 Remediation Plan

Nowadays, according to national and regional legislation (following indications of European Union) contaminated sites have to be rehabilitated in order to prevent health risks and to preserve natural resources (water, air and soil). The aim of rehabilitation plans should be restoring environmental conditions at a level not far from that before mining operations. The abandoned mine site of Baccu Locci represents the first example of application of a rehabilitation plan on a large area in Sardinia, aimed to reduce the potential risk caused by As and heavy metals diffusion.

Site characterisation confirms that effects of mining on water system are still present in the environment up to several kilometres from the mine area, on account of the wide diffusion of contamination sources. The system have the natural capacity to control effectively the metal contamination in surface water and groundwater; anomalous values with respect to the natural baseline are limited to the mine area. Because of its peculiar behaviour, arsenic escapes this natural control and the level of contamination tends to increase in surface water downstream the mine area, and to be significantly high in some wells of the coastal plain. The point contamination sources of As in the plain of Quirra are spatially related to the flooded areas covered by contaminated wastes transported from the mine area of Baccu Locci. The randomly distribution of As in wells makes challenging the full comprehension of its pathway in groundwater and individuation of possible attenuation mechanisms.

Since As is documented to be present in very high concentrations in the contamination sources and associated with slowly soluble mineralogical phases, As contamination is presumed to persist for a very long time if adequate remediation activities are not undertaken.

Rehabilitation plan started in autumn 2008 and is still in progress. It includes the reclamation of the waste rock dumps, and treatment of acid waters from adits [1]. About 42,000 m^3 of waste rocks and tailings will be moved to a save dump in the upper part of the mine, and on-site treatment of acid waters from adits will be obtained with a treatment through an adsorbent filters system.

The location of the Baccu Locci mine in an impervious territory challenged geologists and engineers when designing the rehabilitation plan. The top of the area turned up to be unsuitable to support the dump, due to a network of superficial mine tunnels, and a fractured substrate. Therefore, the company decided to locate the dump in the mountainside, and to model the steep slope with steps to increase the stability of the dump; specific materials will be used to protect the dump from water infiltration. The difficult road network in the area suggested not to move materials from the old waste-rock dumps in the lower area, but to reclaim them in situ. With respect to the acid water from adits, problems arise from the different geochemical behaviour of contaminants involved and require two treatments: an effective immobilization of As will be obtained through an adsorbent filters system at low pH values, while the same process for heavy metals will be applied at high pH values [1].

13.6 Conclusions: What to Learn from the Abandoned Mine in the Baccu Locci Catchment

The example of Baccu Locci showed that we are still suffering heavy environmental degradation from the waste legacies and unwise water management in mined areas, with a negative impact on economy and public health. On the other hand it also showed that thanks to an accurate characterisation plan including geochemical, mineralogical and hydrological investigations it has been possible to individuate sources, paths and targets of arsenic contamination in the environment, and to arrange adequate countermeasures. Though BLS surface water is expected to improve quality in the next years due to the positive effect of the remediation plan, nothing has been undertaken to mitigate the risk in the coastal plain, where As pollution of groundwater is documented. New international policy on environmental security assessment should take in account that the lack of environmental prevention actions in mining, inevitably demands remediation actions extremely expensive, which, however, may be not satisfactory in recovering completely water resources.

In the future it will be mandatory for all countries not to spoil natural resources. Demographic boom, global climate changes, diffuse industrialization are some of the reasons for a constant increasing of water demand in the Mediterranean area. The lack of a correct water supply (in terms of quality and quantity) will inevitably drive us toward serious social conflicts. Scientific community should play a key-role in indicating the way to preserve environment and peace.

Acknowledgments This article has been written thanks to the contribution of many scientists, that in the past years studied, under different aspects, the site of Baccu Locci, which has been an exceptional natural laboratory to improve the knowledge on arsenic geochemistry and mineralogy.

References

1. Angeloni A, Bavestrelli A, (2009) Bonifica e recupero ambientale dell'area mineraria di Baccu Locci nei territori dei comuni di Villaputzu e San Vito – Provincia di Cagliari. In: Proceedings Sardinia 2009, Twelfth international waste management and landfill symposium, S. Margherita di Pula, Cagliari, Italy, 5–9 Oct 2009
2. Ardau C (2002) Mineralogy and geochemistry of arsenic from the dismantled mine area of Baccu Locci. Plinius 27:39–43
3. Ardau C, Frau F, Dadea C, Lattanzi P, Mattusch J, Wennrich R, Titze K (2001) Solid-state speciation of arsenic in waste materials and stream sediments from the abandoned mine area of Baccu Locci (Sardinia, Italy). In: Cidu R (ed) Proceedings WRI-10, Villasimius, Italy, 10–15 June 2001. A.A. Balkema Publishers, Rotterdam, ISBN: 9026518242, 1173–1176
4. Asante KA, Agusa T, Subramanian A, Ansa-Asare OD, Biney CA, Tanabe S (2007) Contamination status of arsenic and other trace elements in drinking water and residents from Tarkwa, a historic mining township in Ghana. Chemosphere 66:1513–1522
5. Bakos F, Carcangiu G, Fadda S, Mazzella A, Valera R (1990) The gold mineralization of Baccu Locci (Sardinia, Italy): origin, evolution and concentration processes. Terra Nova 2:234–239
6. Cidu R, Biddau R, Fanfani L (2009) Impact of past mining activity on the quality of groundwater in SW Sardinia (Italy). J Geochem Explor 100:125–132
7. Concas A, Ardau C, Zuddas P, Cristini A, Cao G (2006) Mobility of heavy metals from tailings to stream waters in a mining activity contaminated site of Sardinia (Italy). Chemosphere 63:244–253
8. Conti P, Funedda A, Cerbai N (1998) Mylonite development in the Hercynian basement of Sardinia (Italy). J Struct Geol 20(2/3):121–133
9. Duker AA, Carranza EJM, Hale M (2005) Arsenic geochemistry and health. Environ Int 31:631–641
10. Fanfani L, Pilia A (2007) Arsenic mobilization in groundwater at Quirra Plain (Sardinia, Italy). In: Bullen TB, Wang Y (eds) Proceedings WRI-12, Kunming, China, 31 July–5 Aug 2007, pp 1227–1230

11. FAO, UNICEF, WHO, WSP (2010) Towards an arsenic safe environment in Bangladesh. Executive summary of the WORLD WATER DAY 2010. http://www.unicef.org/media/files/Towards_an_arsenic_safe_environ_summary(english)_22Mar2010.pdf
12. Frau F, Ardau C (2003) Geochemical controls on arsenic distribution in the Baccu Locci stream catchment (Sardinia, Italy) affected by past mining. Appl Geochem 18:1373–1386
13. Frau F, Ardau C (2004) Mineralogical controls on arsenic mobility in the Baccu Locci stream catchment (Sardinia, Italy) affected by past mining. Min Mag 68(1):15–30
14. Frau F, Ardau C, Lorrai M, Fanfani L (2001) Geochemistry of waters in the dismantled mine area of Baccu Locci (Sardinia, Italy): the arsenic contamination. In: Cidu R (ed) Proceedings WRI-10, Villasimius, Italy, 10–15 June 2001. A.A. Balkema Publishers, Rotterdam, pp 1229–1232, ISBN: 9026518242
15. Frau F, Biddau R, Fanfani L (2008) Effect of major anions on arsenate desorption from ferrihydrite-bearing natural samples. Appl Geochem 23:1451–1466
16. Frau F, Ardau C, Fanfani L (2009) Environmental geochemistry and mineralogy of lead at the old mine area of Baccu Locci (south-east Sardinia, Italy). J Geochem Explor 100:105–115
17. Liu C-P, Luo C-L, Gao Y, Li F-B, Lin L-W, Wu C-A, Li X-D (2010) Arsenic contamination and potential health risk implications at an abandoned tungsten mine, southern China. Environ Pollut 158:820–826
18. Pilia A (2005) Mobilizzazione e trasporto dell'arsenico nell'acquifero alluvionale della Piana di Quirra. Ph.D thesis, Università di Cagliari, pp 174
19. Progemisa (2004) Piano della Caratterizzazione area Baccu Locci – Quirra – Piano di Investigazione iniziale– Report
20. Ravenscroft P, Brammer H, Richards K (2009) Arsenic pollution: a global synthesis. Wiley-Blackwell, Chichester, UK, pp 616
21. Regione Autonoma Sardegna (1998) Il Parco geominerario della Sardegna. Grafiche Sainas, Cagliari, pp 79
22. Yáñez L, García-Nieto E, Rojas E, Carrizales E, Mejía J, Calderón J, Razo I, Díaz-Barriga F (2003) DNA damage in blood cells from children exposed to arsenic and lead in a mining area. Environ Res 93:231–240
23. Zucchetti SC (1958) The lead-arsenic-sulfide ore deposit of Bacu Locci (Sardinia-Italy). Econ Geol 53:867–876

Chapter 14
Acid Mine Drainage Migration of Belovo Zinc Plant (South Siberia, Russia): A Multidisciplinary Study

Svetlana Bortnikova, Yuri Manstein, Olga Saeva, Natalia Yurkevich, Olga Gaskova, Elizaveta Bessonova, Roman Romanov, Nadezhda Ermolaeva, Valerii Chernuhin, and Aleksandr Reutsky

Abstract The distribution of chemical elements (Zn, Cu, Fe, Pb, Cd, As, Sb, Be) in the water and bottom sediments of the Belovo swamp-settler was investigated in our integrated study of geochemistry, geophysics, and hydrobiology. This swamp collects drainage escaping from clinker heaps made up of wastes of pyrometallurgical smelting of sphalerite concentrate. Water in the swamp has high TDS (Total Dissolved Solids) values, with extremely high contents of toxic elements. Bottom sediments in the swamp are a mixture of hydrogenic secondary Cu, Zn, Fe and other elements minerals. High metal concentrations lead to drastic changes in biota: phytoplankton, zooplankton, and bacteria communities. The species richness and the composition of plankton reflect the chemical composition of water of the swamp-settler and allow considering it as an extreme habitat. About 90% of zooplankton individuals have a genetic mutation expressed in morphological deformations. Infiltration of the high TDS swamp water into groundwater was detected by vertical electric sounding. Contouring of settler volume and preliminary estimation of useful component resources were done; as a result, the settler could be considered as a secondary deposit. By laboratory

---

S. Bortnikova (✉) • Y. Manstein • O. Saeva • N. Yurkevich
Trofimuk Institute of Petroleum Geology and Geophysics SB RAS, Novosibirsk, Russia
e-mail: bortnikovasb@ipgg.nsc.ru

O. Gaskova • E. Bessonova
Sobolev Institute of Geology and Mineralogy SB RAS, Novosibirsk, Russia

R. Romanov
Central Siberian Botanic Garden SB RAS, Novosibirsk, Russia

N. Ermolaeva
Institute for Water and Ecological Problems SB RAS, Barnaul, Russia

V. Chernuhin • A. Reutsky
SibEnzyme Ltd, Novosibirsk, Russia

experiments a principal possibility of solution processing is shown. In addition, extraction of some metals from solutions and sediments can decrease remediation costs for the area.

Keywords Acid mine drainage • Phytoplankton • Zooplankton • Electromagnetic sounding

14.1 Introduction

Some concentrated drainage streams interacting with tailings strongly altered by sulfide can be considered not only as a dangerous matter for the environment, but also as a secondary raw material for the extraction of useful components. These streams can be called "liquid ore" due to high concentrations of dissolved metals. One of the examples of such streams is the drainage escaping from clinker heaps of the Belovo zinc processing plant (Kemerovo region, South-Western Siberia, Russia). This drainage is the product of seasonal streams and clinker interaction; it flows into a swamp through a special ditch. This swamp is bounded by an earth-fill dam and allows accumulation and settling of the drainage. The mineral composition of the clinker heaps, zonality, content of metals and their speciation (water-soluble, exchangeable) were described in detail [4, 20].

The aims of this study were: (i) to estimate the element concentrations in the water of the drainage streams and swamp-settler; (ii) to determine chemical and mineral composition of the bottom sediments; (iii) to identify the toxicity of the solution for the biota; (iv) to detect the dissemination depth of the solution in the settler; (v) to process solutions for copper extraction.

14.2 Study Area

The clinker heaps are located in the area of Belovo zinc processing plant in the town of Belovo (Fig. 14.1). The plant extracted zinc by smelting from a sphalerite concentrate which has been obtained from barite-sulfide ores mined at the Salair ore field. The sphalerite concentrate was of low quality and contained a high amount of impurities, because the Salair ore consists of a fine-grained intergrowth of different sulfide minerals. Since the mid-1990s the plant has ceased operation. About one million tons of clinker containing significant amount of sulfuric acid were left in the plant area in the form of heaps. Large amounts of fine-grained coke dust (15–25%) were also present in the waste. The clinker was stored in ~15 m high heaps, with flat top and a steep slope.

This waste was affected by spontaneous ignition of the coke dust and subsequent burning of waste in the heaps, thus causing an acceleration of oxidizing processes. Oxidation rate is shown by the occurrence of abundant secondary minerals on

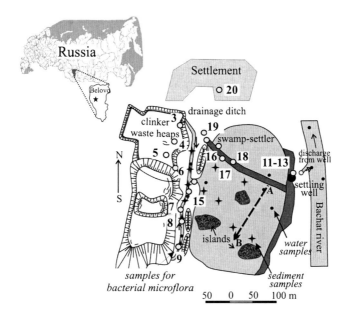

Fig. 14.1 Schematic map of the study area with location of the sampling points

the waste surface. The western border of the waste is bounded by the swamp-settler in which the drainage from the waste discharges. Water from the swamp flows firstly into a settling well and then in the Bachat river. Water of the drainage stream is a bright – dark blue solution, and bottom sediments have white, blue, green and yellow colors.

14.3 Methods

14.3.1 Field Sampling

Various sampling campaigns were carried out in 10 years (1999–2008). A total of 22 sampling points were monitored; they comprise surface water (drainage ditch, swamp-settler, discharge into the Bachat river), groundwater (settling well, wells and water pumps in the neighboring settlement, underground drainage into the Bachat river) and bottom sediments from ditch and swamp (Fig. 14.1). Sampling was carried out in summer and autumn, and in winter 2008 under the ice. The water samples were filtered through 0.45 μm filters using an all plastic equipment and collected into acid-cleaned polyethylene bottles. Values of pH, Eh, T were measured at the sampling site. Bottom sediments (hydrogenic flocs) were sampled into polyethylene packages and then were dried.

Phytoplankton was sampled in summer 2008 from surface layer by filtration of water through membrane filter with pores of 0.55–0.65 μm and was processed in formalin fixed and nonfixed state.

Samples of zooplankton were collected in winter 2008 by filtration of 300–400 l of swamp-settler water through Dzhedy net with cells of 62 μm and were conserved with formalin (4%).

Samples (bottom sediments, efflorescents, water) for bacterial microflora investigation were taken in 1.7 ml Eppendorf test tubes from different parts of the explored area: the drainage ditch, the dam, the swamp-settler, and the heap slope.

14.3.2 Laboratory Analyses

14.3.2.1 Inorganic Components

Anion species in solutions were determined by ion chromatography. Metals were analyzed by ICP-AES. In addition, dry salt residues were received by evaporation and analyzed by the Synchrotron X-ray fluorescent (SR-XRF) method, allowing us to decrease the detection limits. Bottom sediments were analyzed by SR-XRF method. Mineral composition was determined by X-ray powder diffraction.

14.3.2.2 Plankton Study

Phytoplankton and zooplankton were studied using optical microscope, according to standard methods. The samples were processed by countable-weight method in the Fux-Rosental and Bogorov's chamber. The cells and individual number of phytoplankton were counted separately; the latter is the number of solitary cells, temporal aggregates, colonies, coenobia, filaments, trichomes etc. irrespective to cell number in these units. The separate weight of Cladocera and Rotatoria were determined on the length of an organism's body using the equation of relationship between these parameters.

14.3.2.3 Calculation Method of Bacteria Colonies Quantity

Fifty micro liters of the sample were placed (soil, clay, sand, water etc.) in a 1.7 ml Eppendorf test tube, and distilled autoclaved water up to a volume of 1 ml was added. Each sample was initially carefully stirred with a sterile pipette tip, then with a shaker for 1 min. Two micro liters of mixture was shown on two Petri dish and placed at a temperature of 25°C for 5 days. Then we counted the amount of growing colonies. The strains were identified according to Berge's manual [3].

14.3.3 Geophysical Investigation

Estimation of solution extent at depth in the swamp volume was carried out in winter 2008 by geoelectric sounding. The sounding was carried out by using a multielectrode resistivity meter. The electrodes penetrated through the holes drilled on the ice. The physical prerequisite for the use of the electric sounding in this case is the high mineralization of the swamp-settler solutions. The problem of contouring area contaminated by the solution in the swamp is reduced to the detection of the electrolyte with relatively low specific electric resistivity in the accommodating media [12]. Electric resistivity tomography was performed using a Schlumberger array, with 48 electrodes spaced by 5 m.

14.3.4 Solution Processing

The concentration and extraction of metals (Cu, Zn, Cd, Pb, Fe) and associated elements (Se, Be) from drainage can be performed by a variety of ways: chromatography, precipitation and co-precipitation, extraction, and electrolysis. Electrolysis is the means that allows pure metal to be recovered as Me°. Industrial scale electrolysis was initially applied in the 1970s, and the study of these processes started just 20 years ago [17, 19, 23].

14.3.4.1 Cell-Electrolysis of Drainage Water

The experimental cell was assembled as follows: the cathode was a copper plate, and the inert anode was a graphitic rod, connected to a constant-current source model B5-50 (ISTOCHNIK Ltd., Russia). Electrodes were placed in 300 ml of drainage (sampled in winter 2008) with pH 3.5-4.0 and containing 3.5 g/l of copper and a total dissolved metal concentration of approximately 5 g/l (Fig. 14.2).

While current is applied to the electrodes, the following reactions occur:

– on the cathode, copper reduction

$$Cu^{2+} + 2e^- \rightarrow Cu^0,$$

– on the anode, oxygen formation and protons regeneration

$$2H_2O \rightarrow O_2 + 4H^+ + 4e^-.$$

The optimal conditions were selected according to the rate of increasing plate mass within a solution temperature range of 50–60°C, which raised the process rate. The duration of electrolysis was 3 h.

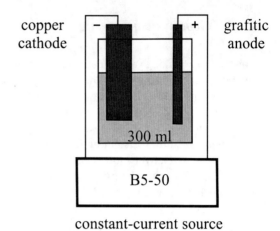

Fig. 14.2 Scheme of the electrolytic cell

14.3.4.2 Ion-Changing Copper Extraction

In order to improve the yield of metallic copper obtained from electrolysis, concentration methods were applied, obtaining a concentration level of the feed solution reaching the range 25–35 g/l. Ion-exchange resins (artificial high-molecular organic polyelectrolytes, which possess ion-exchanging properties) were used for the removal and extraction of metals from waste water [10].

Elution was carried out in an ion-exchanging column filled by 100 g of ion-exchange resin KU-2-8 (cation exchange capacity 4.4–4.8 eq-mg/g by 0.1-n *CaCl$_2$* solution). The resin was activated using 10% *H$_2$SO$_4$*. A 2000 ml drainage solution with an initial pH of ~4.0 and a copper concentration of 3.5 g/l was dripped down through the column at a rate of 5 ml/min (Fig. 14.3). Elution of the elements and column regeneration were carried out by using 300 ml of 20% H$_2$SO$_4$. The concentrated solution derived in the previous stage was used as feed liquid for electrolysis processing as described in sect. 3.4.1.

14.3.4.3 Method Based on Addition of Metallic Aluminum

In terms of electrochemical activity, aluminum is located left of the base metals: K, Ca, Na, Mg, Al, Mn, Zn, Fe, Co, Ni, Pb, H, Cu, Ag. In other words, aluminum has more negative potential than all of these metals. In this case, the interaction of metallic aluminum and a solution of salt of one of the less negative potential metals will lead to electron transition from aluminum to the metal cation. Therefore, the metal cation is reduced and metallic aluminum is oxidized:

$$Me^{n+} + ne^- \rightarrow Me°, Al° - 3e^- \rightarrow Al^{3+}.$$

Fig. 14.3 Scheme of elution

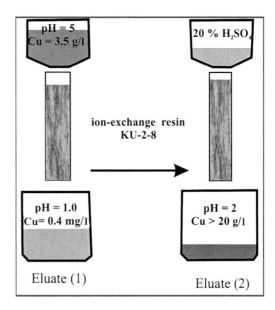

The experiment was performed according to the following procedure:

– Addition of 0.3 g of aluminum foil to 100 ml of sample
– 7 days wait, until about half of aluminum foil dissolved producing a discolored solution (pH=3.9), and 0.5 g of copper powder is precipitated
– Separation of the copper powder by filtration (aluminum hydroxides form at pH>4, but our current pH 3.9 solution allows us to separate the copper powder easily)
– Neutralization of the filtrate by ammonia to pH 7.5 leading to the precipitation of aluminum hydroxide. The resulting white flakes were also separated by filtration.

Samples of drainage water and filtrates from each step were analyzed by ICP AES.

14.4 Results

14.4.1 Inorganic Components

14.4.1.1 Solution Composition

Water of the drainage ditch from the clinker heaps shows high salinity and extremely high metal concentrations (Table 14.1). Composition of the solution varies with time, but the main trend is an increase of metal concentrations from 1999 to 2008.

Values of pH in the ditch range from 3.3 to 5.8. It is important that metals (Zn and Cu) account for about 50 equ-% of the cationic composition. In the swamp-settler

Table 14.1 Solution composition in the drainage ditch and the swamp-settler, mg/l

Periods	October 1999 Ditch	October 1999 Settler	June 2000 Ditch	June 2000 Settler	June 2005 Ditch	June 2005 Settler	February 2008 Ditch	February 2008 Settler
pH	4.27	5.05	4.36	3.56	5.80	4.81	3.30	3.40
SO_4^{2-}	3.5[a]	2.6	4.2	6.8	1.7	3.7	20	23
Cl^-	180	220	190	230	120	190	42	70
K^+	150	83	51	80	21	76	120	110
Na^+	150	120	450	670	210	530	920	910
Mg^{2+}	150	10	130	260	110	360	700	670
Ca^{2+}	5.3	4.8	440	490	250	510	280	390
Fe_{tot}	2.2	0.48	3.0	25	0.18	0.40	15	8.8
Zn	0.37[a]	0.22	0.54	1.0	0.15	0.37	1.7	1.6
Cu	0.19[a]	0.06	0.27	0.57	0.15	0.46	3.0	2.5
Al	5.4	8.6	7.5	40	2.5	6.7	200	130
Mn	4.3	12	18	32	4.6	22	83	75
Cd	2.5	1.3	4.6	7.4	1.5	4.2	9.2	9.4
Co	0.7	0.9	1.0	2.5	1.0	3.9	18	16
Ni	0.8	1.1	1.5	2.2	1.2	4.5	21	18
Be	<0.1[b]	<0.1	<0.1	<0.1	1.1	2.1	13	17

[a]SO_4^{2-}, Zn, Cu in g/l
[b]Be in µg/l

pH values of the solutions vary from 3.4 up to 4.8. Total dissolved solids measurement (TDS) is not so high, but metals Zn and Cu account for the greater share of the cations. In 2008, when the sampling occurred in the winter time, TDS of the solutions were higher than in summer solutions and concentrations of metals were sharply increased: Mg, Na, K 4–6 times, Fe and Al by 2 orders, Zn, Cu, Co and Ni by more than one order.

Furthermore, extremely high concentrations of very toxic elements (Cd and Be) were observed. An obvious principal reason to this was the freezing and concentration of metals under the ice. Nevertheless, occurrence of such high concentrations of dissolved metal species indicates more serious danger for the ecosystem, with respect to what was observed in the analysis of the summer solutions.

Discharge into the river Bachat carried out through settling well is not completely cleaned from metals (Table 14.2).

As a result, metal concentrations are significantly higher in the downstream water than in the upstream water and Cd in water from column and wells in the settlement located in 100 – 200 m from the clinker heaps is three times above the Russian water quality standards of 0.01 mg/l (Maximum Allowable Concentration).

14.4.1.2 Bottom Sediments

Bottom sediments in the drainage ditch and swamp-settler are hydrogenic secondary minerals formed by supersaturation of the solution. Sediments are in the form of

14 Acid Mine Drainage Migration of Belovo Zinc Plant

Table 14.2 Concentration of some elements in surface and groundwaters as a result of the waste inflow, mg/l (Zn, Cu, Fe) and μg/l (Pb, Cd, As)

	Zn	Cu	Fe	Pb	Cd	As
Run-off into the Bachat river						
Surface	6.8	0.19	1.0	5.2	230	17
Underground	11	0.32	2.7	11	430	30
Water from the Bachat river						
30 m upstream	0.76	0.087	0.4	3.5	4.6	4.1
300 m downstream	3.8	0.18	6.4	1,200	91	17
Water in the settlement						
Pipeline	0.04	0.003	< 0.05	2.4	0.13	< 1
Column in 100 m	0.56	0.032	0.04	< 1	1.0	< 1
Wells in 100 m	0.76	0.17	0.12	2.0	3.0	3.0
in 150 m	0.1	0.0063	<0.05	0.52	0.33	<1
MAC[a]	1	1	0.5	10	1	10

[a]Maximum Allowable Concentrations [21]

Table 14.3 Averaged metal content in bottom sediments from drainage ditch and swamp-settler, ppm and % (Cu, Fe, Zn)

	Blue-green layers		Yellow-orange layers	
Elements	Ditch	Settler	Ditch	Settler
Cu	19	14	2.4	5.3
Fe	7.3	6.2	20	12
Zn	1.6	0.97	0.76	0.59
Pb	470	180	1,400	300
Cd	90	46	32	47
Cr	74	81	350	230
Ag	79	25	190	40
As	100	67	610	130

stratified substance. Generally, the upper layer is blue-green and the lower one is yellow-green or yellow-orange; thin black lenses occur in both layers. Blue-green color is caused by high concentration of Cu-compounds, concentration of this element reaches up to 27% (Table 14.3). Higher concentrations of Ca, Zn, Cd, Ni were determined in blue-green layers in comparison with yellow-orange ones where Fe concentration reaches up to 19% and Sb, As, Cr are elevated. A wide spectrum of chemical elements was determined both in blue-green and in yellow-orange sediments.

Regarding the mineral composition, sediments consist of Cu, Zn, Fe, Ni – water hydroxylsulfates. Reduced minerals such as native Cu deposited on Fe-scrub, sulfides and arsenides of Ni, Cu, and Fe (rammelsbergite, maucherite semseyite, cubanite) which can be formed at solution-coke dust interaction were also found in the ditch.

Carbonates like malachite, azurite, rosasite, were also determined in the swamp-settler. They can be formed under drainage neutralization by the bed rocks of swamp. Stratification of sediments and physico-chemical modeling indicate that

Al, Pb, and *Ca* sulfates are precipitated from solutions earlier than *Cu* and *Zn* sulfates. In the settler solutions metal species and their ratios are the same as in the ditch solutions.

14.4.2 Biota State

Extremely high level of metal concentrations in the habitat results in a drastic biota state, as described in the following paragraphs.

14.4.2.1 Phytoplankton

Fourteen species of algae and cyanobacteria were found in the plankton sampled during summer: Cyanoprokaryota – 1, Chlorophyta – 5, Streptophyta – 2, Heterokontophyta – 1, Bacillariophyta – 3 (including one species identified by the empty valves), Cryptophyta – 1 and nonidentified autotrophic flagellates (Chrysophyceae or Cryptophyta) – 1 species. The few other autotrophic eukaryotes with low abundance or featureless were non-identified. The teratoid forms were not found.

The phytoplankton abundance was maximal near the drain mouth (168.2 10^6 cell/l, 167.5 10^6 ind./l, 27.2 g/m^3), while in the drain and other parts of swamp-settler it was significantly lower (2.2-13.8 10^6 cell/l, 1.8-13.8 10^6 ind./l, 0.1–3.5 g/m^3). The chlorophytes *Chlamydomonas acidophila* Negoro sensu [6] (up to ~100% of phytoplankton abundance) and cf. *Stichococcus bacillaris* Näg. (up to 14% of cell and individual number, 11% of biomass), streptophyte *Koliella* cf. *sigmoidea* Hind. (up to 51% of cell number, 50% of individual number) and non-identified autotrophic flagellates (up to 42% and 43% respectively, 48% of biomass) were the most abundant species in phytoplankton in different parts of the swamp-settler.

The low species richness of algae and cyanobacteria and the simple composition of phytoplankton of Belovo swamp-settler are characteristic for different extreme habitats [18], including those with low pH values [11, 14]. One of the most abundant species of summer phytoplankton of the Belovo swamp-settler, the *C. acidophila*, may grow at pH 2–6 ([7]; Cassin, cit. by: [22]). This species was commonly found in extreme acid habitats [14], it adapted to low pH values, high concentrations of heavy metals (Nishikawa, Tominaga, cit. by: [13]) and low insolation [8].

The algae and cyanobacteria of acid mine drainage may influence their environment by different ways [5]. Despite its presence and sometimes high abundance, there is little evidence that algae and cyanobacteria directly contribute to improve perceived water quality. Probably one of the most important effects of it on the acid mine drainage is its biomass and extracellular products, that serve as a carbon source for the sulfate-reducing bacterial community in a predominantly anaerobic environment, which in turn produces alkalinity [5].

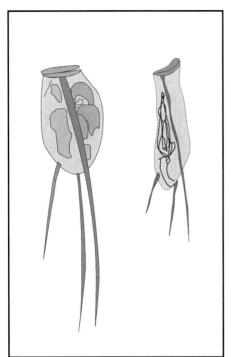
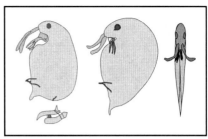

Bosmina longirostris (O.F.Mull.)

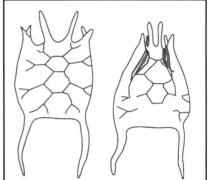

Filinia longiseta (Ehrb.)　　　*Keratella quadrata* (O.F.Mull.)

Fig. 14.4 Variations in morphology of some zooplankton organisms from the swamp-settler. The left figures show normal forms, the right – teratoide

14.4.2.2 Zooplankton

Seven groups of zooplankton have been found in winter samples: cladoceran *Bosmina longirostris* (O.F. Mull.); rotifers *Filinia longiseta longiseta* (Ehrb.), *Keratella quadrata* (O.F. Mull.), *Keratella cochlearis* (Gosse), *Br. angularis angularis* (Gosse), *Brachyonus quadridentatus* (Hermann), *Testudinella patina* (Hermann). The 90% of the detected zooplankton organisms have morphological deformities which considerably complicate specific identification (Fig. 14.4). Judging by attributes *Bosmina longirostris* (O.F.Mull.) resembles *Bosmina obtusirostris arctica* Lill., but has typical for *B. longirostris* postabdomen. The majority of rotifers have considerably deformed armours.

Obviously, the influence of high metal concentrations leads to genetic mutation as long as all individuals are deformed identically. In the samples taken in summer and autumn zooplankton was absent. It seems that temperature increase and acceleration of metabolism lead to accumulation of lethal dose of toxic elements from the water.

The zooplankton in the Belovo swamp-settler develops exclusively in the subglacial period and is presented by forms which are found in fresh reservoirs of Western Siberia the whole year round [16]. At the expense of diapaused eggs Rotatoria and

Table 14.4 Taxonomic divergence of bacterial microflora

Samples, see Fig. 14.1	Amount of colonies	Taxonomic groups
3[a]	2	*Microbacteriacaea* – 2
4	1	*Micrococcus* – 1
5	7	*Arthrobacter* – 3, *Microbacteriaceae* – 4
6	3	*Microbacteriaceae* – 1, *Bacillus* – 1, *Micrococcus* – 1
7	21	*Microbacteriaceae* – 12, *Bacillus* – 7, *Micrococcus* – 2
8	3	*Microbacteriaceae* – 3
9	5	*Microbacteriaceae* –5
11	9	*Bacillus* – 3, *Microbacteriaceae* – 2, *Micrococcus* – 4
12	24	*Bacillus* – 13, *Micrococcus* – 8, *Planococcus* – 2, *Microbacteriaceae* – 1
13	13	*Micrococcus* – 5, *Microbacteriaceae* – 8
15	7	*Micrococcus* – 6, *Microbacteriaceae* – 1
16	26	*Bacillus* – 5, *Micrococcus* – 18, *Microbacteriaceae* – 3
17	20	*Microbacteriaceae* – 4, *Arthrobacter* – 3, *Bacillus* – 8, *Micrococcus* – 5
18	12	*Microbacteriaceae* – 6, *Micrococcus* –6
19	7	*Microbacteriaceae* – 5, *Micrococcus* – 2
20	13	*Microbacteriaceae* – 6, *Planococcus* – 2, *Micrococcus* – 5

[a]Samples No. 3–6 – efflorescents; 7–15 – bottom sediments; 16–20 – soil

Cladocera are capable to avoid adverse conditions, while Copepoda are constantly exposed to toxic influence of metals.

During the summer period water temperature rises, speed and intensity of metabolism of zooplanktonic organisms increase. High concentration of heavy metals appear at first sublethal (and call bisexual reproduction and occurrence of diapaused eggs), and then lethal. Thus, in summer all zooplankton community of the swamp-settler is in a status of diapause, it waits till adverse conditions are over. With the fall of temperature its metabolism level decreases, Cladocera and Rotatoria (most likely already presented by both male and female) appear from resting eggs and have time to give new generation. Copepoda, which as a rule experience diapause in ontogenetic of stage II – III, are not present in the community.

14.4.2.3 Taxonomic Divergence of Bacterial Microflora

It is important to note that only ~1% of bacterial species are cultivated [2]. In the present analysis only saprophytic eubacteria may be isolated. Results of taxonomic analysis are shown in Table 14.4.

All isolated bacterial strains are Gram-positive, not capable to anaerobic growth, and catalase-positive. They belong to four families: *Bacillaceae, Micrococcaceae, Microbacteriacea* and *Planococcaceae*.

The family *Bacillaceae* includes species of one isolated bacterial genus *Bacillus*. The family *Micrococcaceae* includes species of two isolated bacterial genera

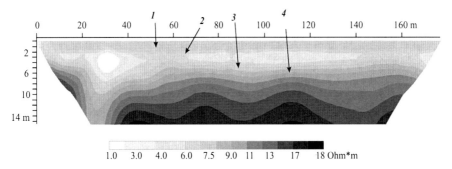

Fig. 14.5 Vertical zoning of Belovo swamp-settler according to the data of geophysical sounding

Arthrobacter and *Micrococcus*. The family *Planococcaceae* includes species of one isolated bacterial genus *Planococcus*. In the family *Microbacteriacea* genera was not identified.

It is notable that there are not any Gram-negative species including halophilic representatives of Flavobacterium and Pseudomonas genera. Taxonomic richness of isolated bacteria is relatively poor and its representatives are typical for the soil samples. But Streptomyces (largest genus of Actinobacteria is characteristic of majority of soil samples) are absent.

The most heterogeneous and diverse bacterial Bacillus genus is characterized by the wide spectrum of environment conditions [9]. They form spherical endospores which allow to survive in wide spectrum of unfavourable conditions. Considerable part of Bacillus genus representatives are oligotrophic organisms because they can live in an environment that offers very low levels of nutrients.

Representatives of *Micrococcaceae*, *Microbacteriacea* families often adapt to halophilic conditions, low temperature and also may be oligotrophic organisms [3]. The representatives of *Planococcus* genus are halophilic and often associated with cyanobacterial mat and are typical not only for soil, but for water (fresh and sea) samples.

14.4.3 Electric Tomography

The result of geophysical sounding was the geoelectric cross-section up to a depth of 14 m, on which several zones with different specific resistivity of media were revealed (Fig. 14.5). Each of them corresponds to particular hydrogeochemical conditions:

- Horizon with specific resistivity 3.5–4 Ω ×m. It is a surface layer represented by concentrated water in which metal concentrations (Cu, Zn, Cd, Ni, Co, and others) reach high toxicity levels. Solution composition is caused by discharge of the drainage under the clinker heaps.

- Bottom sediments formed by secondary water Cu, Zn, Fe, Ni (and other metals) sulfates with combination of silt and organic matter; pore space is filled by solutions of the same composition as the surface layer; specific resistivity is 4.2–6.8 $\Omega \times$ m.
- Horizon of groundwater; according to values of specific resistivity, the infiltration of high toxic solutions into groundwater can be inferred.
- Impermeable clay horizon having resistivity of 8 – 15 $\times$ m is located at 6.5–10 m depth.

Infiltration of the concentrated solutions into groundwater is confirmed by the analysis of water samples taken from the wells and water pumps situated on the area of the settlements in 60–100 m far from the waste heaps and swamp-settler (Table 14.2). On the base of integrated results it can be arguable that the waste heaps and the swamp-settler are the sources of large-scale environmental pollution, and metal dissemination occurs uncontrolled and in different directions.

At the moment, the control of these processes is a practically impossible issue. In comparison with similar drainage solutions described in the literature, the Belovo swamp-settler corresponds to one of the most mineralized reservoir. Even in comparison with well-known drainage within the Richmond mine (Iron Mountain), where solutions had pH values ranging to negative [1, 15], solutions within Belovo waste heap have higher metal concentrations in some cases. While in the Richmond mine the main share of TDS in the solutions belongs to iron, in the Belovo swamp-settler leading metals are Cu and Zn. The solutions in the ditch and in the settler can be considered as potential row material for economic extraction of useful components which can decrease remediation costs.

Estimated resources of the swamp-settler for an area of 100 × 100 m (which corresponds only to the south part of the settler) are: 40 tons of Zn and 50 tons of copper as dissolved species in the settler solution, plus 800 tons of Zn and 8,000 tons of Cu as secondary soluble minerals in bottom sediments.

Total estimation of settler resources will show larger amounts of metals, which are in easily extracted species. This task can be addressed by comparing reclamation and exploitation costs with respect to benefits for the ecological system. As a first step we tried to process solutions from the drainage system, to extract copper as main useful component.

14.4.4 Solution Processing

14.4.4.1 Results of the Ion-Exchange Copper Extraction

During the passage of the solution through the column, pH fell to a value of 1.0. The main reason is the replacement reaction of H^+ for cations from a solution on the resin. The copper concentration in the eluate (1) was 0.4 mg/l, reflecting the 99.99% of copper extraction from the solution on the resin. Also, more then 98% of Zn, Al,

Table 14.5 Results of ion-exchanging elements extraction, mg/l

Elements	Settler solution	Eluate (1)	Eluate (2)
pH	4.0	1.0	2.0
Cu	3,500	0.4	25,000
Zn	1,400	0.2	3,200
Mg	1,000	0.8	2,700
Al	270	0.3	680
Mn	90	0.02	220
Co	26	0.01	69
Ni	20	0.1	56
Cd	13	0.005	33
Pb	7.5	0.1	20
As	1.2	0.1	8.4
Mo	1.0	0.2	2.3
Sb	0.5	0.1	1.0
Tl	0.5	0.1	1
Be	0.03	0.0005	0.062

Cd, Co, Ni, Pb, Mn, Mg, Be and more then 80% of As, Bi, Mo, Sb, Tl, W were removed (Table 14.5).

Eluate (1) concentrations of only Tl, As, Be, Sb, Ni, Pb, Cd (as against 18 toxic elements of drainage) exceeded Maximum Allowable Concentration of chemical substances in terms of drinking and community use.

After column regeneration, eluate (2) pH was 2 and eluent copper concentration was ~25 g/l, consequently septuple Cu concentration and almost 100% resin wash-out of copper occurred. The first resin wash-up leads to 100% As, and Ba removal, whereas other metals were concentrated just two-three times (resin wash-out of these metals was about 30–40%) and full removal for these requires additional H_2SO_4.

In summary, we conclude that ion-exchange resins can be used not only for copper concentration but also for its specific separation from different elements that complicate the electrolysis.

14.4.4.2 Experiments with Solution Electrolysis

Results of copper electroextraction from drainage and concentrated solutions are presented in Fig. 14.6 as a function of the copper cathode mass increase and electrolysis time.

The dependence shows that, after 2 h of electrolysis of the drainage solution (3.5 g/l), the mass increase of the copper cathode slows, reaching a plateau after the 3rd hour. In the experiment with the drainage solution 0.42 g of metallic copper were produced and in the experiment with a concentrated solution (25 g/l) 1.6 g metallic copper were produced during the same time (3 h).

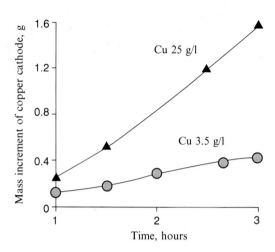

Fig. 14.6 Mass increment of copper cathode derived from electrolysis

Table 14.6 The composition of the drainage solution and the resulting filtrates, mg/l

Elements	Settler solution	Filtrate 1	Filtrate 2	MAC
pH	3.5	3.9	7.5	6.5–8.5
Al	320	1,900	0.5	0.5
Cu	5,500	10	0.2	1.0
Zn	4,000	4,100	220	1.0
Fe	10	40	0.5	1.0
Mn	130	170	100	0.1
Co	28	32	6.2	0.1
Ni	30	40	10	0.02
Pb	5.5	<0.01	<0.01	0.01
Se	1.2	0.9	< 0.01	0.01
Be	0.033	<0.0002	<0.0002	0.0002

The slowing of the cathode mass increment was not observed in the second experiment. This shows again that the electrolyte concentration is a controlling and limiting factor during electrolysis. There are more metal ions at the cathode surface in the more concentrated solutions, so less time is needed for the ion transfer to the electrode, with the result of an increased metal extraction rate.

The multistage cell electrolysis method requires time (ion-exchange), and energy (heating), plus sulfuric acid. Due to these facts, the method based on the addition of metallic aluminum is much simpler to perform and has lower ecological risks.

14.4.4.3 Experiments with the Addition of Aluminum

The experimental results listed in Table 14.6 (Filtrate 1) demonstrate that the introduction of drainage and aluminum foil leads to reduce the concentrations of Cu to

10 mg/l and Pb to <0.05 mg/l, with concentrations of other metals remaining unchanged. One of the reasons for the Cu and Pb discrimination is the maximal potential difference between them and Al.

The Al concentration became equal to MAC (Maximum Allowable Concentration) level after ammonia neutralization of the filtrate (Table 14.6, Filtrate 2). Due to co-precipitation of Se and Be on aluminium hydroxide flakes, their concentrations dropped to MAC levels also; and Fe and Cu concentrations dropped to less than MAC levels.

The final filtrate solution contained 6% Zn, 20% Co, 60% Ni, and 77% Mn of their initial concentrations. Moreover, resulting copper powder could be melted and hydroxide aluminum could be used in industrial production of coagulant and fire-retardants.

These results allow us to propose a recommended drainage remediation method based on addition of metallic aluminum followed by neutralization with ammonia, to significantly reduce effluent concentrations of toxic elements and increase the pH to natural levels, in order to prevent environmental contamination.

14.5 Conclusions

Water of the Belovo swamp-settler is concentrated in strong toxic solutions with extreme content of hazardous elements. Both the level of concentration and mobility of the elements are determined by ore processing as well as waste storage conditions. The richness of species and the composition and seasonal changes of plankton reflect the chemical composition of water of the Belovo swamp-settler, and allow considering it as an extreme habitat.

Cultivated bacterial strains belong to four families: *Bacillaceae, Micrococcaceae, Microbacteriacea and Planococcaceae*. It is notable that there are not any Gram-negative species including halophilic representatives of *Flavobacterium* and *Pseudomonas* genera, nor Gram-positive strains from *Streptomyces* genus, that are typical of different soils.

Remediation actions are urgently required because the heap and the settler are located within the range of the city and houses are in close proximity to these sources of pollution.

The depth of the settler solutions spreading is indicative of intensive dispersion of pollutants into groundwater, which significantly complicates the choice of environmental measures. The first experimental trials to process this solution testify a possibility of extraction of Cu and other metals by electrolyses. As a result, costs of minimizing the environmental damage can be reduced by processing the easily extracted Zn and Cu compounds from the settler solutions and bottom sediments.

Acknowledgements The researches were financially supported by RFBR (grant No. 08-05-00688). We are grateful to the operators of analytical devices – S. Nechepurenko, Yu. Kolmogorov, N. Palchic.

References

1. Alpers CN, Nordstrom DK, Burchard JM (1992) Compilation and interpretation of water-quality and discharge data for acidic mine waters at Iron Mountain, Shasta County, California, 1940–91. U.S Geological Survey Water-Resources Investigations Report 91-4160: 173
2. Amann RI, Ludwig W, Schleifer K-H (1995) Phylogenetic identification and *in situ* detection of individual microbial cells without cultivation. Microbiol Rev 59(1):143–169
3. Holt JG et al (ed) (1997) Bergey's manual of determinative bacteriology, 9th edn in 2 vols (Rus. transl. by ed. akad). Russian Academy of Science G.A. Zavarzin (in Russian)
4. Bortnikova SB, Gaskova OL, Airijants AA (2003) Waste lakes: origin, development and influence on the environment. Geo, Novosibirsk (in Russian)
5. Das BK, Roy A, Koschorreck M, Mandal SM, Wendt-Potthoff K, Bhattacharyaa J (2009) Occurrence and role of algae and fungi in acid mine drainage environment with special reference to metals and sulfate immobilization. Water Res 43:883–894
6. Fott B (1956) Flagellata extrémně kyselých vod. Preslia 28:145–150
7. Fott B, McCarthy AJ (1964) Three acidophilic volvocine flagellates in pure culture. J Protozool 11(1):116–120
8. Gerloff-Elias A, Spijkerman E, Schubert H (2005) Light acclimation of Chlamydomonas acidophila accumulating in the hypolimnion of an acidic lake (pH 2.6). Freshw Biol 50(8):1301–1314
9. Graumann P (ed) (2007) Bacillus: cellular and molecular biology, 1st edn. Caister Academic Press, Peter Graumann University of Freiburg, Norfolk
10. Lebedeva KB (1975) Ionite in the nonferrous metallurgy. Publishing house Metallurgy, Moscow (in Russian)
11. Lessmann D, Fyson A, Nixdorf B (2000) Phytoplankton of the extremely acidic mining lakes of Lusatia (Germany) with pH ≤ 3. Hydrobiologia 433:123–128
12. Manstein AK, Epov MI, Voevoda VV, Sukhorukova KV (2000) RF Patent Certificate no. 2152058, G 01 V 3/10 24.06.98. Byull. Izobret. 18 (in Russian)
13. Nishikawa K, Yamakoshi Y, Uemura I, Tominaga N (2003) Ultrastructural changes in Chlamydomonas acidophila (Chlorophyta) induced by heavy metals and polyphosphate metabolism. FEMS Microbiol Ecol 44:253–259
14. Nixdorf B, Fyson A, Krumbeck H (2001) Review: plant life in extremely acidic waters. Environ Exp Bot 46:203–211
15. Nordstrom DK, Alpers CN (1999) Negative pH, efflorescent mineralogy, and consequences for environmental restoration at the Iron Mountain superfund site, California. Proceed Nat Acad Sci 96:3455–3462
16. Polukhina NI, Dvurechenskaya SYa, Sokolovskaya IP, Baryshev VB, Anoshin GN, Vorotnikov BA (1998) Some toxic microelements in Novosibirsk reservoir's ecosystem (data XRF SR and AAS techniques). Nucl Instrum Meth Phys Res A 405:423–427
17. Prasad PB (1994) Seeding of electrolysis copper waste for copper removal and recovery as elemental copper. Mater Sci Lett 13:15–16
18. Seckbach J, Chapman DJ, Garbary DJ, Oren A, Reisser W (2007) Algae and cyanobacteria under environmental extremes: final comments. In: Seckbach J (ed) Algae and Cyanobacteria in extreme environments. Springer, Dordrecht, Netherlands, pp 465–485
19. Seryanov YV, Kvyatkovskaya LM (1994) Electrolytic recovery of precious metals from dilute. Russ Electrochem 30(3):364–369 (in Russian)
20. Sidenko NV, Giere R, Bortnikova SB, Cottard F, Palchik NA (2001) Mobility of heavy metals in self-burning waste heaps of the zinc smelting plant in Belovo (Kemerovo Region, Russia). J Geochem Explor 74(1–3):109–125
21. State Standard GN 2.1.5.1315-03 (2003) Maximum allowable concentrations (MAC) of chemical substances in object of drinking and community use (in Russian)
22. Visviki I, Palladino J (2001) Growth and cytology of chlamydomonas acidophila under acidic stress. Bull Environ Contam Toxicol 66:623–630
23. Zhang HH, Coury LA (1992) Copper recovery from copper or its alloy treating solutions. Anal Chem 65(11):1552–1561

Chapter 15
Distribution and Origin of Boron in Fresh and Thermal Waters in Different Areas of Greece

Elissavet Dotsika, Dimitrios Poutoukis, Wolfram Kloppmann, Brunella Raco, and David Psomiadis

Abstract World Health Organisation (WHO) guideline value for boron is 0.5 mg/l. The drinking-water standard of the European Community [11] (EC Directive, 1983) is twice the value recommended by WHO, but still boron concentrations in many ground- and surface waters in Greece exceed this value, rendering such water unacceptable according to the European standards. Boron is biologically an essential element, but at high concentrations is toxic to plants (above approximately 1 mg/l in irrigation water) and probably to humans. Because of this potential toxicity and the need of implementation of EU regulation on national level, the study of the boron levels in both ground- and surface water is of great significance for water management. In Greece, a significant number of thermal, mineral and superficial water springs, especially in Northern Greece and in islands, present high boron values. Nevertheless, such ground waters or borehole water with high temperature and high boron content are frequently used for irrigation and drinking purposes, and could therefore have an antagonistic effect on crop yield and health. In order to study the boron contamination and to elucidate the origin of B, we collected a number of hot and fresh waters all over Greece. The relatively high concentrations of boron in groundwater became a major problem in the study areas. Some million m^3 of groundwater are rendered unusable due to the high boron content. The need of significant quantities of high quality water is an important issue due to rapid

---

E. Dotsika (✉) • D. Psomiadis
Inst. of Materials Sciences, National Center for Scientific Research "Demokritos",
Aghia Paraskevi, Athens, Greece
e-mail: edotsika@ims.demokritos.gr

D. Poutoukis
General Secretariat for Research and Technology, Messogion 14–18, Athens, Greece

W. Kloppmann
BRGM, BP 6009, Orleans Cedex 2 F 45060, France

B. Raco
CNR Institute of Geosciences and Earth Resources, Via G. Moruzzi 1, Pisa 56124, Italy

urbanization and increasing touristic development of the study areas. Thus, the investigation of the boron problem is crucial, with regard to B spatial distribution and its possible sources. In all sampled waters, the boron concentration exceeds the limit of 0.5 mg/l, which is the former recommended WHO limit. Moreover, in the irrigation waters examined, the boron concentration exceeds the value of 0.75 mg/l, which is the limit for sensitive plants (for plants of moderate and high tolerance, these values vary between 0.75–3 mg/l and >3 mg/l respectively). In all cases, elevated boron and salinity could be attributed to geothermal activity, anthropogenic sources and/or seawater intrusion into the aquifers. This finding has important implications for water management: In a setting of high natural geochemical background values, control of the pollution source is not possible and water managers have to cope with a local to regional geochemical anomaly that implies boron specific water treatment or mixing with low-boron water resources to bring concentrations down.

Keywords Stable isotopes • Origin of boron • Pollution • Hydrochemical methods • Greece

15.1 Introduction

Boron has been identified as a natural and anthropogenic element of waters that in many cases has caused water quality degradation. The presence of boron in surface and groundwater could pose a potential threat to their use for drinking and irrigation purposes. Entering a period of special attention on water resources and reserves, the control of water contaminants is highly significant. Boron is not found as a free element in nature and is classified as moderate to high toxic element in the aquatic environment. It usually occurs as orthoboric acid in some volcanic spring waters and as borates like borax and colemanite. In the earth's crust, the average boron concentration is 10 ppm, a value that characterizes boron as a lithophilic element.

Natural sources include weathering of igneous rocks and leaching from sedimentary deposits of marine and non-marine origin. Anthropogenic sources include landfill leachate, drainage from coal mines and leaching of mining related wastes, flash and relevant products, agricultural runoff because boron is minor constituent in fertilizers or pesticides, and sewage effluents due to presence of boron in detergents [17, 49, 55].

The geochemical cycle of boron is more easily understood if we consider a basaltic magma, from which boron can be obtained by slow cooling and deformation of a range of rocks such as gabbro, diorite, tonalite, granodiorite [21]. During its crystallization, boron is concentrated in magmatic residues and forms boron silicate minerals. The creation of borate and boron silicate raw materials can therefore be regarded as the products of the deformation and crystallization of basaltic magma. Furthermore, boron is present in volcanic environments where it appears at the surface either in the form of soluble compounds or in liquid or as gas phase [16]. Boron compounds may either dissolve and discharge into the sea or concentrate in closed basins; if boron appears in gaseous form, it either remains in the atmosphere or goes back to the biosphere through the rains.

The gaseous boron of the atmosphere makes up 97% of total boron in the troposphere. Summarizing, boron in the atmosphere is mainly derived from sea-spray degassing and fumaroles gases with a boron residence time of approximately 1 month.

In aqueous solutions, boron is present as $B(OH)_4$ ion, undissociated boric acid $B(OH)_3$, polyborate ions and borates [$(Na-Ca-Mg)B(OH)_4$]. The distribution of these species is controlled by the pH, salinity and specific cation concentrations.

In general, B is relatively easily mobilized into aqueous fluids during water–rock reactions. However, B speciation is not a conservative property in groundwater. B contents change as a consequence of water–rock interactions, mixing with waters of different origin, and addition of contaminants (natural or anthropogenic). Additionally, the isotopic compositions of B vary considerably in environment and have distinctive ranges that are useful in groundwater, and other crustal fluid studies.

At temperatures higher than 150°C, seawater leaches boron from sediments and volcanic rocks [15, 16, 32, 35]. At lower temperatures, boron can be removed from solution by absorption onto clay minerals and incorporation into secondary minerals [23, 40].

Boron is an essential element for both plants and animals, but when it exceeds a certain threshold, specific for each species, it can be rather toxic for plants and presumably also for humans. Boron concentration <1 mg/l in irrigation water is required for sensitive crops. Moreover a long-term trigger value in irrigation water of 0.5 mg/l has been proposed [1]. Despite the fact that boron is one of the seven most important trace elements for the growth of plants, vegetation is reduced or even absent in areas where borates are present at high concentrations. Moreover, many crops are sensitive to high boron levels in irrigation water. Boron can also be detected in human and animals tissues, in which it plays an important physiological role. However, high boron levels in drinking water are considered to bear risks for human health [6]. In animal experiments, high doses of boron affect fertility and pregnancy.

Occasionally observed high concentrations (boron concentration in most groundwater is less than 1 mg/l) in potable and irrigation waters are attributed to hydrothermal action [18], evaporative concentration, intrusion of seawater, residual seawater, dissolution of evaporites [18], mineral weathering [29], human pollution [3, 50] and the process of sorption and de-sorption of B to mineral surfaces [37].

The World Health Organisation guideline value for boron in drinking water is 0.5 mg/l [54]. The drinking-water standard of the European Community [11] is two times this recommended value but still, in many water resources worldwide, boron concentrations in ground- and surface water exceed this value rendering such water unacceptable according to the European standards. Although there are limited toxicological data of boron on human health, a tolerable daily intake of 400_g/kg body weight per day has been proposed. Drinking water and food contribute about 30% and 65% of boron mean daily intake for an adult, respectively [53].

Boron compounds are widely used in industrial applications such as electronics, nuclear reactors, metallurgy, glass making, pharmaceuticals, and plastics. However, the main industrial applications of boron that apparently affect ground water systems are washing powders and agricultural applications of boron-fertilizers and boron-pesticides. Boron contamination is clearly not a major issue in the context of diffuse pollution due to agricultural activities dominated by the problem of nitrates, pesticides and herbicides. Nevertheless, the increase of the concentration of boron in water is many times indirectly correlated with the development of agriculture; exploitation of high amounts of groundwater often results in significant drop of piezometric levels and saline intrusion. High levels of boron are also detected in coal and landfills. The main source of boron contamination of German landfills is the washing powder. A major boron problem has been identified in Greece. Recently, the results from an EU project entitled BOREMED showed that the occurrence of boron in groundwater in various Mediterranean countries is a serious threat to the future use for drinking and irrigation purposes [4, 9, 10, 24]. Greece was among the countries that were under study in this project. The research goal of the project was to thoroughly investigate the presence of boron in groundwater and to find out the origin of this element in the study area. High boron concentrations in Greece water resources make them unusable for irrigation and drinking water purposes. The need for significant quantities of high quality water is an important issue due to rapid urbanization and increasing touristic development, making the investigation of the spatial distribution and possible source a priority.

Greece as a coastal country with a strong geothermal activity combines a variety of potential natural sources of boron and represents an interesting context for the investigation of the provenance of the contaminant. In Greece, we find geothermal fields of low, middle and high enthalpy, a high geothermal gradient, which is due to the volcanic activity of the area and to the active tectonics, which allow the circulation of water in great depths. In these fields or close to them, high concentrations of boron are present. The study of boron geochemistry is not only an academic interest, but also a necessity for water management in a country that is largely under water stress due to an increasing demand for drinking and irrigation water. The rapid urbanisation of some areas, the tourist industry and the agricultural activities lead to a lowering of water tables and consequent deterioration of water quality by the intrusion of seawater.

The main goal of this paper is to present the boron distribution in water resources of Greece and distinguish natural sources of boron (geothermal activity, seawater intrusion, evaporate dissolution, mixing with sedimentary brines) from anthropogenic sources (boron contaminated surface and ground waters). Chemical and isotopic fingerprints have been used to investigate the origin of boron. Statistical treatment of the results was also employed in order to clarify similarities among samples and the case of West Chalkidiki area was chosen for thorough investigation.

15.2 Sampling and Analysis

In total 113 samples of natural waters, spring and boreholes, were collected all over Greece. Figure 15.1 shows locations of water sampling areas. These areas cover a wide range of typical aquifers. Many samples come from Ikaria, Lesvos (Mytilini)

15 Boron Contents in Greek Fresh and Thermal Waters

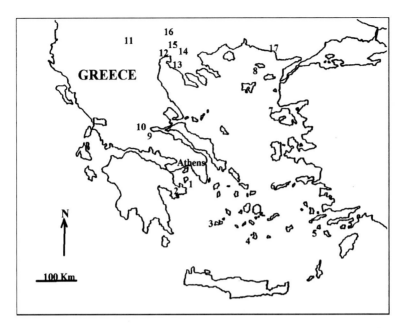

Fig. 15.1 Location of water sampling sites in Greece: 1-Egina, 2-Methana, 3-Milos, 4-Santorini, 5-Nisyros, 6-Ikaria, 7-Lesvos, 8-Samothraki, 9-Thermopyles, 10-Ypati, 11-Aridea, 12-Thessaloniki (Sedes), 13-Chalkidiki (Nea Appolonia, N. Triglia, Eleochoria, Nea Tenedos, Petralona), 14-Mygdonia basin (Lagadas, Volvi, Nymfopetra), 15-Nigrita, 16-Sydirokastro, 17-Traianoupoli

and Samothrace islands in eastern Greece, from the volcanic arch (Methana, Milos, Santorini, Nisyros, Egina) and from the continental Greece (Thermopiles, Ypati, Trajanoupolis, Aridea, Nigrita, Sidirokastro, Lagada, Nymphopetra, Nea Appolonia, N. Triglia, Eleochoria, Nea Tenedos, Sedes, Petralona).

Temperature was measured directly in the field. The water samples were filtrated through 0.45_m membrane filters and stored at 4°C. The samples were analyzed for B, Ca, Mg, Na, K, HCO_3, Cl, Br, SO_4, NO_3 and As. Alkalinity (CO_3, HCO_3) was measured titrimetrically with HCl. Ca and Mg were measured by EDTA titrimetric method. The ions Na, K, SO_4, Cl, Br and NO_3 were determined by ion chromatography. Arsenic was determined by flame atomic absorption/hydride generation.

The B content was determined photometrically using the curcumin method. The samples were acidified and evaporated to dryness (at 55°C) in the presence of curcumin. The red precipitate can be dissolved in ethyl alcohol. Boron in the red alcoholic mixture was photometrically determined at $\lambda = 540$ nm (404A Method in Standard Methods for the Examination of Water and Wastewater, APHA-AWWAWPCF, 16th edition, [2]).

Samples (not acidified) were also subjected to isotopic analysis of $\delta^{11}B$, $\delta^{18}O$, δ^2H. The $\delta^{18}O$ compositions of water in the samples were determined from CO_2 equilibrated with the water [13]. The δ^2H compositions of water were determined from the H_2 generated by the Zn-reduction method [7].

$\delta^{11}B$ was determined by positive ion thermal ionization mass spectrometry [41] on a Finnigan MAT 261 single collector solid source mass spectrometer. Boron isolation was obtained by using the Amberlite IRA-743 boron selective resin according to a method adapted from Gaillardet and Allègre [14]. The sample was then loaded on Ta single filament with Cs and mannitol, and run as $Cs_2BO_2^+$ ion. Values are reported on the δ scale relative to NBS951 boric acid standard where $\delta^{11}B$ (in ‰) is defined as $[\{(^{11}B/^{10}B) \text{ sample } / (^{11}B/^{10}B) \text{ standard}\} -1] \times 10^3$. The $^{11}B/^{10}B$ value obtained for the NBS951 boric acid standard after oxygen correction was 4.04491 ± 0.00136 (2δ, number of determinations = 13). The long-term external reproducibility based on replicate analysis is ± 0.34‰. The internal error is often better than 0.2‰.

15.3 Origin of B and $\delta^{11}B$

The B and Cl⁻ concentrations can be used for the comprehension of the provenance of boron contents in water resources. Cl⁻ is regarded as a conservative ion even in geothermal environments. Its behavior is thus particularly useful for investigating the origin of salinity. In seawater boron concentrations range between 4.5 and 5 mg/l, and B/Cl ratio is 2×10^{-4}. In fossil brines, of seawater origin, boron concentrations range between 47.6 and 1,379 mg/l, and B/Cl values range between 5.6×10^{-4} and 9.9×10^{-3} [19, 46]. In rainwater, boron concentration is approximately 0.01 mg/l, while in fresh and uncontaminated surface water this concentration is in general less than 0.05 mg/l with a B/Cl ratio of 1.3×10^{-3}. In waters contaminated by domestic waste waters, and other anthropogenic sources (derived from Na or Ca-borates), the concentrations of boron are significantly elevated. In these waters B/Cl ratio is much higher than that of seawater.

The concentrations of boron in oil fields waters are greater than 100 mg/l. Connate waters are also enriched in boron, with values that often exceed 350 mg/l [38] and B/Cl ratios that range from 5×10^{-3} to 4×10^{-2}.

In thermal waters, which contain juvenile water, boron is present at various concentrations and the ratio of B/Cl ranges from 2×10^{-2} to 0.4 [52]. In general, the distribution of boron among hydrothermal waters depends on the provenance, temperature, vapour pressure and lithology of aquifers.

In order to facilitate the discussion of $\delta^{11}B$ variations, typical ranges of boron isotopes in different environments from literature are illustrated in Fig. 15.2. Each of these sources presents distinctive boron isotopic compositions: Granites, fresh oceanic basalts and non-desorbable (fixed) boron in clay minerals have $\delta^{11}B$ value of 0‰ (−5–5‰), altered oceanic basalts have $\delta^{11}B$ values from 0‰ to 9‰, and desorbable boron, leached from marine clay minerals, has $\delta^{11}B$ from 13.9‰ to 15.8‰ [20, 44, 49]. Modern marine carbonate sediments have $\delta^{11}B$ from 14.2‰ to 32.2‰ [48].

Boron isotopic compositions in fluids have also different signature: $\delta^{11}B$ in the ocean is 39‰ [43, 48], $\delta^{11}B$ values in volcanic gases range (Fig. 15.2) between 1.5‰ and 6.5‰ [22]. Representative studies of $\delta^{11}B$ in rain (0.8–35‰) became

15 Boron Contents in Greek Fresh and Thermal Waters

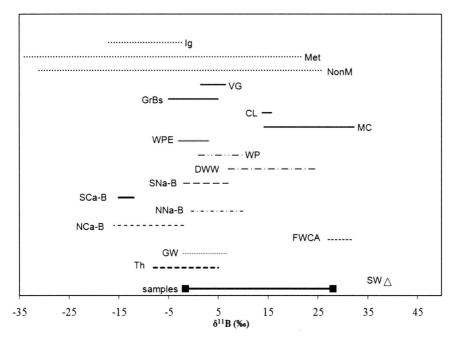

Fig. 15.2 $\delta^{11}B$ values and ranges from literature. *SW* sea water [43, 48], *GW* geothermal waters [33, 34], *FWCA* fresh waters-coastal aquifer, *NCa-B* natural Ca-borate, *NNa-B* natural Na-borate, *SCa-B* synthetic Ca-borate, *SNa-B* synthetic Na-borate [49, 50], *DWW* domestic waste-water, *WP* washing powder, *WPE* washing powder (Europe) [49, 50], *MC* marine carbonates [49], *CL* clays [20, 44, 49], *GrBs* granites, basalts [20, 44, 49], *VG* volcanic gases [22], *NonM* non-marine B, *Met* metamorphic B, *Ig* igneous B [36, 49], *Th* thermal waters

available only recently after some early studies [41], estimating a wide range of $\delta^{11}B$. Chetelat et al. [5] found a range of 33.9–44.8‰ for coastal rains in French Guiana with a weighted mean of 41 ± 3.5‰ and concluded on marine origin of boron and a slight enrichment in ^{11}B due to kinetic fractionation during evaporation. Mather and Porteous [28] measured comparable values (38–42‰) in Southern England (Guildford) but a much larger range for rains on the Isle of Wight (−13–24‰). Rose-Koga et al. [39] measured two coastal rain samples from California (19.8 and 26‰) and of continental rains and snows (−10–+34‰). Rainwater from Germany was reported to have $\delta^{11}B$ equal to 13‰ [12]. In the hydrothermal fluids the $\delta^{11}B$ is close to 0‰ (−10–10‰) [34]. In continental waters, the upper limit of $\delta^{11}B$ is 59‰, observed in Australian crater lakes [47] and the lower limit is −27‰ observed in Cecina River, Italy [36]. The $\delta^{11}B$ of fresh groundwater depends on contributions of aqueous boron from weathering (low $\delta^{11}B$) and marine aerosols (high $\delta^{11}B$). Subsurface brine has a large range of $\delta^{11}B$ values according to the origin of brine and the interaction with sediments [49]. In general, the ^{11}B values (Fig. 15.2) of the main B reservoirs are 4–58‰ for marine boron; −31–26‰ for non-marine B; −17–−2‰ for igneous B and −34–22‰ for metamorphic B [36, 49].

Boron compounds are also used as boron-fertilizers and boron-pesticides, which results in an increase of B contents in groundwater. The $\delta^{11}B$ in boron fertilisers is between $-15‰$ and $7.5‰$ [49, 50]. In particular, $\delta^{11}B$ of synthetic Na-borate is $-0.4-7.5‰$ while the $\delta^{11}B$ of natural Na-borate minerals is $-0.9-10.5‰$ [49, 50]. In contrast, $\delta^{11}B$ values of synthetic Ca-borate products are low, -15 to $-12‰$ and overlap with those of natural Ca-borate minerals [49].

15.4 Geochemical and Isotopic Characterization of Waters

The quality of waters collected from the Greek islands (islands of the volcanic arc and islands of the NE Greece), has been degraded by both the elevated salinity and the high boron levels (B = 0.1–2.6 mg/l). Moreover, the thermal waters from the same areas show very high concentrations of B (B = 1–12 mg/l), a fact that is probably related to the relatively high temperature ($T = 30-55°C$). The waters of the continental Greece have also high temperature, B/Cl ratio much higher than that of seawater (10^{-2}), and Cl⁻ concentration less than 200 mg/l.

In the diagram B versus Cl⁻ (Fig. 15.4), two linear correlations appear between B and Cl⁻, suggesting a different origin for both elements. The trend with the lower slope, defined by all waters sampled in the islands, represents an ideal mixing between seawater and fresh water. This suggests that Cl⁻ and B derive from seawater.

The trend with higher slope is a mixing line between cold water and deep thermal water. The hot member is water with temperature less than 62°C, coming from Nigrita, Sidirokastro, Aridea and Chalkidiki and the cold water coming from the same areas with non-marine origin of the B. For the water of the continental Greece the B content of fresh and hot water is higher than seawater and increases with temperature.

In the diagram B/Cl *versus* Cl⁻ (Fig. 15.4), representing the values of all the samples, two main groups can be distinguished: the Cl⁻ concentration of the first group is less than 200 mg/l and its B/Cl ratio less than 200. Waters with Cl⁻ concentrations exceeding 700 mg/l and a B/Cl ratio higher than 800 fall in the second group.

Waters sampled in areas of continental Greece belong to the first group. Most of these waters have high temperature, B/Cl ratio much higher than that of seawater (2×10^{-4}) and are freshwaters with Cl⁻ concentrations less than 200 mg/l. The measured Cl⁻ concentration and the relationship between $\delta^{18}O$ and $\delta^{2}H$ values (Fig. 15.5) indicate a meteoritic origin of water and thus the possible boron contamination by seawater intrusion cannot be excluded.

Boron originated from B-fertilizers, B-pesticides and industrial and urban wastewater would result in waters rich in B and other ions such as Ca^{2+}, Na^+ and NO_3^- compared to the local unpolluted fresh waters, a fact that was not observed [10]. Therefore, the above results indicate a geothermal origin for boron, meaning that the boron content derives from interaction between water and rock.

15 Boron Contents in Greek Fresh and Thermal Waters

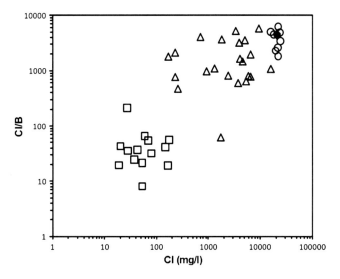

Fig. 15.3 B concentrations versus chloride. *Squares*: Thermopiles, Ypati, Trajanoupolis, Aridea, Nigrita, Sidirokastro, Lagada, Nymphopetra, Nea Appolonia, N. Triglia, Eleochoria, Nea Tenedos, Sedes, Petralona. *Triangles*: Methana, Milos, Santorini, Nisyros, Egina. *Circles*: Ikaria, Mytilini Therma, Methana Therma. *Black Rhombus*: Seawater

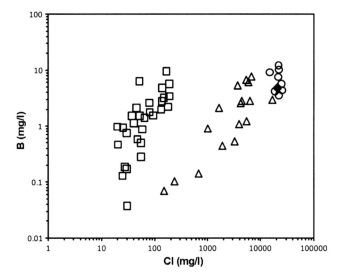

Fig. 15.4 Cl/B ratio versus Cl⁻. For symbols see Fig. 15.3

The waters of the second group come mainly from the islands and the geothermal fields of the volcanic arc. The positions of the points (Fig. 15.3) representing these waters, show the participation of seawater in the deep hydrothermal system indicating a marine provenance of boron. In fact in Fig. 15.4 the position of the samples coming from the islands is very close to that of mean

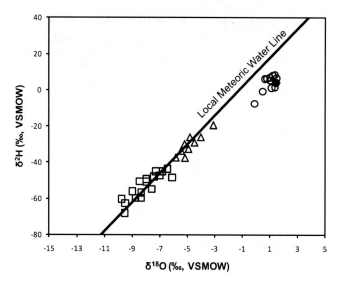

Fig. 15.5 δ^2H versus δ^{18}O diagram. *Line*: local meteoric water line. For symbols see Fig. 15.3

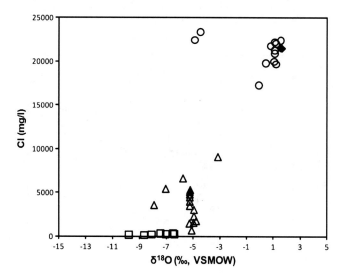

Fig. 15.6 δ^{18}O versus Cl⁻. For symbols see Fig. 15.3

seawater confirming its purely marine origin. The position of some samples under the local meteoric water line (Fig. 15.5) could reflect mixing between seawater and fresh water.

In the diagram δ^{18}O *vs* Cl⁻ and in a δ^2H *vs* Cl⁻ (Figs. 15.6 and 15.7) the positions of the spots, which are related to waters with B/Cl ratio lower than that of seawater, are distributed along different lines, between seawater and geothermal water of meteoric origin, corresponding to different sampling zones.

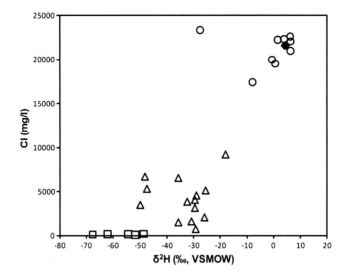

Fig. 15.7 δ^2H versus Cl⁻. For symbols see Fig. 15.3

The high boron content of the waters from the volcanic arc can be mainly attributed to the temperature, which is mostly responsible for the formation of saline geothermal water. In Ikaria (B/Cl = 2 × 10⁻⁴), the provenance of boron is clearly related to the intrusion of marine water [8]. On the contrary, the high boron contents of water from Lesbos [30], Samothrace and Egina, are attributed to both the elevated temperatures and the intrusion of marine water.

15.5 The Case of West Chalkidiki

15.5.1 The Study Area

The study area is located in the west part of Chalkidiki, S-SE of the Katsika mountain up to the coastal area (Fig. 15.8). The area is under rapid urbanization and tourist development with increasing demand of drinking water supply. Agricultural activities in the area also demand high quantities of irrigation water. Due to overexploitation the water table has been lowered and the water quality deteriorated. The presence of elevated concentrations of arsenic is also a significant problem in the major area [26]. The study area belongs geologically to the Peonia Unit [25, 27, 45]. Mesozoic limestone is dominant in this region, along with granites-granodiorites, which are located in the north-western part [31]. In particular, the karst system Katsika – Petralona (Fig. 15.8) consists of extensively fissured and karstified Jurassic limestone under thin sandstone layers. Several bauxite horizons, slightly metamorphic, are also intercalated. This karst system develops in the sedimentary cover on top of granitic bedrock. A discontinuity exists on the north, where the limestone overlies the granites.

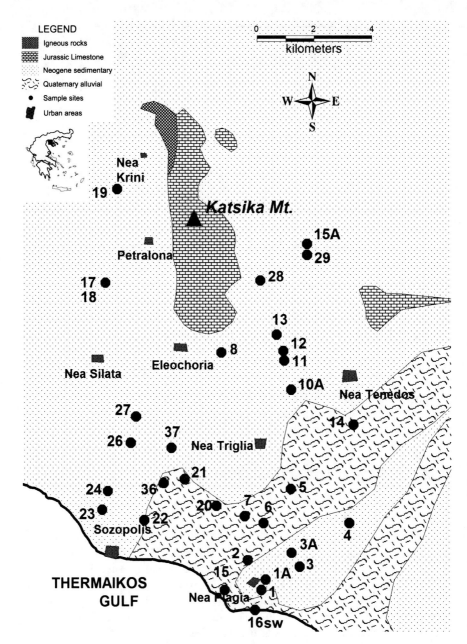

Fig. 15.8 Schematic geological map of West Chalkidiki area, showing the location of water sampling sites

Potential sources of salinity and boron in the aquifers of West Chalkidiki area could be the following:

- Seawater intrusion. Many boreholes are located in coastal areas and seawater intrusion is possible due to overexploitation
- Boron contamination from hydrothermal waters resulting from interactions with the hydrothermal reservoir matrix. The presence of geothermal field in the study area suggest this assumption
- Anthropogenic sources
- Combination of the above sources.

15.5.2 Geochemistry

Regarding the water samples' temperatures, these can be divided into fresh waters (13–20°C), semi-thermal waters (20–25°C) and thermal waters (31–42°C). Bicarbonate was the dominant anion in groundwater, followed by chloride. Calcium and magnesium were the dominant cations although sodium exhibited significant contribution in certain wells. Thus, the waters sampled are classified into two water types: HCO_3 and Cl groundwaters. Thermal spring waters (Ca-HCO_3 type), semi thermal waters (Mg-Na-HCO_3), fresh spring waters (Ca-HCO_3 and Mg-HCO_3 type) and almost all sampled borehole waters (Mg-Ca-HCO_3 type) are included in the HCO_3 group. Only two samples (15-Nea Plagia and 24) are included in the Cl -group.

Fig. 15.9 shows the relationship between Cl$^-$ and major cations; the chemical signature of all waters is characterized mainly by the mixing between hydrothermal water and fresh spring water (fresh waters are plotted close to the axes' intersection), but water-sample 15 does not follow this trend. This water has Cl$^-$/Br ratio similar to that of seawater suggesting that the supply of Cl$^-$, Na$^+$ and K$^+$ ions by seawater intrusion is not negligible. Also, this water shows about 160 mg/l nitrate, indicating that part of its chemical composition is controlled by human activities.

Hot water springs have the highest concentration of Ca^{2+} and low Mg^{2+} (Fig. 15.9), as well as high HCO_3^-, indicating that their chemical composition has been modified by water-rock interaction processes at high temperature and pCO_2. These thermal waters circulate in carbonate rocks and the enrichment in Ca^{2+} is attributed to dissolution of carbonates. It is possible that part of Ca^{2+} may come from the dissolution of detrital Ca-bearing plagioclase, which occurs in the aquifer matrix.

The B concentrations in Chalkidiki waters are from minimal (0) to maximum 0.6 mmol/l. In the diagram Cl/B versus Cl$^-$ (Fig. 15.10), where B addition from rocks and seawater is also shown, two members are distinguished: fresh water and thermal water. The lowest Cl/B values are observed in semi thermal and thermal waters. The boron excess with respect to Cl$^-$ can be explained by water-rock interaction with boron-bearing mineral phases, and eventually subsequent mixing with seawater.

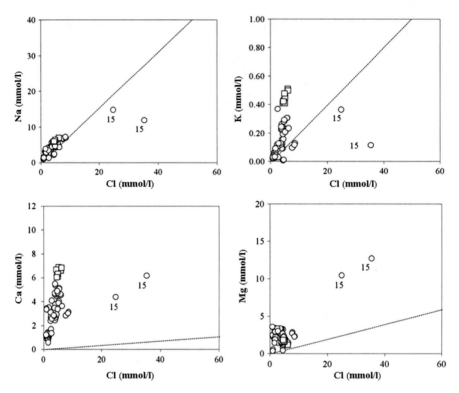

Fig. 15.9 Plots of Na$^+$, K$^+$, Ca^{2+} and Mg^{2+} versus Cl$^-$ concentrations in the Chalkidiki waters. *Circles*: fresh waters, *squares*: thermal waters, *line*: sea water dilution

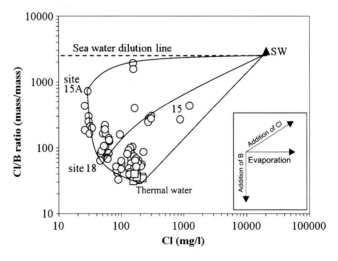

Fig. 15.10 Plot of Cl/B ratio versus Cl$^-$ concentrations in the Chalkidiki waters. *Circles*: fresh waters, *squares*: thermal waters

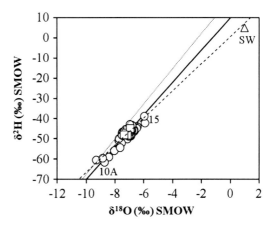

Fig. 15.11 Values of δ^2H versus $\delta^{18}O$ in the Chalkidiki waters. *Circles*: fresh waters, squares: thermal waters. *solid line*: global meteoric water line; *dash-line*: regression line of sampled waters; *dot-line*: local meteoric water line; *SW* seawater

Thus, the potential source of boron in the study area is related mainly to the interaction of thermal water with the reservoir matrix. A minor source of boron contents might be attributed to seawater intrusion and anthropogenic pollution.

15.5.3 Isotopic Composition

The stable isotope contents of all sampled waters show variations of $\delta^{18}O$ ranging from −9.31‰ to −5.9‰ and δ^2H from −61.5‰ to −42.1‰ (Fig. 15.11). The maximum value is observed for two samples due to seawater contribution. The most negative value was measured in water sample 10A. The thermal waters (samples 8 and 14) have approximately the same isotopic signature with respect to fresh spring waters (samples 9, 15A, 17 and 19), indicating their purely meteoric origin. The $\delta^{18}O$ and δ^2H values of the rest of waters range between the saline samples and the sample 10A. Therefore, if sample 15 is excluded, the position of all samples in this diagram confirms their meteoric origin. Minor variations observed in thermal springs could reflect different recharge altitudes.

In all waters sampled the $\delta^{11}B$ range from −1.62‰ to 28.96‰. In order to determine the origin of B, the relationships between the B/Cl ratio and $\delta^{11}B$ contents of the water sampled are illustrated in Fig. 15.12. In this diagram, the poles highlighted by the water chemistry are easily recognized also from δ11B of water subgroups: fresh spring water, sample 9, thermal spring water, samples 8 and 14 (end-member), semi thermal water, samples 10, 12, 13 and 11 (end-member), and finally the saline member, sample 15.

The highest B contents (around 6 mg/l) associated with a low $\delta^{11}B$ (0.50‰) were observed in the thermal waters. In the karst hot system, the $\delta^{11}B$ value in thermal water is 0‰. This $\delta^{11}B$ value indicates the non carbonate source of B. The $\delta^{11}B$ values of fresh spring water, sample 9, are similar to those of thermal fluids suggesting the same B source. Most borehole waters present $\delta^{11}B$ similar to that of thermal waters, suggesting that B in these waters has geothermal origin.

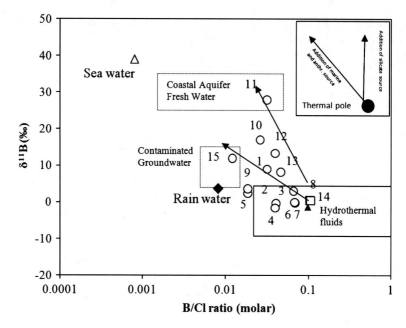

Fig. 15.12 Values of $\delta^{11}B$ versus B/Cl in the Chalkidiki waters. *Arrows* represent theoretical mixing lines ([48, 50]). *Circles*: fresh waters, *squares*: thermal waters

Starting from the thermal pole, all the waters are plotted inside a mixing field defined by two-members (Fig. 15.12, inset). One trend is clearly related to geothermal input. This trend is mainly related to sample 15, which shows an increase of $\delta^{11}B$ and B/Cl ratio compared to geothermal water. For this sample, the seawater intrusion is indicated by Cl/Br ratio (Fig. 15.10) and by the $\delta^2H–\delta^{18}O$ relation (Fig. 15.11).

The second trend is related again to geothermal input and also to water from the Neogene sediments that are characterized by the highest $\delta^{11}B$ (sample 11). This trend is mainly related to samples that are enriched in ^{11}B with respect to the thermal waters. It is also remarkable that these waters found in the area of Eleochoria and Nea Tenedos have high ^{11}B content in contrast to those in close vicinity, where low ^{11}B concentrations were found. The model that could explain this phenomenon is that the meteoric water goes deep into the limestone and the granodiorite, where water temperature and B concentration increase with increasing depth. This hydrothermal karstic water is pumped at Eleochoria and Nea Tenedos (boreholes 8 and 14). On the other hand, the semi thermal borehole waters, 10, 11, 12, 13 (their borehole depth is approximately 150 m) are recharged from the Neogene sediments. However, due to the great thickness of these sediments, the borehole waters do not reach the karstic limestone. Prevalent circulation of these waters in the Neogene sediments, sand and carbonates is confirmed by the enrichment of $\delta^{11}B$.

15.6 Conclusions

Generally, all sampled waters across Greece present boron concentration that exceeds the former WHO limit of 0.3 mg/l and in many waters in regions with high geothermal activity, boron is higher than 1 mg/l, which is the acceptance value for the human consumption in the EU. Moreover, in some irrigation waters the boron concentration exceeds the value of 0.75 mg/l, which is the limit for sensitive crops. About the origin of B, both boron concentrations and the B/Cl ratios of waters indicate that their enrichment in boron is due to their vicinity of geothermal fields and/or to the intrusion of seawater. Especially the elevated B concentrations in some water samples from the N. Greece may be related with the geothermal field of Central Macedonia. The B contents of water samples from the islands can be attributed mainly to the high temperature at depth, and to the intrusion of marine water.

The chemical and multi isotope approach for the case of West Chalkidiki showed high spatial variation of Boron contents, ranged from 0.04 to 6.5 mg/l. The boron excess with respect to Cl^- contents might reflect interactions with the hydrothermal reservoir matrix. The saline component, with high Cl^- and B contents is due to seawater intrusion caused by overexploitation of the aquifers. The influence of human activities on the water chemistry, shown by the increase of NO_3, suggests that fertilisers or sewage could contribute to the boron budget. In the borehole waters, the $\delta^{11}B$ values exhibit in general three increasing trends in relation to the thermal water, reflecting different B sources: One is clearly related to geothermal inputs with $\delta^{11}B = 0‰$ and B/Cl ratio equal to 0.1. The second one is related to the seawater intrusion and has a value of $\delta^{11}B > 10‰$ and B/Cl ratio = 0.01, while the third one is related to "sand and carbonates" of Neogene input and has a value of $\delta^{11}B \gg 10‰$ and B/Cl > 0.01. In conclusion, the B dissolved in groundwater of Chalkidiki area is geogenic, i.e. it does not derive from anthropogenic sources, but it is released from the aquifer sediments. The overpumping and uncontrolled drilling of the aquifers for new wells in the major area could lead to an increase risk of appearance of boron in uncontaminated groundwaters posing a potential threat to their use for drinking and irrigation purposes. Moreover, the distinctive isotopic composition of boron sources makes boron isotope a potential environmental tracer for distinguish the origin of dissolved constituents (natural versus anthropogenic), and hence, sources of contamination in groundwater.

References

1. ANZECC (2000) Australian and New Zealand guidelines for fresh and marine water quality, Chapter 9
2. APHA, AWWA, WPCF (1985) Standard methods for the examination of water and wastewater, 16th edn. APHA, AWWA, WPCF, Washington, DC
3. Barth S (1998) Application of boron isotopes for tracing sources of anthropogenic contamination in groundwater. Water Res 32:685–690

4. BOREMED (2004) Boron contamination of water resources in the Mediterranean Region: distribution, sources, social impact and remediation, Contract EVK1-CT-2000-00046. Final Report
5. Chetelat B, Gaillardet TJ, Freydier TR, Négrel P (2005) Boron isotopes in precipitation: experimental constraints and field evidence from French Guiana. Earth Planet Sci Lett 235:16–30
6. Cidu R, Frau F, Tore P (2010) Drinking water quality: comparing inorganic components in bottled water and Italian tap water. J Food Compos Anal, March 2011, 24(2):184–193
7. Coleman ML, Shepard TJ, Durham JJ, Rouse JE, Moore GR (1982) Reduction of water with zinc for hydrogen analysis. Anal Chem 54:993–995
8. Dotsika E, Michelot JL (1992) Origine et temperatures en profondeur des eaux thermals d'Ikaria (Greece). C R Acad Sci Paris t 315(Serie II):1261–1266
9. Dotsika E, Kouimtzis Th, Kouras A, Poutoukis D, Voutsa D (2002a) Hydrochemical and isotopical study for the origin of boron in the west Chalkidiki area. In: Proceedings of the 6th Hellenic hydrogeological conference, Xanthi, pp 347–352
10. Dotsika E, Kouimtzis Th, Kouras A, Poutoukis D, Voutsa D (2002b) Deep temperature of hot karstic water of the west Chalkidiki area. In: Proceedings of the 6th Hellenic hydrogeological conference, Xanthi, pp 377–384
11. EC (1998) Directive related with quality of water intended for human consumption 98/83/EC
12. Eisenhut S, Heuman KG (1997) Identification of groundwater contaminations by landfills using precise boron isotope ratio measurements with negative thermal ionization mass spectrometry. Fresenius J Anal Chem 359:375–377
13. Epstein S, Mayeda T (1953) Variation of ^{18}O content of waters from natural sources. Geochim Cosmochim Acta 4:213–224
14. Gaillardet J, Allègre CJ (1995) Boron isotopic compositions of corals: seawater or diagenesis record? Earth Planet Sci Lett 136:665–676
15. Harder H (1970) Boron content of sediments as a tool in facies analysis. In: Walker CT (ed.) Geochemistry of boron. Halsted, Stroudsburg, pp 105–127
16. Harder H (1973) Boron. In: Wedepohl KH (ed.) Handbook of geochemistry II-1. Springer, Berlin
17. Helvaci C (2005) Borates. In: Selley RC, Cocks LRM, Plimer IR (eds.) Encyclopedia of geology, vol 3. Elsevier, Amsterdam, pp 510–522
18. Hem JD (1985) Study and interpretation of the chemical characteristics of natural water, 3rd edn. U.S. Geological Survey Water-Supply Paper 2254, pp 263
19. Herrmann AG, Knake D, Schneider J, Peters H (1973) Geochemistry of modern seawater and brines from salt pans: main components and bromide distribution. Contrib Miner Petrol 40:1–24
20. Ishikawa T, Nakamura E (1993) Boron isotope systematics of marine sediments. Earth Sci Planet Lett 117:567–580
21. Jensen ML, Bateman AM (1979) Economic mineral deposits, 3rd edn. Wiley, New York, pp 593
22. Kanzaki T, Yoshida M, Nomura M, Kakihana H, Ozawa T (1979) Boron Isotopic composition of fumarolic condensates and sassolites from Satsuma Iwo-Jima, Japan. Geochim Cosmochim Acta 51:1939–1950
23. Karen R, Mezuman V (1981) Boron adsorption by clay minerals using a phenomenological equation. Clays Clay Miner 29:198–204
24. Kloppmann W, Bianchini G, Chalalambides A, Dotsika E, Guerrot C, Klose P, Marei A, Pennisi M, Vengosh A, Voutsa D (2005) Boron contamination of Mediterranean groundwater resources: extent, sources and pathways elucidated by environmental isotopes. Geophys Res Abstr 7:10162
25. Kockel F, Mollat H, Walther H (1977) Erlanterungen zur geologischen Karte der Chalkidiki and angrenzender Gebiete 1:100.000 (Nord-Griechenland). Bund Geowiss Rohst, Hannover, 119 p
26. Kouras A, Katsoyiannis I, Voutsa D (2007) Distribution of arsenic in groundwater in the area of Chalkidiki, Northern Greece. J Hazard Mater 147:890–899

27. Lalechos N (1986) Correlation and observation in molassic sediments in on shore and off shore area of Northern Greece. Miner Wealth 42:7–34
28. Mather JD, Porteous NC (2001) The geochemistry of boron and its isotopes in groundwaters from marine and non-marine sandstone aquifers. Appl Geochem 16:821–834
29. McArthur JM, Ravenscroft P, Safiullah S, Thirlwall MF (2001) Arsenic in groundwater: testing pollution mechanisms for aquifers in Bangladesh. Water Resour Res 37:109–117
30. Michelot JL, Dotsika E, Fytikas M (1993) A hydrochemical and isotopic study of thermal waters on Lesbos Island (Greece). Geothermics 22(2):91–99
31. Monod O (1965) Etude géologique du massif du mont Chortiatis (Macédoine). Inst Geol Geoph Res Athens X(4):221–279
32. Mosser C (1983) The use of B, Li, Sn in determining the origin of some sedimentary clays. Chem Geol 38:129–139
33. Musashi M, Oi T, Ossaka T, Kakihana H (1990) Extraction of boron from GSJ rock reference samples and determination of their boron isotopic ratios. Analytica Chimica Acta 231:147–150
34. Palmer MR, Sturchio NC (1990) The boron isotope systematic of the Yellowstone National Park (Wyoming) hydrothermal system: a reconnaissance. Geochim Cosmochim Acta 54:2811–2815
35. Palmer MR, Spivack AJ, Edmond JM (1987) Temperature and pH controls over isotopic fractionation during adsorption of boron on marine clay. Geochim Cosmochim Acta 51:2319–2323
36. Pennisi M, Gonfiantini R, Grassi S, Squarci P (2006) The utilization of boron and strontium isotopes for the assessment of boron contamination of the Cecina River alluvial aquifer (central-western Tuscany, Italy). Appl Geochem 21:643–655
37. Ravenscroft P, McArthur JM (2004) Mechanism of regional pollution of groundwater by boron: the examples of Bangladesh and Michigan, USA. Appl Geochem 19:1255–1291
38. Rittenhouse G, Fulton R, Grabowsk R, Bernard J (1969) Minor elements in oil field waters. Chem Geol 4:189–209
39. Rose-Koga EF, Sheppard SMF, Chaussidon M, Carignan J (2006) Boron isotopic composition of atmospheric precipitations and liquid–vapour fractionations. Geochim Cosmochim Acta 70:1603–1615
40. Seyfried WE, Janecky DR, Motti MJ (1984) Alteration of the oceanic crust: implications for geochemical cycles of lithium and boron. Geochim Cosmochim Acta 48:557–569
41. Spivack AJ (1986) Boron isotope geochemistry. PhD thesis, Massachusetts Institute of Technology/Woods Hole Oceanographic Institution, 184 p
42. Spivack AJ, Edmond JM (1986) Determination of boron isotope ratios by thermal ionization mass spectrometry of the dicesium metaborate cation. Anal Chem 58:31–35
43. Spivack AJ, Edmond JM (1987) Boron Isotope exchange between seawater and oceanic crust. Geochim Cosmochim Acta 51:1033–1042
44. Spivack AJ, Palmer MR, Edmond JM (1987) The sedimentary cycle of the boron isotopes. Geochim Cosmochim Acta 51:1939–1950
45. Syridis G (1990) Lithostromatografical, biostromatografical and paleostromatografical study of Neogene-Quaternary formations of Chalkidiki Peninsula (in Greek). Ph.D. of Aristotle University of Thessaloniki, 243p
46. Valyashko MG (1956) Geochemistry of Bromine in the processes of salt deposition and the use of the bromine content as a genetic and prospecting criterion. Geochemistry 6:570–589
47. Vengosh A, Chivas AR, McCulloch MT, Starinski A, Kolodny Y (1991) Boron isotope geochemistry of Australian salt lakes. Geochim Cosmochim Acta 55:2591–2606
48. Vengosh A, Starinsky A, Kolodny Y, Chivas AR, Raab M (1992) Boron isotope variations during fractional evaporation of sea water: new constraints on the marine vs. non marine debate. Geology 20:799–802
49. Vengosh A, Heumann KG, Juraske S, Kasher R (1994) Boron isotope application for tracing sources of contamination in groundwater. Environ Sci Technol 28:1968–1974

50. Vengosh A, Kolodny Y, Spivack AJ (1998) Ground-water pollution determined by boron isotope systematic. IAEA. In: Application of isotope techniques to investigate groundwater pollution, IAEA-TECDOC-1046, pp 17–38
51. Vengosh A, Weinthal E, Kloppmann W, the BOREMED team (2004) Natural boron contamination, Geotimes. (www.geotimes.org)
52. White DE, Warning GA (1963) Volcanic emanations. Data of geochemistry, chapter k. U.S. Geological Survey, Paper no. 440-k
53. WHO (1998) Environmental health criteria 204. Boron, Geneva
54. WHO (2006) Guidelines for drinking-water quality, 4th edn, Annex 4. World Health Organization, Geneva, ISBN 92 4 154696 4
55. Wyness AJ, Parkman RH, Neal C (2003) A summary of boron surface water quality data throughout the European Union. Sci Total Environ 314–316:255–269

Chapter 16
Environment and Water Resources in the Jordan Valley and Its Impact on the Dead Sea Situation

Wasim Ali

Abstract The hydrologically and the environmentally degraded situation of the Dead Sea is reflected by the hydrogeological situation in the Upper and Lower Jordan Valley and the surrounding basin of the Dead Sea. The Dead Sea, with a drainage area of around 42,200 km², is mainly fed by the Jordan River from the north and from side wadis in the east and west. With a surface runoff from 1,000 mio m³/year (before 1950) to less than 200 mio m³/year nowadays, it is clear why the water level in the Dead Sea is lowering and its surface is declining. Due to the continuous decrease of water runoff into the Jordan Valley and the side wadis, the surface water level of the Dead Sea is declined by around 1 m a year, and this caused more than 25 m decline of the Dead Sea level in the last 30 years. Geoelectric sounding measurements by Jordanian research teams showed that areas underlying the coastal aquifers formerly occupied by the Dead Sea water are gradually becoming flushed and occupied by fresh water. The latter is becoming salinized due to the residuals of Dead Sea salted water in the aquifer matrix, showing a lower salinity than that of the Dead Sea water. The salt dissolution from the Lisan Marl formation through the surface and groundwater movement is causing collapses along the shorelines in form of sinkholes. These sinkholes are tens of meters in diameter and depth and cause damages to roads, parks and agriculture areas. The sinkholes are threatening the increased urban and touristic activities in the Dead Sea area. The activities of the mineral extraction industry and the increasing of needed fresh water supply for the tourism and agriculture and urban development exploit the resources of the Dead Sea without consideration of the natural needs of the Dead Sea. Without any dramatic change of the present unsustainable development policies of the Dead Sea area, a tragedy ending by drying of the Dead Sea will be the consequence in less than a century. The Planned Red – Dead Sea project can be a solution to ease the deteriorating situation of the Dead Sea.

W. Ali (✉)
Department of Hydrogeology, Karlsruhe Institute of Technology,
Karlsruhe, Germany
e-mail: wasim.ali@kit.edu

Keywords Water resources • Jordan Valley – Dead Sea • Red Sea • Dead Sea canal project

16.1 Introduction

The hydrology and the environmental situation of the Dead Sea area are reflected by the hydrogeological situation in the Jordan Valley and the surrounding basin of the Dead Sea.

The Dead Sea, with a drainage area of about 42,200 km^2, is mainly fed by the Jordan River from the north and from side wadis in the east and west. With a surface runoff larger than 1,000 mio m^3/year before the year 1950 and smaller than 200 mio m^3/year nowadays, it is clear why the water level in the Dead Sea is lowering and its surface is declining. Only during rainfall and flood seasons the Dead Sea receives a considerable quantity of water otherwise the underground water flows from the fresh ground water bodies of the surrounding aquifers towards the Dead Sea.

16.2 Climate, Geology and Hydrogeology of the Dead Sea Area

The prevailing climate in the surrounding area of the Dead Sea is the typical semi-arid mediterranean climate with rainfall occurring during the winter months from October until April. Whereas along the ridge of the Mountains west and east of the Dead Sea (around Jerusalem and Amman) the annual rainfall is up to 700 mm, the precipitation decreases to 150 mm in Jericho and Sweimah area in the Jordan Valley. The average annual temperature increases from 18°C in the Mountains by Amman and Jerusalem to around 26°C in Jericho and Swiemah.

The geological formations of the surrounding of the Dead Sea in the eastern side range from the Triassic Zerqa Ma'in Group to the Quaternary Lisan Marl Formation and young basaltic flows. On the western side the formations are divided into three sedimentary groups: the Cenomanian-Touronian group consisting of limestones, dolomites and some marls, the Senonian-Paleocene group with prevailing chalks and the Dead Sea Group (Pliocene-Pleistocene), representing the sedimentary fill of the Jordan Valley composed of fan-deposits, clays, marls, chalks and some gypsum and halite. The Dead Sea area is structurally intensively affected by the development of the Jordan Rift Graben System.

Hydrogeologically the eastern side comprises three important regional aquifer complexes of Jordan. The Deep Sandstone Aquifer Complex, the Upper Cretaceous Aquifer Complex and the alluvial deposits and the basalts build the Shallow Aquifer Complex. The main recharge of the aquifers takes place along the highlands and the plateau. Due to the climate there are no year-round flowing streams in the area, even some of the springs are just perennial in dry years. Only in 20% of the years the surface runoff reaches along the wadis to the Jordan Valley itself and is dependent on very intensive rainfalls in the surrounding mountains. Two main aquifers exist in the western side, the deep Cenomanian–Turonian aquifer of mainly limestone and the shallow aquifer consisting of young sediments like sand, conglomerates of wadi sediments and lisan formation of gypsum, marl and sand.

16.3 The Effects of Lowering of Dead Sea Level on Water Resources

The population of the surrounding area of the Dead Sea due to the lack of surface water depends mainly on some springs and wells and only in rainy periods on floodwater to cover their urban and agriculture demand. The water quality of these wells are mostly brakish to saline and only few have fresh water quality. Figure 16.1 shows the discharge quantity of the major springs and shallow wells in the lower Jordan Valley north of the Dead Sea [7].

Most of the springs have a discharge of less than 5 MCM (Million Cubic Meters) per year and only some springs near Jericho in the west, like Ein Sultan, have more than 5 MCM per year.

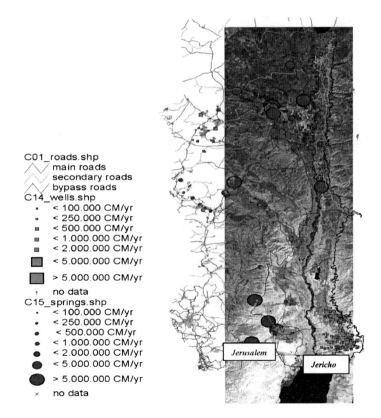

Fig. 16.1 Wells and springs discharge of the lower Jordan Valley north of the Dead Sea, deep wells are not included after *Issac and Sieber* [7]

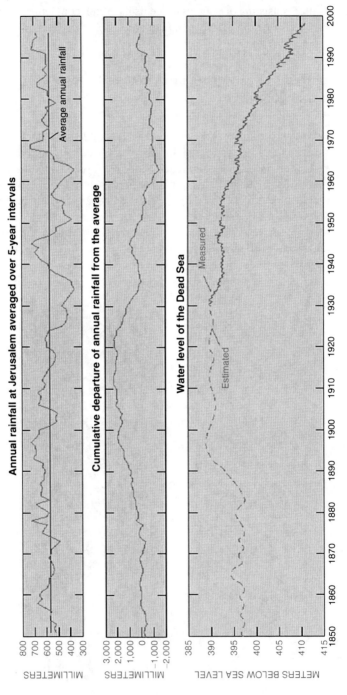

Fig. 16.2 The influence of rainfall and water-resources development on Dead Sea water levels [2]

Figure 16.2 shows the relation between the annual rainfall and the Dead Sea level [2]. Until around 1970, Dead Sea water levels and rainfall showed a correlation. After 1970 although rainfall generally increased during this period, water levels declined steeply, corresponding to decreased inflows from the Jordan River. Although the effects of rainy years in the -1980s, and especially 1992, are still evident, their influence on Dead Sea water levels is moderate. Development of water resources exploitation have a direct impact on the lowering of the Dead Sea water level.

16.4 Negative Effects of the Dead Sea Lowering

16.4.1 Fresh Water Losses

The most negative effect of the lowering of the Dead Sea level is the fact that after the calculation of Salameh et al. [8], 370 Mio m^3/y of freshwater of surrounding aquifers are lost to the Dead Sea through the interface readjustment mechanisms, as a result of their over exploitation of waters which formerly fed the Dead Sea. Geoelectric sounding measurements after Salameh and El-Nasser [8] and Ali et al. [1] and other hydrogeological tests showed that the areas underlying the coastal aquifers, which formerly were occupied by Dead Sea water, are gradually becoming flushed and occupied by freshwater. That fresh water becomes salinized due to the residuals of Dead Sea water in the aquifer matrix. The present salinity of that matrix is lower than that of the Dead Sea water. At the same time salt dissolution from the Lisan Marl formation is causing collapses along the shorelines in the form of sinkholes, which are tens of meters in diameter and depth.

16.5 Sinkholes

The sinkholes appear in the Dead Sea shore since more than 10 years. They differ in form and can reach more than 20 m in length and depth. The mechanism of the sinkholes phenomena hast been explained after many studies, by the fact that the salt dissolution from the Lisan Marl young formation surrounding the Dead Sea is causing collapses along the shorelines of the Dead Sea in the form of sinkholes. Figure 16.3 shows series of sinkholes near the village Alhadith and the Lisan island on the east side of the Dead Sea in Jordan. These sinkholes cause a lot of troubles for the farmers and the touristic activities around the Dead Sea shore.

16.6 Construction of Small Dams on the Side Wadis

To cover the rapidly increasing demand of water for the urban and agriculture demand in the Jordan Valley, small dams on the side wadis of the Dead Sea were constructed, to catch the wadi water before it flows to the Dead Sea. Figure 16.4 shows a small dam constructed during year 2005 on the mouth of Wadi Moujib, on

Fig. 16.3 Several sinkholes on the Jordanian side of the Dead Sea shoreline near Alhaditha village and Lisan Island (Photo [1])

Fig. 16.4 A small dam constructed 2005 on the mouth of Wadi Moujib (Photo [1])

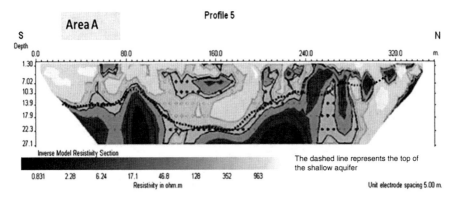

Fig. 16.5 Geoelectrical resistivity profile (nr. 5) measured in the Sweimah area on the Dead Sea shore line, Jordan. The dasched black line represent the depth of the shallow aquifer in the area [1]

the eastern side of the Dead Sea shore at the south of Sweimah, Jordan. This water is used partly for irrigation and partly transferred to Amman City.

16.7 Some Results of the Investigations on the Eastern Side, Sweimah Area

Following are some results of the joint research activities of Al Balqa Applied University (Salt, Jordan) and KIT (Karlsruhe Institute of Technology, former Karlsruhe University, Germany) in the years 2005–2006 in the Sweimah area on the north eastern Dead Sea shore line, Jordan. The research activitis include remote sensing interpretations, landuse mapping and geophysical measurements. Figure 16.5 shows a geoelectrical resistivity profile as an example for this activity. The dashed black line represents the depth of the shallow aquifer in the area [1].

16.8 Some Results of the Investigations on the Western Side, Ein Feschcha Area

Another research activity conducted as a joint research activity of Al Quds University, West Bank, Palestine and KIT (Karlsruhe Institute of Technology, Germany) in the years 2005–2008 in the Ein Feshkha springs area, northeastern shore of the Dead Sea, West Bank, Palestine. Ein Feshkha springs are considered the major outlit of the deep and shallow aquifer of the mountain area between Beitlehm and Ramallah. Figure 16.6 shows a geohydrological profile from west (Beitlehm) to east (Ein Fechkha-Dead Sea) after [3–5]. This model shows that fresh water can be bumped before its mixing with saline water on the north-western shore of the Dead Sea.

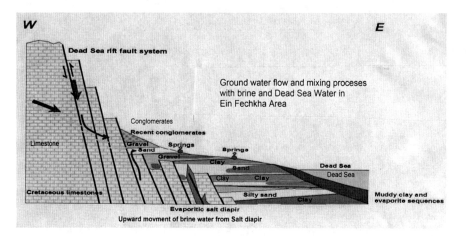

Fig. 16.6 Geohydrological flow model of Ein Fechkha area as a result of geophysical, hydrochemical investigations. In the model the groundwater flow and mixing processes with brine and Dead Sea water are presented after [4, 5]

Previous studies estimated the discharge of Ein Feshkha springs up to 60 MCM/year. As a result of the drop in the level of the Dead Sea, this leaded to the migration of the fresh/saline water interface, and the fresh water is becoming salinized due to the residuals of Dead Sea water in the aquifer matrix, the mixing with old water bodies and the dissolution from Lisan formation. Nevertheless it is believed that the water receives its salinity only in the direct vicinity of the springs, being of better quality in the Cretaceous mountain reservoir. This finds its expression in EC values, from 0.7 to >15 mS/cm and Cl-concentrations of 137–>11,048 mg/l. Examination of the composition of the saline water in Ein Feshkha, may lead to the source and the distribution of the saline water and its mixing points with fresh water. Hence, this may allow capturing the fresh water before its mixing with the brines [4–6].

16.9 Projects to Rehabilitate the Dead Sea, the "Red Dead Sea Canal"

The Dead Sea Canal is a proposed project of building a canal from either the Mediterranean Sea (MDSC) or the Red Sea (RSDSC) to the Dead Sea, taking advantage of the 400-m difference in water levels between the seas. The water flowing through the canal will to stop the drop in the level of the Dead Sea observed in recent 40 years. As an advantage the canal can also be used to generate hydroelectric power because of surface difference and maybe by salinity gradient power, and desalinate water by reverse osmosis.

On 2008, the World Bank was ready to support a feasibility study for the Dead Sea – Red Sea Canal project and chose Coyne et Bellier to carry out the feasibility study for this canal project. The feasibility study needs 24 months, and costs nearly seven million USD.

16 Water Resources in the Jordan Valley 237

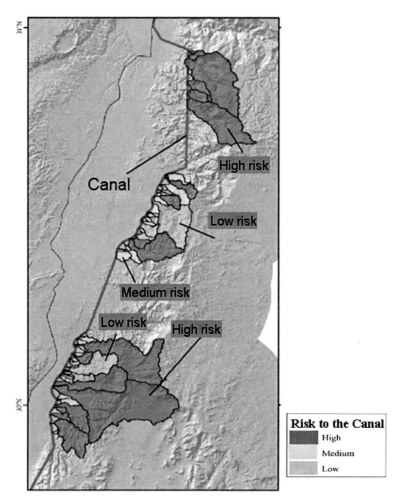

Fig. 16.7 A satellite photo of Wadi Araba and the surrounding region with the Red-Dead Canal, on the left and the risk zones through faults and earth quakes on the right [9]

The project will consist of transporting, via a 200 km long canal, around two billion m³ of water per year with the aim of restoring the water level of the Dead Sea, a natural, historical and economic site of foremost importance. It also provides for drinking water production, in particular for the urban agglomeration of Amman, and electricity generation. It will provide the beneficiary countries of Jordan, Israel and the Palestinian Authority with an opportunity to cooperate on a major project, thereby promoting the efforts for peace in the region. This study will be overseen by a Technical Committee consisting of representatives from the World Bank and the countries concerned: Jordan, Israel and the Palestinian Authority. The financing for it will be provided by France, Japan, the United States of America, the Netherlands and Greece. Figure 16.7 left, shows the path of the Dead Sea – Red Sea project with its surrounding, and the map on the right side shows the risk zones of the canal path, mainly through faults and earthquake activities.

Around 20 international groups had expressed their interest in this study and 6 of them were selected to present a tender. After a strongly contested process, the proposal from the group led by Coyne et Bellier was judged by the World Bank to be the most appropriate. Coyne et Bellier will work in association with Tractebel Development Engineering, a Belgian subsidiary of Tractebel Engineering (SUEZ), and KEMA Nederland BV in the Netherlands (Tractebel Engineering, GDF SUEZ, Brussels, Belgium).

Acknowledgments The author is thankful to the Ministry of Research & Education (BMBF) in Germany for the financial support of the research activities documented in this paper. He is thankful to for his colleagues in Morocco and Italy for Organizing this NATO supported Advanced Research Workshop on water security, management and control.

References

1. Ali W, Al-Zoubi A, Al-Ruzouq R, Abuladas A, Hötzl H (2006) Hydrogeological and geophysical investigations in the Sweimah Area, Dead Sea, Jordan. In: The 3rd symposium of the German-Arab society for environmental studies, Environmental Protection in the Middle East and North Africa, Frankfurt am Main, Germany
2. EXACT (1999) Overview of middle east water resources – water resources of Palestinian, Jordanian and Israeli Interest. Compiled by U.S. Geological Survey for Executive Action Team, Washington, ISBN 0-607-91785-7, 44
3. Guttman Y, Flexer A, Hötzl H, Bensabat J, Ali W, Yellin-Dror A (2004) A 3-D hydrogeological model in the arid zone of Marsaba-Feshchah region, Israel. In: The 5th international symposium on Eastern Mediterranean geology. Thessaloniki Proceedings 3, pp 1510–1513
4. Hannich D, Hötzl H, Ali W (2008) New geophysical insights concerning the interfaces between fresh, brackish and saline water in the Ein-Fashkha area at the north-western Dead Sea shore. In: EGU 2008, Vienna, 13–18 Apr 2008
5. Hassn J (2008) Hydrogeology and hydrochemistry of the brackish Ein Feshckha spring group (Dead Sea area). Ph.D. thesis, University of Karlsruhe, Karlsruhe
6. Hötzl H, Ali W, Rother M (2001) Ein Feshkha springs as a potential reservoir for fresh water extraction, Dead Sea Area. Le premier colloque national d'hydrogeologie et environment, Fes, Morocco
7. Issac J, Sieber (2000) Developing sustainable water management in The Jordan Valley. Report EU project 1907–2000, Graz and Bethlehem
8. Salameh E, El-Nasser H (2000) Changes in the Dead Sea level and their impacts on the surounding groundwater bodies. Acta Hydrochim Hydrobiol (Karlsruhe) 28:24–33
9. Shirav-Schwartz M, Calvo R, Bein A, Burg A, Ran Nof (Novitsky), Baer G (2006) Red Sea – Dead Sea conduit geo-environmental study along the Arava Valley, (MERC) program, Bureau for Global Programs. US Agency for International Development, Jerusalem, Report GSI/29/2006

Chapter 17
Hydrochemistry and Quality of Groundwater Resources in Egypt: Case Study of the Egyptian Southern Oases

Anwar A. Aly, Abdelsalam A. Abbas, and Lahcen Benaabidate

Abstract Due to the arid climate of the southern part of Egypt, oases groundwater is a most precious natural resource, providing reliable water supplies for the population of these oases. This study aims to assess the quality of groundwater resources in the southern Egypt oases; Dahkla, Kahrga, and Uweinat, taking part from the Nubian Sandstone geological complex. Groundwater, which is a major source of water supply in the southern Egypt oases, is facing severe quantity and quality problems. The large concentrated agriculture constitutes the main source of its contamination. Water scarcity combined with the typically arid climate and the excessive use of soils for agriculture causes severe declining of the groundwater piezometric head levels, in addition to water salinity problems. However, groundwater remains acceptable for irrigation in the studied area.

Keywords Dahkla • Kahrga • Uweinat • Groundwater • Contamination • Salinisation

17.1 Introduction

Egypt covers an area of slightly over one million square kilometers. Ninety-nine percent of the population (65 million) lives in a small band alongside the Nile River in the Nile valley, Nile delta and coastal areas on about 4% of the Egyptian

A.A. Aly (✉) • A.A. Abbas
Soil and Water Sciences Department, Faculty of Agriculture,
Alexandria University, Alexandria, Egypt
e-mail: anwarsiwa@yahoo.com

L. Benaabidate
Faculty of Sciences and Technology, University of Sidi Mohamed Ben Abdellah, Fez, Morocco

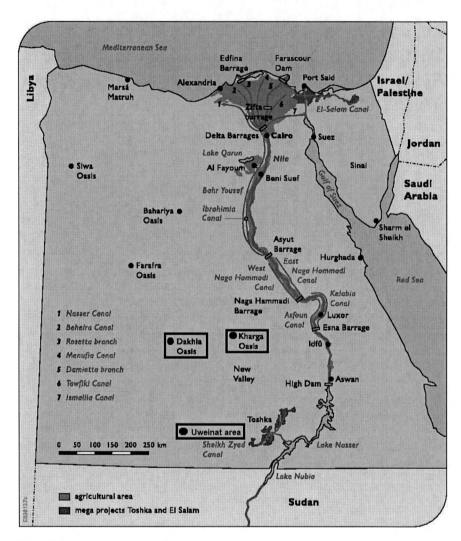

Fig. 17.1 Situation of the study area

land. This is because Egypt depends on Nile River as the main source of water resources for irrigation, drinking, and other different activities. In the recent years a great deal of attention has been directed toward expanding agricultural project depending on groundwater as the population increases and the Nile water supply is limited.

The southern oases in the western desert of Egypt are considered the most important Agriculture areas depend on groundwater. These oases are south of Lat. 26° in the western desert of Egypt including: Kharga, Dakhla, and oases east of Uweinat Mountain, as shown in Fig. 17.1 [17].

17.2 Studied Oases Landform

Kharga and Dakhla Oases occupy part of a great natural excavation in the southern section of the Western Desert of Egypt. They belong to the Nubian geological Complex. This excavation includes also the slightly elevated plain (140 m above sea level) between them. The depression is open towards the south and southeast, altitude rises gradually to the southwest, reaching 400 m in the direction of Uweinat Mountain.

The long axis of Kharga Oasis (about 200 km west of the Nile) is in a north-south direction. It is bounded on the north and east by steep loftily escarpment. To the south and southwest there is no definite boundary, consequently it is difficult to estimate the area of the depression precisely. It is long and narrow, 185 km from north to south and between 15 and 30 km from east to between 300 and 400 m below the surrounding plateau. The lowest point of the oases floor is almost at sea level whereas the highest is at 115 m above the sea level.

Dakhla Oasis is about 120 km west of Kharga oasis. Its long axis is in W-NW-E-SE direction. Its length is about 55 km and its width varies between 10 and 20 km. Altitude is in range 100–400 m above sea level [14, 18].

Uweinat is isolated in an area of extreme aridity located to the south of Dahkla of about 300 km, convergence of the borders of Libya, Egypt and Sudan. At the centre of the Libyan Desert (Eastern Sahara), the western granitic part is located entirely in Libya, while the eastern sandstone part, a series of high plateaus, lies mainly in Sudan, the northern flanks jutting into Egypt [18].

17.3 Climate

The Kahrga, Dakhla, and Uweinat are located in a dry rainless part of the Great Sahara. Rainfall is almost null whereas the mean annual relative humidity is lower in summer (26–32%) than in winter (53–60%). The mean annual evaporation is 18.4 mm/day and range from 25.1 to 9.5 mm/day in summer and winter respectively. Temperature is moderate in winter, but rises very high in summer to record extreme maxima of about 50°C in both oases.

The Mean annual rainfall for Kharga is less than 1 mm. However, on rare occasions, torrential storms occur. The hottest month is July, with daily maximum and minimum of about 40°C and 20°C respectively. Maximum temperature occasionally approaches 50°C, with January being the coolest month. Mean monthly relative humidity ranges from 30°C during the summer to 56% during the winter. Evaporation on bare soil ranges from about 6 mm/day in December to 23 mm/day in June.

17.4 Water Resources

The water resources in southern oases are underground. In Kharga Oasis, there are two distinct strata separated by a 75 m band of impermeable grey shale. The upper bed is exposed at the surface and forms the true artesian water sandstone from which the flowing wells of the Kharga Oasis derive their supply. In Dakhla Oasis, the water supply comes from a bed of white sandstone which corresponds to the surface sandstone water of Kharga.

In 1956, major agricultural development programs were started at Kharga and Dakhla Oases, where a large number of deep wells were drilled. At the beginning of the projects, water originally flowed under artesian pressure, at an annual discharge rate of about 200 million m^3, with the expansion of facilities; piezometric head levels started declining as much as 1 m/year. This caused the production from individual wells to decrease or, in the case of some shallow wells, it required the drilling of new deep wells in order to maintain overall production. The declining head has also caused at least half of the deep wells in Kharga Oasis to stop flowing freely; consequently these artesian wells have been transformed to pumping wells [1].

The water demand in the area has been met by pumping both shallow and deep wells. The total annual extraction from deep wells has fluctuated over the years; however, the annual withdrawal from deep wells has exceeded extraction from shallow wells. About 17 billion m^3 of water was withdrawn from the combination of shallow and deep wells in the period 1960–1980 [13].

17.4.1 The Nubian Aquifer System

Water in the southern oases is supplied by deep wells that reach depths ranging between 600 and 1,000 m. Some of these wells are artesian, while others are operated by motorized pump. As the rainfall in Egypt's western desert is virtually zero, the oases depend exclusively on the water supply from these deep wells, which tap the Nubian Sandstone Aquifer System.

The Nubian Aquifer spans across Sudan, Chad, Libya, and Egypt and reaches Sinai in the North. The aquifer system has a subsurface extent of over 630,000 km². The thickness of the sandstones that contain the artesian flow system ranges from 400 m depth near Kharga Oasis to a depth of over 2 km at Siwa Oasis. The aquifer system is stratified and consists of several different horizons at Dakhla and Kharga Oases, there are two layers: an "upper aquifer, reported at the depth of about 400 m, and a lower aquifer, reported at a depth of about 650 m from the surface" [12].

The Nubian Aquifer System is a source of fossil water, as it was accumulated over thousands of years and experiences "negligible recharge in the present arid conditions". This means that the water is a non-renewable resource. It is not clear how long such groundwater will be available for extraction from the aquifer, especially as extraction rates are intensifying in most North-African countries that tap the aquifer system. In Egypt about 0.6 billion m^3 of water are pumped from deep aquifers [11].

Shallow well production declined slowly but continuously subsequent to 1956 in Kharga Oases. According to the FAO report of [8], the decline in discharge from shallow wells may be regarded as a measure of the change in the water balance in the upper aquifers due to the development of deep wells that have intercepted upward flowing water. Heads and discharges continue to decline in both the shallow and deeper parts of the system, and hydrodynamic conditions are still transient with water being withdrawn from storage.

17.4.2 Dynamics of the Nubian System

Kharga and the other oases of the Western Desert have been centers of groundwater discharge from springs and wells for many thousands of years. These long periods of discharge have influenced the regional flow system. Components of vertical and horizontal flow must be understood in the overall development of groundwater from the area. Many previous workers have attempted vertical divisions of the aquifer system on the basis of the impermeable shale layers; however, studies of the subsurface geology of the area indicate that the shale layers are lenticular and they act as local confining layers. Further, because of faulting, jointing, and folding of these beds, they have developed a very extensive preferential vertical and horizontal leakage between the water bearing sandstones of the Nubian system. Philip et al. [13] also indicated that the groundwater flow in Kahrga was from the Southwest, and that groundwater discharge was creating distinct cones of depression in the piezometric surface. They also concluded that there is some flow into the northern Kharga depression from the southeast under a low gradient. Kharga Oases appears to capture water entering the area from the south and southeast, while Dakhla intercepts flow entering from the west and northwest. Hydrographs of changes in water levels from wells tapping deep aquifers in Kharga Oases for the period 1965–1980 show that the rate of decline in head for the deep aquifer in the northern part of the southern oases is 17 m, while the head changes in the central part of the oases, where heavy extractions have been made from the deep aquifers, show declines as much as 32 m [13].

17.4.3 Challenges to Efficient and Sustainable Water Use in Southern Oases

A research was undertaken by the American University in Cairo in Abu Minqar [2], Farafra Oasis, has shown that the primary concern of the population is not yet the long-term sustainability of water availability. Rather, people worry more about water inefficiencies of the current agricultural system. First, available irrigation water is significantly limited in the summer months. The vast majority of landholders leave a portion of their land fallow due to the water shortages from heightened evaporation and plants water needs. For example, farmers who own six feddan (2.52 ha) generally only plant crops on two to four feddan (0.84–1.68 ha) in summer, because the received water is not plentiful enough for all their agricultural needs, making the inefficient irrigation practices the main issue rather than water scarcity. In the southern oases generally about 25% – in some cases even up to 50% – of water are wasted because of poor irrigation infrastructure: the water running through unlined and uncovered canals evaporates quickly and seeps into the ground.

Additionally, the lack of drip irrigation hoses and sprinklers forces farmers to use flood irrigation, the least efficient way to water a field.

17.5 Social and Economic Benefits

As one purpose of the government is to improve the social and economic conditions, a policy may be considered for the development of groundwater resources in Kharga. Such development can be logically and effectively planned so that mining of groundwater is highly beneficial and constitutes a rational step in orderly optimum development of water resources.

Future development in the Kharga, Dakhla, or other oases of the Western Desert, however, must consider the short-term, as well as the long-term impact of that development on the rest of the desert areas. Development in the Western Desert of Egypt is a complicated socioeconomic problem, and will require the mining of groundwater. Many different types of development schemes must be considered: irrigation, duck farms, tourism, and fish culture, requiring different amounts of water, involving different economic parameters, and resulting in different types of benefits to the re-settled people. There are large remaining reserves of groundwater in the Western Desert of Egypt that can be developed effectively for the future but they must be developed cautiously. The Nubian is a geologic complex unit comprising a very large storage system containing good quality water that moves through the sandstone beds very slowly, generally from South to North toward points of discharge. This slow movement in a way comprises a safety factor; even though humans have been extremely wasteful of this water in the past, and artesian heads have been reduced drastically in some local areas, much water remains for future development.

Hellstrom [9] estimated the rate of water movement under Kharga and Dakhla at 15 m/year. He concluded that 50,000 years would be required to traverse the distance between the nearest rainfall areas (recharge) and the oases, a period later generally confirmed by carbon dating, which indicated 25,000 years for Dakhla Oasis and 50,000 years for Siwa Oasis. The rate of water movement has been calculated from 100 to 200 m/year. Storage beneath Kharga and Dakhla Oases is sufficient to provide about 1,000 million m^3 of water annually for 200 years. Even if recharge is insignificant, there obviously is a great amount of water available in such large volume of water-bearing earth material that can be withdrawn under practical pump lift conditions.

17.6 Farmer's Knowledge, Attitudes and Practices (KAP) Survey

17.6.1 Kahrga and Dahkla Oases

Kahrga and Dahkla oases farmers depend mainly on planting date palm, Clover, Wheat, Beans with some intercrops. The farmer uses the surface irrigation and depend on pumping machine (controlled by government) to irrigate their farm. Each well irrigates around five acres. The yields of the oases land (sandy soil) were low generally due to soil salinity, limitation of irrigation water, and soil fertility problems which are considered the most important agricultural problem in the two oases ecosystem.

17.6.2 Uweinat

The Uweinat area was considered a new reclaimed area located in the southern part of Egypt near Sudan border. The agriculture areas in Uweinat were not as the previous two oases (Kahrga and Dahkla), where farms are represented as small areas. In Uweinat the large agricultural investors cultivate large areas depending on a central pivot for irrigation; each central pivot irrigates around 240 acres. The main agriculture activities in Uweinat depend on planting Wheat, Barly, Onions, Potato, and Soya bean.

17.7 Water Quality Evaluation of Southern Oases

17.7.1 Materials and Methods

17.7.1.1 Sampling Sites and Dates

In this study, water samples were collected from 14 different geo-referenced locations that cover three areas of Egyptian western desert, in southern part, in attempt to capture the spatial variations in the water resources quality of the studied areas (Table 17.1).

Table 17.1 Contents of major elements (mg/l)

N	Name	Cu	Fe	Mn	Zn
1	Kahrga Elnagda 23	0.007	3.660	0.404	0.124
2	Kahrga Company 55	0.016	0.996	0.095	0.058
3	Kahrga Company 5	0.012	2.165	0.155	0.194
4	Kahrga Hebes	0.007	1.042	0.195	0.045
5	Kahrga Monera ElBalad	0.016	1.550	0.130	0.124
6	Karga 1	0.005	3.194	0.120	0.094
7	Kahrga Nadora	0.016	2.580	0.238	0.123
8	Dahkla Ziat	0.012	9.100	0.121	0.044
9	Uweinat ElGeash 2	0.248	0.377	0.091	0.276
10	Uweinat	0.025	0.604	0.025	0.520
11	Uweinat ElGeash 1	0.012	0.397	0.296	0.121
12	Uweinat ElGeash	0.012	1.027	0.036	0.448
13	Uweinat ElGeash 7	0.004	0.186	0.027	0.093
14	Uweinat ElGeash 8	0.006	0.118	0.023	0.149

Once collected, water samples were transported immediately to the laboratory in iceboxes and chemical analyses were carried out to assess the water quality according to Klute [10].

17.7.1.2 Water Quality Measurements

The water pH, EC, soluble ions, Boron, and heavy metals were determined as follows:

- The water reaction (pH) was determined using a pH meter (pH meter – CG 817).
- The total soluble salts were measured by using electrical conductivity meter (EC) in dS/m at 25°C (Test kit Model 1500_20 Cole and Parmer).
- The Soluble Potassium and Sodium were determined by using flame photometer apparatus.
- The Soluble Calcium and Magnesium were determined by Versenate titration method (EDTA), Soluble carbonates and Bicarbonates were determined by titration with sulfuric acid using phenolphthalein as indicator for the former and methyl orange for the latter, and Soluble Chlorides were determined by titration with silver nitrate solution and potassium chromate as indicator as described by Klute [10].
- The boron was determined calorimetrically by using carmine methods as described by U.S. Salinity Laboratory staff [17].
- The heavy metals: The water samples were acidified (to pH<2) with concentrated nitric acid and filtered if necessary, and a Flame Atomic Absorption Spectrometer (Perkin Elmer A.A. model, AAS Varian 6083) was used for measuring iron, zinc, manganese and copper.

17.7.2 Results and Discussions

17.7.2.1 Water Quality Evaluation

The assessment of the suitability of the southern oases groundwater resources for irrigation has been carried out by the FAO method [3]. The obtained data (Tables 17.1 and 17.2) showed that there are generally no considerable spatial differences between water qualities obtained from different locations and depths within the Oases. In general, water obtained from three locations was the same in its quality, and it was also observed that the water salinity as one primary indicator for water quality for irrigation was in the acceptable level for irrigation in all places. Generally, the studied wells for the three studied locations were within the acceptable levels for salinity, chloride and alkalinity problems that make this water suitable for irrigation. Specific toxic elements such as boron were in the acceptable safe limit in most wells.

The heavy metals and trace elements levels generally were within the safe and acceptable levels for surface irrigation except Bir Dahkla Ziat which contain Iron concentration more than 9 mg/l, this mean that this well's water may contribute to the loss of availability of essential phosphorus and molybdenum. Most of groundwater

Table 17.2 Contents of trace elements (mg/l)

No	Name	pH	EC dS/m	Ca++	Mg++	Na+	K+	HCO$_3^-$	Cl-	SO$_4^-$
1	Kahrga Elnagda 23	7.73	0.40	24.0	14.6	26.45	30.498	115.9	53.25	43.2
2	Kahrga Company 55	7.91	0.47	20.0	9.8	56.81	22.678	176.9	71	0
3	Kahrga Company 5	7.88	0.34	20.0	7.3	32.89	19.55	146.4	39.05	0
4	Kahrga Hebes	8.16	0.35	20.0	22.0	27.37	29.325	164.7	49.7	28.8
5	Kahrga Monera El Balad	8.25	0.59	20.0	9.8	80.27	24.633	219.6	71	0
6	Karga 1	7.42	0.21	16.0	7.3	12.19	12.512	79.3	35.5	0
7	Kahrga Nadora	7.97	0.41	20.0	14.6	34.5	43.01	183	63.9	0
8	Dahkla Ziat	7.24	0.14	12.0	17.1	7.36	5.865	54.9	42.6	19.2
9	Uweinat ElGeash 2	7.62	0.16	16.0	7.3	10.35	0.391	61	28.4	19.2
10	Uweinat	7.71	0.61	48.0	17.1	46.92	0.391	97.6	67.45	115.2
11	Uweinat ElGeash 1	7.75	0.59	52.0	9.8	43.47	0.391	97.6	74.55	76.8
12	Uweinat ElGeash	7.42	0.30	28.0	9.8	20.47	0.391	103.7	46.15	4.8
13	Uweinat ElGeash 7	8.08	0.70	40.0	22.0	72.91	2.737	140.3	120.7	62.4
14	Uweinat ElGeash 8	7.75	0.42	36.0	14.6	26.45	0.391	42.7	60.35	86.4

has a manganese concentration that may be toxic to a number of crops, but this usually happens in acid soils, and not in the oases alkaline soil (Fig. 17.2).

In case of using drip or sprinkler irrigation systems, the water suitability might be different, due to the expected chemical and physical blocking processes that might occur through the irrigation nets by the heavy precipitated materials of iron and manganese. Overhead sprinkling may result in unsightly deposits on plants, equipment and buildings as shown in the following equation:

$$4\ Fe^{2+}\ (soluble) + 3O_2 \rightarrow 2\ Fe_2O_3\ (precipitate)$$

The iron concentrations in Kahrga and Dahkla generally more than Uweinat, this explain why Kahrga and Dahkla uses surface irrigation without problem, on the other hand Uweinat uses sprinkler irrigation (central pivot) successfully.

17.7.2.2 Effect of SAR on Irrigation Water

The quality of the irrigation water salinity is usually expressed by classes for convenience. Most classifications reflect the conductivity and most of the sodium content. The concentration of sodium is important in the classification of irrigation water because it reacts with the soil and reduces its permeability [7]. Salinity in commonly reported as electrical conductivity (EC) in µS·cm^{-1} of the solution. The Sodium Adsorption Ratio (SAR), defined as $Na/(Ca+Ma)^{0.5}$ in solution, where concentrations are expressed in meq·l^{-1}, is a useful parameter that it is closely related to the exchangeable sodium percentage in the soil. Water quality criteria for irrigation must consider the direct impact on crop yields and the indirect impact related to the effects on the soil chemical and physical properties [16].

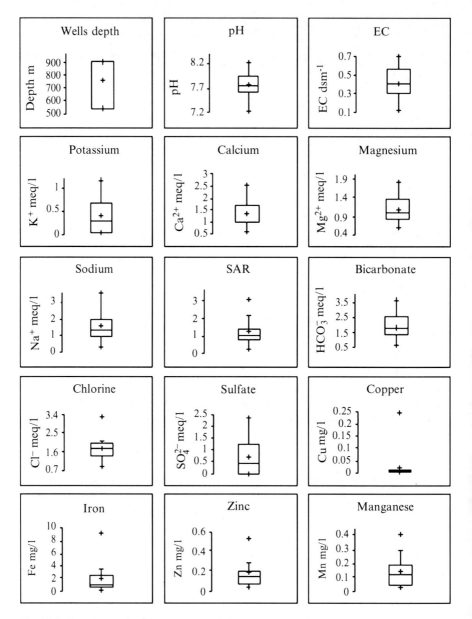

Fig. 17.2 Box plot graph of the hydro-chemical compositions of studied wells

The plot of SAR average values based on the values of electrical conductivity (Fig. 17.3) has identified the following classes of salinity tolerance:

– Class 1: represented by samples 6, 8 and 9, revealed that those waters have a low risk of salinisation and alkalinization

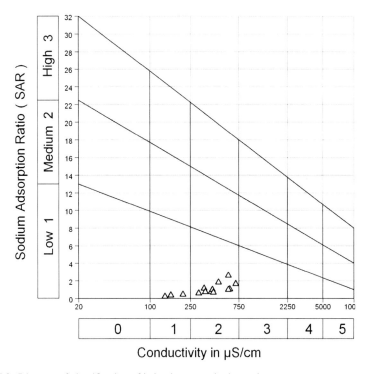

Fig. 17.3 Diagram of classification of irrigation water in the study area

- Class 2: containing remain samples showed that waters have a low risk of alkalinisation but a medium risk of salinisation.

In general, and according to the classification of the Riverside scale, the studied waters are classified as "good" for irrigation.

17.7.2.3 Hydrochemical Classification of Studied Waters

The determination of the chemical types of the studied water has been carried out by the Piper diagram (Fig. 17.4). This figure shows that these waters can be classified into three groups: (i) calcium (magnesium) – chloride which is represented by samples n° 1, 9, 10, 11 and 13; (ii) calcium (magnesium) – bicarbonate with samples n° 4, 6 and 12; (iii) sodium – bicarbonate, represented by samples n° 2, 3, 5 and 7.

This graphical representation of Piper, although it allows a global visualization of samples, does not adequately discuss the geographical distribution of studied waters. Obtained chemical types are mixed with a predominance of waters rich in calcium, magnesium and bicarbonates. This suggests that these waters are issued from deep reservoirs composed of carbonate rocks, with minerals rich in calcium and magnesium such as calcite, aragonite and magnesite.

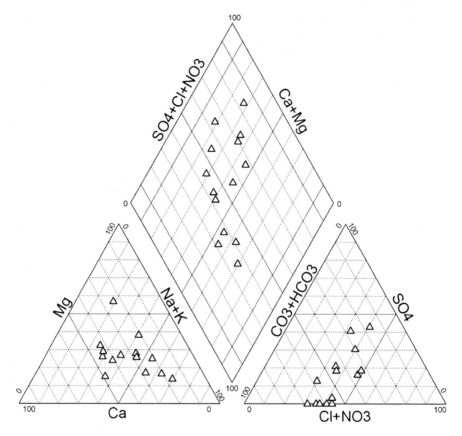

Fig. 17.4 Studied water chemical types

17.7.2.4 Test of Equilibrium with Carbonates

The equilibrium diagram of carbonate minerals was carried out by Carpenter [5, 6] and incorporated by Stumm and Morgan [15]. The projection of the two parameters of pCO_2 and the logarithmic ratio of calcium and magnesium activities relating to the studied waters, computed by the Solmineq Model, on the diagram (Fig. 17.5) shows that the these waters are in equilibrium with dolomite in the field of stable minerals and with aragonite in the unstable minerals field. This balance means that the dolomite is the main carbonate mineral in deep reservoirs. However, this equilibrium diagram considers only pure mineral species. Accordingly, a highly magnesian calcite can be likened to a dolomite [4].

17 Hydrochemistry of Egypt Southern Oases

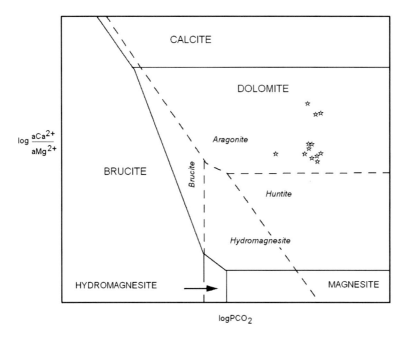

Fig. 17.5 Water equilibrium with carbonates

The saturation index (SI) is the form most commonly used for groundwater. Water is in equilibrium with a mineral when the SI of this mineral is equal to zero. It is under saturated if this index is below zero and it is over saturated when the SI is above zero. However, the inaccuracy on the pH measurements due to measuring devices, the variation of this parameter when the water flow toward surface and the error that could occur during chemical analysis, these factors result in an inaccuracy in the calculation of the saturation index. Therefore, it is recommended to consider that the saturation is obtained in a wider area such that $-1 < SI < +1$ [6].

The test of the saturation state of studied water with respect to certain carbonate minerals such as calcite, dolomite and aragonite was obtained graphically by calculation of the saturation index of these minerals (Fig. 17.6).

The use of the SI showed that almost all studied water are saturated with respect to aragonite, calcite and dolomite, with the exception of sample 8 for calcite, 5 and 8 for the aragonite and 4, 7, 8, 12 and 13 for dolomite.

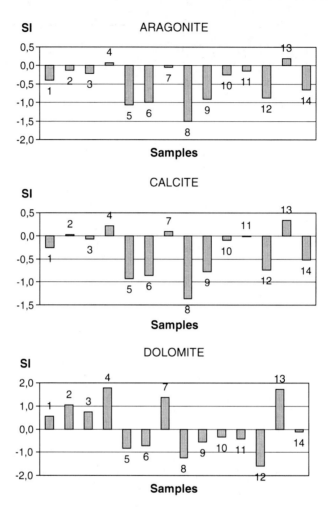

Fig. 17.6 Water saturation with respect to some minerals

17.8 Conclusion

Water resources in Egypt are becoming scarce. Surface-water resources originating from the Nile are now fully exploited, while groundwater sources are being brought into full production. Egypt is facing increasing water needs, demanded by a rapidly growing population, by increased urbanizations, by higher standards of living and by an agricultural policy which emphasizes expanded production in order to feed the growing population. The Egyptian western desert is considered an important area for agriculture expansion depending on groundwater, on the other hand this groundwater needs more studies regarding its quality and sustainability. This paper has focused on the quality and hydrochemistry of groundwater in the Egyptian southern oases, i.e. Dahkla, Kahrga, and Uweinat. The result concluded that the groundwater quality for irrigation was in the acceptable level for irrigation in all places. In case of using drip or sprinkler irrigation systems, suitability might be different, due to the expected chemical and physical blocking processes that might occur through the irrigation nets by the heavy precipitated materials of iron and manganese. The results of hydrochemistry suggests that the southern oases groundwater are issued from deep reservoirs composed of carbonate rocks, with minerals rich in calcium and magnesium such as calcite, aragonite and magnesite.

Acknowledgements The authors are indebted to Agric. Eng. Soilman Mohamed (Uweinat Agriculture Company) and Mr. Mohamed Soilman Mohamed (Undergraduate student in Faculty of Agriculture, Alexandria University, Egypt and a Native citizen of Kahrga Oasis) for their help in KAP surveys and water samples collection.

References

1. Ahmed MU (1977) Digital computer model for designing well. Fields in the Libyan sahara. In: Proceedings of the international conference on computer applications in developing countries, Vol. 1 & 2, sponsored by Asian Institute of Technology, 13 Bangkok
2. American University in Cairo (2008) Community-based water demand management. http://www.aucegypt.edu/ResearchatAUC/rc/ddc/research/WDM
3. Ayers RS, Westcot DW (1985) Water quality for agriculture. Irrigation and Drainage Paper 29 (rev.1). FAO, Rome
4. Benaabidate L (2000) Caractérisation du bassin versant de Sebou: hydrogéologie, qualité des eaux et géochimie des sources thermales. Thèse Es-sc. F.S.T. Fès, 250 p
5. Carpenter AB (1962) Carbonate system. In: HH Schmitt (ed), Equilibrium diagrams for minerals. Geological Club of Harvard. Cambridge, Mass, 199 p
6. Daoud D (1995) Caractérisation géochimique et isotopique des eaux souterraines et estimation du taux d'évaporation dans le bassin de Chott Chergui (zone semi-aride) Algérie. Thèse univ. Paris sud. Centre d'Orsay
7. Djabri L (1987) Contribution à l'étude hydrogéologique de la nappe alluviale de la plaine d'effondrement de Tebessa. Essai de modélisation. Thèse univ. Franche-Comté, UFR. Sc. Tech., Besançon, 171 p
8. FAO (1977) Groundwater pilot scheme, New Valley, Egypt. Agricultural development prospects in the New Valley: Rome, EGY 71/561, Technical Report no. 4
9. Hellstrom B (1940) The subterannean water in the Libyan Desert: Sartryek Ur Geografiska Annater. Stockholm 34:206–239
10. Klute A (1986) Methods of soil analysis, Part 1, 2nd edn, Agronomy Monograph 9. ASA and SSSA, Madison
11. National Water Resources Plan (NWRP) (2005) Water for the future, NWRP 2017 Project, final report. Ministry of Water Resources and Irrigation, pp 2–12
12. Osmond JK, Dabous AA (2003) Timing and intensity of groundwater movement during Egyptian Sahara pluvial periods by U-series analysis of secondary U in ores and carbonates. Quaternary Res 61:85–94. Available online at www.sciencedirect.com
13. Philip EL, Bashir AM, Hussein I (1985) Groundwater development, Kharga Oases, western desert of Egypt: a long-term environmental Concern. J Environ Geol Water Sci 7:129–149
14. Schandelmeier H, Darbyshire F (1984) Metamorphic a magmatic events in the Uweinat – Bir Safsaf Uplift (Western Desert/Egypt). Geol Rundsch 73:819–831

15. Stumm W, Morgan JJ (1981) Aquatic chemistry. An introduction emphasizing chemical equilibria in natural waters, 2nd edn. Wiley, New York, 780 p
16. Suarez DL, Wood JD, Lesch SM (2006) Effect of SAR on water infiltration under a sequential rain-irrigation management system. Agric Water Manage 86:150–164
17. U.S. salinity laboratory staff (1954) Diagnosis and improvement of saline and alkali soil. U.S. Dept. Agric. Handbook No. 60, 1609
18. Zahran MA, Willis AJ (2009) The vegetation of Egypt, 2nd edn. Springer, New York, pp 74–80. http://www.peakware.com/peaks.html?pk=2165

Chapter 18
Means of Mobilization and Protection of Water Resources in Algeria

Mohammed Kadri, Ahmed Benamar, and Brahim Bendahmane

Abstract Increasing demand for water, particularly in arid and semi-arid regions of the world, has shown that the extended groundwater reservoirs are invaluable for water supply, and storage by recharge becomes essential. In many areas of the world, aquifers that supply drinking-water are being used faster than they recharge. The overexploitation of groundwater of Sebaou River (Algeria) for the production of drinking-water and for irrigation, in addition to the lack of aquifer recharge sites, induced lowering of the level of the water table during the recent years. Owing to untreated storm water and wastewater discharged from different sources, Sebaou River undergoes a significant degradation along its course and becomes a true dump for the domestic and industrial wastes. The artificial recharge of the aquifer is an adequate means of storing water and protecting the local environment. It also avoids considerable water loss by evaporation and silting and makes the drinking-water of better quality. Artificial recharge as a means to boost the natural supply of groundwater aquifers is becoming increasingly important in groundwater management. Because the chemical and microbial quality of groundwater is linked to the events occurring above the aquifer, it is imperative to assess the health risks associated with any recharge option. This paper discusses the potential of water resources in Algeria and the integrated approach to groundwater management. A numerical study dedicated to the potential of aquifer recharge from Sebaou River, and based

---

M. Kadri (✉)
University of Boumerdes – LGEA (UMMTO), Boumerdes, Algeria
e-mail: dzkad@yahoo.fr

A. Benamar
University of Le Havre – Laboratoire d'ondes et Milieux Complexes
FRE CNRS 3102, 53 rue Prony, Le Havre, France

B. Bendahmane
University of Bejaia, Bejaia, Algeria

on mathematical modeling of surface and groundwater flows, is presented. The results show the suitability of artificial recharge and the effects on the water table level and the reorientation of the flow direction, avoiding marine intrusion.

Keywords Water resources • Dams • Artificial recharge of aquifers • Modeling

18.1 Introduction

Countries of the south area of the Mediterranean Sea are among the poorest countries in terms of water potential. The Mediterranean region faces high levels of water stress, but only limited water quantity is recycled. In these areas of the world, aquifers that supply drinking-water are being used faster than they recharge. Not only does this represent a water supply problem, it may also have health implications. Moreover, in coastal areas, aquifers containing potable water can become contaminated with saline water if the potable portion is withdrawn faster than it can naturally be replaced. To remedy these problems, some authorities have chosen to recharge aquifers artificially, using either infiltration or injection. Artificial recharge of aquifers by resources drawn from far away is a mobilisation mode adopted especially in arid areas [9].

The main purpose of artificial aquifer recharge technology is to store excess water for later use, while improving water quality (decreasing the salinity level) by recharging the aquifer with better water. The use of artificial recharge to store surplus surface water within the underground can be expected to increase as growing populations demand more water. The Algerian territory covers an area of nearly 2.4 million km^2, but 90% of this area corresponds to arid and semi-arid regions where rainfall is very low. In this part of the territory, the surface water resources are very scarce and are limited only to the portion of the septentrional flank of the Atlas Mountains. In the opposite of this, the groundwater resources are more available, but are very weakly renewable (septentrional Sahara). The potential of renewable water resources is located in northern Algeria, which includes tributary basins of the Mediterranean and closed basins in the highlands.

The water potential of the country is estimated at approximately 16.8 billion m^3, of which only 80% are renewable (70% for surface water and 10% for groundwater). According to the report of the CNES (Conseil National Economique et Social) in 2000, Algeria is facing a situation of water stress. If in 1962 the theoretical water availability per person per year was 1,500 m^3, it was only 720 m^3 in 1990, 680 m^3 in 1995 and 630 m^3 in 1998. Estimated at about 500 m^3 to date, it will be only 430 m^3 in 2020 and would be further reduced with reduced mobilized water resources. This is due to the scarcity of the water resources on one side, the low rate of mobilization on the other side, aggravated by drought and loss by evaporation and seepage.

The means of mobilization, management and conservation of water resources have not been at the level of the development of Algeria. Therefore, the large recurrent deficit affects economic and social development. It is interesting to note the following facts: a very significant reduction of the irrigated area, a decrease of storage capacity and natural regeneration of groundwater due to extraction of the superficial sand layer in the river, and a frequent tension on water between the domestic, agriculture and industry needs.

To solve the problem of balance between mobilization of water resources and the satisfaction of drinking-water demand, industrial and agricultural water requirement, the government launched an integrated project to build a network of interconnected dams enabling easy management and optimum distribution of this resource. To date the total amount of mobilized water in Algeria is about five billion m^3, but the storage capacity is quickly decreased by siltation of dams and the loss by evaporation and seepage. It is therefore important to develop a water management policy that would enhance our water resources. This must be based, on the one hand, on the protection of existing resources and, on the other hand, on increasing storage capacity by using artificial recharge of deep aquifers. In order to effectively manage the water resources, the government required relevant information on the feasibility of the aquifer recharge technology.

Sebaou groundwater in northern Algeria is an important resource, which supplies drinking water for a large region. The overexploitation of this groundwater and the weak rate of natural replenishment of the resource lead to a drastic lowering of the groundwater table. This situation, if not corrected, can lead to long-term depletion of this groundwater resource. A numerical simulation of the replenishment scenarios of the Sebaou groundwater reservoir was performed. The feasibility of artificial recharge of the alluvial aquifer of Sebaou, based on mathematical modeling of flows, is being undertaken. The coupling of mathematical equations of Saint-Venant and groundwater flow allows the estimation of the volume of infiltrated water and the residence time. Artificial recharge leads to increasing available water resources, modifying the chemical and thermal characteristics of water and potentially restoring the disturbed environmental balance by preserving the fauna and flora over thousands of hectares.

18.2 The Potential of Water Resources in Algeria

The Barcelona Convention is part of the legal framework supporting the Mediterranean Action Plan, which was established to protect the Mediterranean region by addressing environmental degradation and linking development with sustainable resource management. In Algeria, the surface flows are primarily concentrated in the septentrional fringe of the country, leading to unequal distribution of water resources as shown in Table 18.1. The northern part is subdivided into four hydrographic regions depending on different hydrographic basins. This zoning is made in order to create an interconnection between the hydrographic basins. The annual average capacity of mobilization of water resources of these four areas, according to the National Agency of Hydraulic Resources (ANRH) is estimated at 12,827 billion m^3 (Table 18.2). The Sebaou alluvial aquifer belongs to the third

Table 18.1 Total capacity of mobilisation of water resources in Algeria

	Surface water (billions m^3)	Ground water (billions m^3)	Total (billions m^3)	%
North of Algeria	12.0	1.9	13.9	82
South of Algeria	1.5	1.4	2.9	18
Total	13.5	3.3	16.8	100

Table 18.2 Annual average capacity of mobilization (ANRH)

	Oranie Chott Chergui	Chélif Zahras	Algérois Soummam Hodna	Constantinois Seybousse Mellègue	Total
Surface km^2	76,000	56,200	50,000	43,000	225,200
Annual average capacity of mobilizations (millions m^3/year)	958	1,974	4,300	5,595	12,827

Table 18.3 Capacity of large dams by hydrographic basin

	Oranie Chott Chergui	Chélif Zahras	Algérois Soummam Hodna	Constantinois Seybousse Mellègue	Total
Number of dams	12	13	12	15	52
Capacity (million m^3)	658	1,950	820	1,530	5,200

basin (Algérois, Soummam and Hodna) in terms of area and the second one in terms of potential of water mobilization.

A significant interest was dedicated to the dam's construction. Indeed, with such a significant number of dams (114 large and small dams), Algeria is located in the forefront in the Arab world and occupies the second place in Africa, after South Africa. The total capacity of mobilization of water resources using these dams equals almost six billion m^3 (Table 18.3).

The ANRH has initiated a project on the surface flows for northern Algeria [6] aimed the development of tools for assessing the surface resources in any catchment for the next decades, and 60 potential sites for reservoirs and dams were assessed and evaluated for different uses (irrigation, water supply, etc.).

18.2.1 Water Loss of the Dams Due to Silting and Evaporation

The erosion of the ground in northern Maghreb is very important, creating a direct impact on the rate of silting of dams and consequently on the water storage capacity. In Algeria, 52 large dams receive annually 32 million m^3 of solid materials. The loss by evaporation in 39 storage reservoirs is also considerable. Indeed, the cumulative volume of water lost for the period from 1992 to 2002 is estimated at three billion m^3.

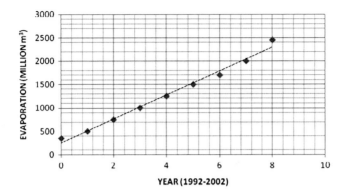

Fig. 18.1 Evolution of the evaporated water volume in 39 dams between 1992 and 2002

The evolution of the cumulative evaporated water volume for this period is shown in Fig. 18.1 [10].

The annual average volume of the evaporation is 250 million m^3 for the 39 dams with a storage capacity of 3.8 billion m^3, i.e. an annual average loss close to 6.5%. According to Fig. 18.1, the total volume of water lost during the period 1992–2002 is about 2.5 billion m^3. By adding the volume of 6.8 billion m^3 discharged into the sea to these losses, we can easily note that the total volume lost is very significant. The comparison between the available resources and the water needs make possible to highlight that by 2020 the Algerian north will be confronted with a water deficit estimated at one billion m^3. On this basis, the capacity of mobilization should be preserved by increasing the storage capacity in dams and using artificial recharge, which involves the increase of the natural movement of surface water into underground formations. The recharge can be either direct or indirect. In direct recharge, water is introduced into an aquifer via injection wells. The injected water is treated to ensure that it does not clog the area around the injection well. In contrast, indirect recharge involves spreading surface water on land so that the water infiltrates through the vadose zone (the unsaturated layer above the water table) downward to the aquifer. An advantage of indirect recharge is that the vadose zone acts as a filter potentially improving the quality of the water percolating through the soil.

18.3 Artificial Recharge of Sebaou Aquifer

18.3.1 Presentation of Sebaou Basin

Located 100 km east from Algiers, the watershed of the Sebaou River occupies an area of 2,500 km^2, with a temperate local Mediterranean climate, and an average annual rainfall of 873 mm. The surface flows represent an annual volume of

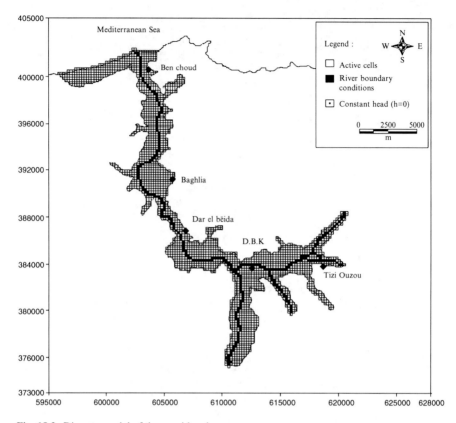

Fig. 18.2 Discrete model of the considered area

627 million m³. The drinking water supply of the region is primarily provided by the alluvial aquifer of Sebaou. The Taksebt Dam (Tizi Ouzou), with a storage capacity of 150 million m³, is the only adequate dam designed to increase the mobilization of surface water. The total water demand of the region is about 100 million m³, a volume equivalent to the exploitable annual resources of the alluvial aquifer of Sebaou (Fig. 18.2).

The overexploitation of the groundwater of the Sebaou River, in order to satisfy the drinking and irrigation water requirements for both wilayas of Tizi-Ouzou and Boumerdes, and the weak level of natural recharge generated a severe lowering of the groundwater level of the aquifer during the last few years. In parallel, the Sebaou River water quality undergoes a significant degradation and becomes a real dump for domestic and industrial wastes. One must also note the anarchistic and uncontrolled extraction of sand and alluvia from the river bed (the protective filter of the aquifer), making the groundwater vulnerable when exposed to the various forms of pollution. Due to this unbalance and irrational use of water resources, the artificial recharge of the aquifer is an effective means to ensure groundwater storage and to avoid water loss. In such an unconfined aquifer, groundwater may be replenished by natural precipitation, irrigation, or artificial recharge. The recharge system may be a group of deep wells, ponds or a combination of them. Their type and location depend on the hydraulic properties of the aquifer. In some cases, water treatment may be required before recharging the water into the aquifer. Aquifer extraction and recharge must be balanced on an annual basis. This will ensure that long-term groundwater levels will be maintained. Also, the quality of extracted water from the aquifer must be suitable for the intended use.

18.3.2 The Aquifer of Sebaou and Pollution

The chemical and microbial quality of groundwater is inextricably linked to events occurring above the aquifer. The many factors that can impact the quality of the groundwater are air deposit of small particles, contaminated rainfall, untreated storm water, polluted agricultural runoff, untreated or partially treated wastewater discharged from municipal and industrial sources, accidental spills and illegal waste dumping. Industrial wastes from local factories (ENIEM, COTITEX, ORLAC, etc.) are directly discharged into the bed of Sebaou River, leading to polluted surface water. The results of chemical analysis (1999–2002) of Sebaou groundwater have shown that water is characterized by an average to high mineralization, the conductivity being ranged between 537 µs/cm and 1,019 µs/cm in high waters and between 504 µs/cm and 1,152 µs/cm in low waters. The pH ranges between 6.8 and 7.9 in high waters and between 7.2 and 7.7 in low waters. According to standards set by WHO, the chemical quality of water is recognized to be of good quality. This situation of uncontrolled discharges can permanently alter the groundwater, precluding the intended use of the aquifer (e.g., irrigation or production of drinking-water) for many decades. An integrated approach to groundwater management in general, and a health impact assessment of the management options in particular, are strongly recommended.

18.3.3 Mathematical Model and Numerical Simulation

The system of exchange between surface and groundwater may be a group of deep wells, rectangular ponds or simply the surface flow after a release from dams, or a combination of them. The type and location depend on the hydraulic properties of the aquifer [4, 5, 12, 13].

18.3.3.1 Surfaces Flows

The surface flows in the rivers are governed by the Saint-Venant equations [1, 7, 11]. These previous results were derived from a vertical integration of Navier-Stokes equations and are given by the following equations:

$$\begin{cases} \dfrac{\partial S}{\partial t} + \dfrac{\partial Q}{\partial x} = q_l + q_n \\ \dfrac{\partial Q}{\partial t} + \dfrac{\partial (Q^2/S)}{\partial x} + g \cdot S \dfrac{\partial h}{\partial x} = g \cdot S \cdot (i - J) \end{cases} \quad (18.1)$$

where:
Q is the flow rate (m³/s);
S is the wetted cross section (m²);
h is the height of water in the section (m);
i is the slope of the river bed;
q_l is the flow of side contribution per unit of length of the river;
q_n is the flow crossing 1 m length of the interface river-aquifer; and:

$$J = \dfrac{Q^2 n^2}{A^2 R^{4/3}}$$

is the slope of the free-water surface of the river. The unknown factors of this system are the flow Q(x,t) and the wetted cross section S(x,t).

18.3.3.2 Groundwater Flows

The flow in a saturated porous medium is governed by the equation of diffusivity [8], obtained from the combination of the equation of continuity (rising from the principle of mass conservation) and the experimental/empirical law of Darcy. In the analysis the recharge wells are considered as point sources. The diffusivity equation is expressed by the following equation:

$$\dfrac{\partial}{\partial x}(K_{xx}\dfrac{\partial h}{\partial x}) + \dfrac{\partial}{\partial y}(K_{yy}\dfrac{\partial h}{\partial y}) + \dfrac{\partial}{\partial z}(K_{zz}\dfrac{\partial h}{\partial z}) = W + S_s \dfrac{\partial h}{\partial t} \quad (18.2)$$

where:
K_{xx}, k_{yy}, K_{zz} are the hydraulic conductivities along the axes, x, y and z (m/s);
H is the hydraulic head (m);
W is the source term or loss per unit volume (l/s);
S_s is the specific storage coefficient (l/s);
t is the time (s).

18.3.3.3 Boundary and Initial Conditions

The boundary conditions of the model can be physical or hydraulic [2]. The physical limits correspond to great sets of surface water like a sea or a lake whereas the hydraulic ones often correspond to flow lines. The boundaries of the Sebaou model correspond to the actual physical limits of the studied area. All the cells of the grid located beyond these limits were taken like not activated. The following boundary conditions were undertaken:

- Northern limit is regarded to be a natural discharge system with null head (sea level).
- Western and eastern limits are regarded to be limits with a null flow.

18.4 Results and Discussion

18.4.1 Steady State

The simulation in steady state was carried out starting from the data (removal, refill, initial loads and points of observation) of the year 2000, the period for which the greatest amount of information is available. This first stage of modeling has several objectives:

- To ensure the correct operation of the model after the modifications of the geometry in the study zone
- To describe the behavior of the tablecloth in the best possible way
- To adjust the hydrodynamic parameters.

Figures 18.3 and 18.4 represent the piezometric map of simulation in steady state and the comparison of the measured and numerical simulation results.

The validation of the model was carried out "manually" by using a "trial and error" approach. The parameters were changed until obtaining sufficiently weak differences between the simulated and measured results.

The results of simulation in steady state lead to a balance of the inlet and total outlet of the modeled area with a flow of 710,775 m^3/day.

18.4.2 Transient State

Simulation in transient state requires the parameters of storage of the aquifer. The storage coefficient ranges between 0.1 and 0.15 [3]. The piezometric values obtained when simulating in transient state the storage on the year 2005 are presented in Fig. 18.5. The model validation in transient state was carried out, and the response of the model to the changes of the recharge and extraction during the period (2000–2006) was simulated and compared with the observed aquifer behavior.

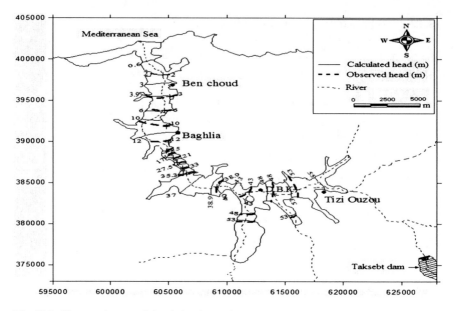

Fig. 18.3 Piezometric map of simulation in steady state

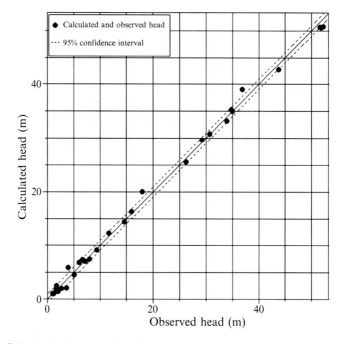

Fig. 18.4 Calculated and measured head

18 Protection of Water Resources in Algeria

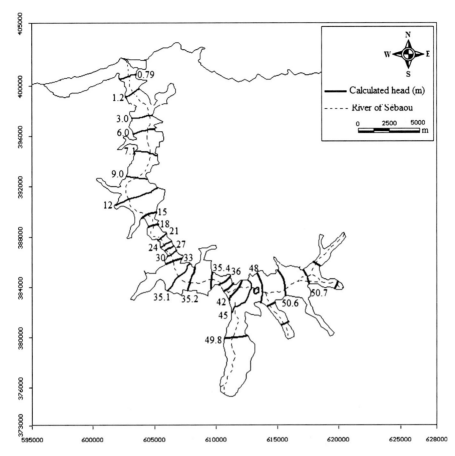

Fig. 18.5 Restored piezometric values of 2005

The comparison of the measured and simulated patterns of water head is shown in Fig. 18.6. One notices that generally the simulation results remain close to the observed ones, and the behavior is characterized by stability of the head over the time. However, a slight overestimation of the head is observed at some points.

18.4.3 Some Scenarios

18.4.3.1 The Over Exploitation

In order to simulate the situation of overexploitation of groundwater, the pumped flow was increased until a value (3Q). The variation of the difference head (ΔH) in the period between 2000 and 2014 shows a decrease in the level of the aquifer (Fig. 18.7) over this time period. We note that in 2014 the direction of groundwater

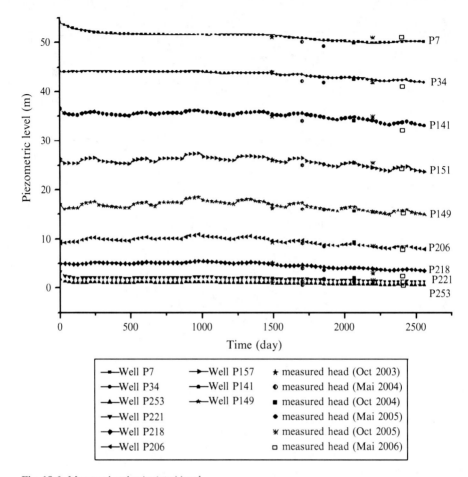

Fig. 18.6 Measured and calculated head

flow will be reversed at the mouth. This implies a seawater intrusion into the region of Ben Chhoud. For better viewing the results of the simulated scenario, these are represented by some checkpoints variations in groundwater level of the Sebaou River during the simulation period (Fig. 18.8).

Therefore, it is advisable to limit the extraction in the downstream sector of the water table and to avoid any implementation of pumping at the mouth of the Sebaou river, in order to avoid the phenomenon of marine intrusion. The increasing salinity will make the water unfit for drinking and also often unfit for irrigation.

18.4.3.2 Artificial Recharge by Injection

The current model offers the possibility to follow the evolution of the piezometric level of the aquifer after the introduction of a flow into the artificial recharge site. Four wells of injection are chosen and installed on the bank of the Sebaou river.

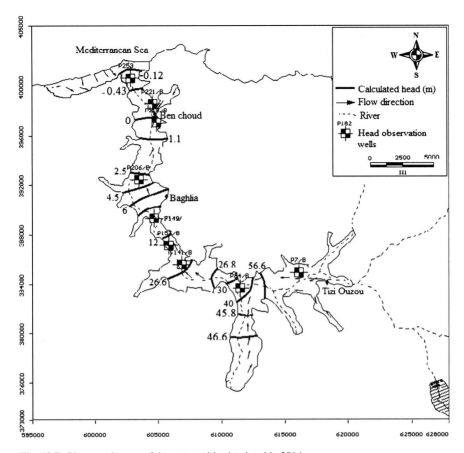

Fig. 18.7 Piezometric map of the water table simulated in 2014

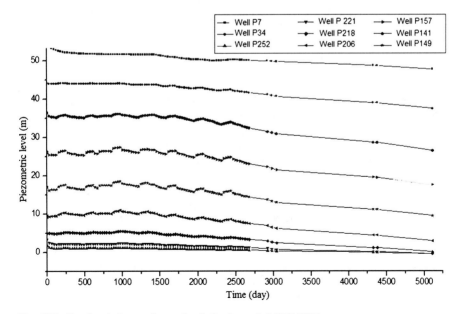

Fig. 18.8 Simulated of groundwater levels for the period 2000/2014

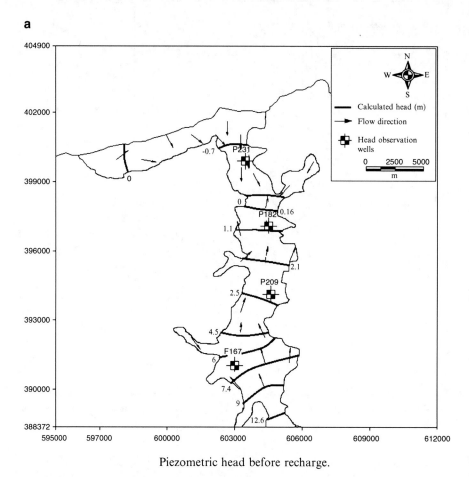

Piezometric head before recharge.

Fig. 18.9 (**a**, **b**, **c** & **d**). Response of the aquifer to the artificial recharge as piezometric head before (**a**) and after recharge (**b**, **c** and **d**) for various flow rates

The flow of the injection is varied for different cases of simulation aiming expecting the response of the water table. The results of the simulations are presented on the Figs. 18.9 and 18.10.

According to the results of Figs. 18.9 and 18.10, we note that the artificial recharge causes a substantial increase of the level of the water table and a reorientation of the direction of the flow, avoiding marine intrusion.

18.4.3.3 Artificial Recharge by Water Release from Taksebt Dam

The artificial recharge may be achieved either by injection wells and direct recharge on permeable zones. This last technique is the most used but present more vulnerability to pollution. The cartography of the groundwater pollution potential of a given area, combined to repartition of favourable zones to artificial recharge, constitute a

18 Protection of Water Resources in Algeria

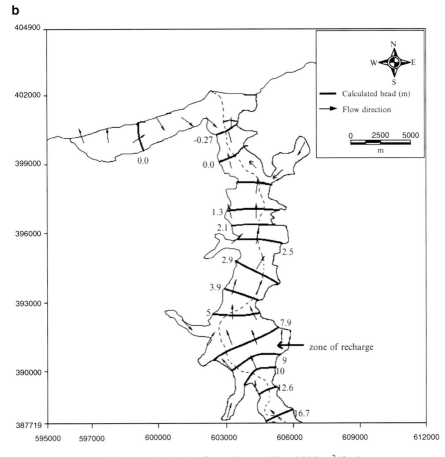

Piezometric head afterrecharge (Q = 2000 m³/day).

Fig. 18.9 (continued)

tool to decision making. The simulation of recharge by infiltration is performed by varying the level of water in the Sebaou River and on its conductance, taking into account the residence time of water at the surface, in order to minimize evaporation loss. The response of the aquifer is presented on Fig. 18.11. One can note that this type of recharge is less efficient compared to the recharge by well injection. A quite weak increase of piezometric level is obtained over the time period simulation.

18.5 Conclusion

The lack of natural fresh water resources has become one of the major concerns of the world today. The valley of Sebaou is experiencing significant demographic and socio-economic development. The continual increasing demand of water and the

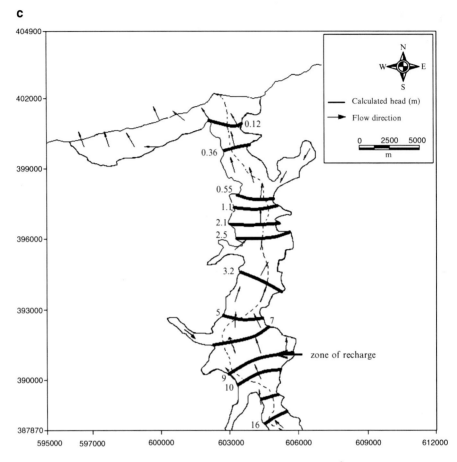

Piezometric head after recharge ($Q = 3000$ m^3/day).

Fig. 18.9 (continued)

irregularity and the reduction of the rainfall allowed a significant reduction of the capacity of water mobilization and a lowering of the groundwater level. At the same time, the Sebaou River is undergoing a significant degradation, leading to a drastic reduction of its filtration capacity. Aquifer recharge is likely to increase in the future because it can restore depleted groundwater levels, provide a barrier to saline intrusion in coastal zones, and facilitate water storage during periods of high water availability.

The feasibility of artificial recharge of the Sebaou aquifer was studied using a mathematical model of surface and groundwater flows. The solution of the coupled equations system was obtained by ModFlow software. The numerical simulation results are in good agreement with the observations and measurement results obtained by the ANRH. For the purpose of analyzing the modification of groundwater table level, different scenarios of recharge are undertaken and the simulation results provide a better understanding of the aquifer behavior in order to plan the recharge,

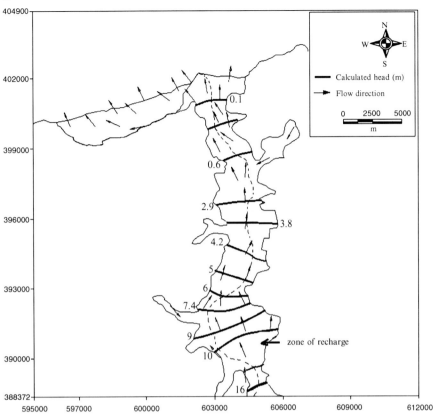

Piezometric head after recharge (Q = 4000 m³/day).

Fig. 18.9 (continued)

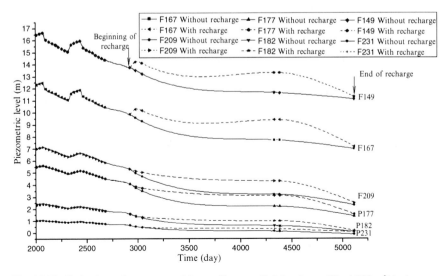

Fig. 18.10 Hydrodynamic response of the aquifer to artificial recharge (Q = 4,000 m³/day)

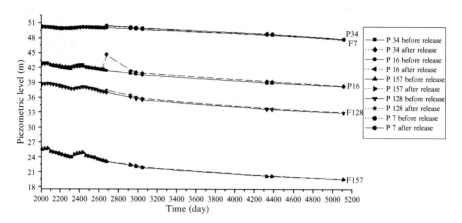

Fig. 18.11 Response of the aquifer to artificial recharge by infiltration of released water from the dam

storage and production management. The recharge by injection was found to be more efficient than water release from a dam. The numerical study aims at carrying out examples of the aquifer recharge and, more specifically, highlighting the important issue of assessing and managing water resources, salination risks and environmental protection of arable lands.

References

1. Ababou R, Bagtzoglou AC (1993) BIGFLOW, a numerical code for simulating flow in variably saturated heterogeneous geologic media, theory and user's manual 1.1, NUREG/CR-6028. Report. U.S. NRC, Government Printing Office, Washington, DC
2. Barone (2000) Modeling the impacts of land use activities on the subsurface flow regime of the upper Roanoake river watershed. Master of Science in Biological Systems Engineering, Faculty of the Virginia Polytechnic Institute and State University, Blacksburg, pp 184
3. Chadrine Y (1975) Étude hydrogéologique de la nappe de l'Oued Sébaou. Document ANRH d'Alger, pp 7
4. Esteves M (1988) Modélisation des relations entre un aquifère alluvial et une rivière: Application au pied de Colmar (Alsace-France), 113ème Congrès national des Sociétés savantes, Strasbourg, pp 41–61
5. Esteves M, Ackerer P (1988) Interaction entre eaux souterraines et eaux de surface, Sweden, Edited by Peter Dahlblom and Gunnar Lindh. In: Proceedings of the international symposium, Department of Water Resources Engineering, Lund University, Lund, 30 May–3 June 1988
6. Laborde JP, Gourbesville P, Assaba M, Demmak A, Belhouli L (2010) Climate evolution and possible effects on surface water resources of North Algeria. Curr Sci 98(8):1056–1062
7. Max A, Saint P (1971) Alimentation artificielle des nappes, Colloque International sur les Eaux Souterraines, Palerme, Italie, 6–8 Décembre 1970
8. McDonald MC, Harbaugh AW (1988) A modular three-dimensional finite difference groundwater flow model. US Geological Survey Techniques of Water-Resources Investigations, Washington, Open-File Report 83-875, pp 588

9. Menani MR (2009) The artificial recharge and the groundwater vulnerability to pollution-case of the El Madher Plain (North East of Algeria). Eur J Sci Res 32(3):288–303
10. Remini B (2005) La Problématique de l'Eau en Algérie. Collection Hydraulique et ransport Solide, Imprimerie Madani, Blida, pp 34–52
11. Roche P, Thiery D (1984) Simulation globale de bassins hydrologiques. Introduction à la modélisation et description du modèle GARDENIA. Rapport BRGM n° 84 SGN 337 EAU
12. Tajjar et al. (1993) Les lâchures des barrages-réservoirs. In: la seine et son bassin: de la recherche à la gestion, E., Fustec et G., d. Marsily (ed) Paris, pp 49–54
13. Tregarot G, Ababou R, Larabi A (1997) Inondations, infiltrations et couplages d'écoulements partiellement saturés et non-saturés, 22èmes journées du GFHN, Meudon, France, 25–26 novembre 1997

Chapter 19
Regional Model of Groundwater Management in North Aquitania Aquifer System: Water Resources Optimization and Implementation of Prospective Scenarios Taking into Account Climate Change

Dominique Thiéry, Nadia Amraoui, Eric Gomez, Nicolas Pédron, and Jean Jacques Seguin

Abstract In the multilayered aquifer system of North Aquitania in South-West France, the Oligocene and Eocene aquifers are pumped intensively for 50 years, mainly for domestic water. The huge pumping, which reaches 290 million m^3 per year, causes a steady decline in groundwater levels. In particular, given the low recharge by rainfall, the Eocene confined aquifer collapsed with a water level decrease exceeding 30 m near the Bordeaux city. The resource is threatened, both in quantity, since the pumping exceeds the renewal and in quality, due to the reversal of flow gradients, with risk of brackish water invasion from the Gironde estuarine area in which the Eocene aquifer outcrops. The Oligocene is also subject to an increased vulnerability to surface pollution. A mathematical model simulating groundwater flows in this multilayer system has been implemented to quantify the water savings to be achieved, and simulate scenarios trend (population growth, industry and agriculture development) and combined scenarios (savings and substitutions). It is a regional multilayer water resources management model with 15 aquifer layers starting from the Plio-quaternaire aquifer at the top, down to the Bajocian aquifer at the bottom. The model, which has a spatial extension varying from 10,000 to 25,000 km^2 according of the aquifer, takes into account 67,000 square cells of size 2 km. It incorporates, at an annual time step, pumping in 3,250 wells and climatic data in five meteorological stations. It is calibrated in transient state using 380 observed time series of water level. The model was used to analyze the possibilities of restoring a balanced and safe state, using simulation of pumping scenarios. These simulations include scenarios incorporating trend forecasts of population growth, scenarios of economy needs, scenarios of pumping alternative

D. Thiéry (✉) • N. Amraoui • E. Gomez • N. Pédron • J.J. Seguin
BRGM, Water Division, 3 Avenue Claude Guillemin BP 36009,
Orléans Cedex 2 45060, France
e-mail: d.thiery@brgm.fr

aquifers and climate change scenarios. The climate change scenario selected is the moderate IPCC Arpege A1B scenario from Météo – France.

Keywords Groundwater management • Multilayered aquifer • Mathematical model • Climate change

19.1 Introduction

The North Aquitania multilayered aquifer system is located in the south-west of France, between the Gironde estuary and the Pyrenees (Fig. 19.1).

Due to the population increase in this region, this aquifer system is heavily pumped. As a result of the intensive pumping, the piezometric level "collapsed" in Bordeaux city region and aquifer zones are dewatered in some locations (Fig. 19.2).

There are threats on the resource in term of quantity because renewal is compromised (for instance in the Eocene formation). There are also threats in term of quality because there are risks of brackish water inflow from the estuarine area (estuary and fossil saline groundwater from the Quaternary) due to flow gradients inversion (Fig. 19.3).

A mathematical model of the aquifer system has been built in order to analyze the risks and study the possibilities of water resource optimization [1–8]. The use of the model must allow actors / decision makers to test scenarios of pumping and

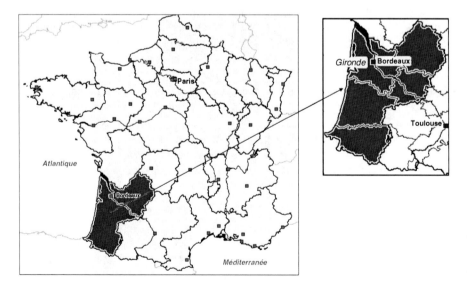

Fig. 19.1 Location of the North Aquitania multilayered aquifer system

19 Regional Model of Groundwater Management in North Aquitania Aquifer System... 277

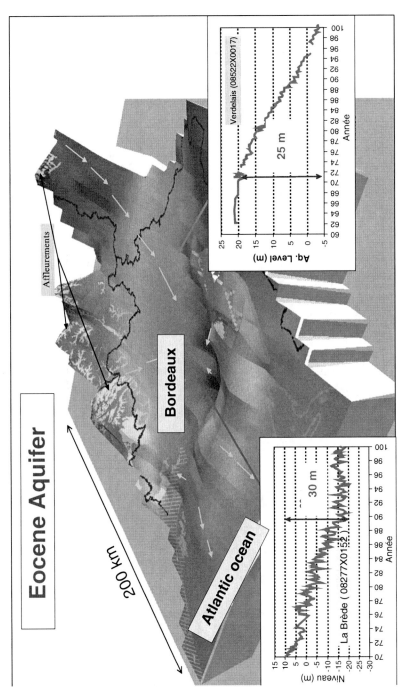

Fig. 19.2 Aquifer collapse by 25–30 m in Bordeaux city region

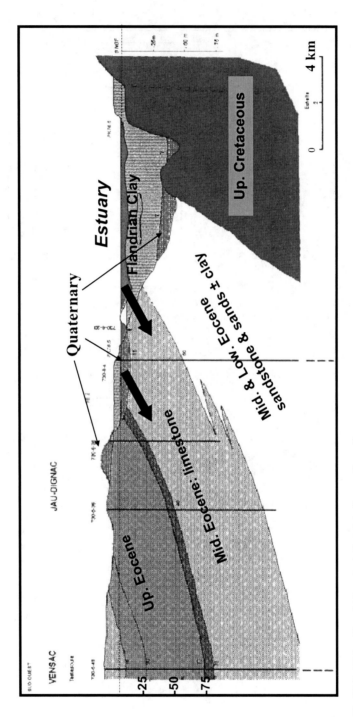

Fig. 19.3 Risks of brackish water inflow into Eocene formation

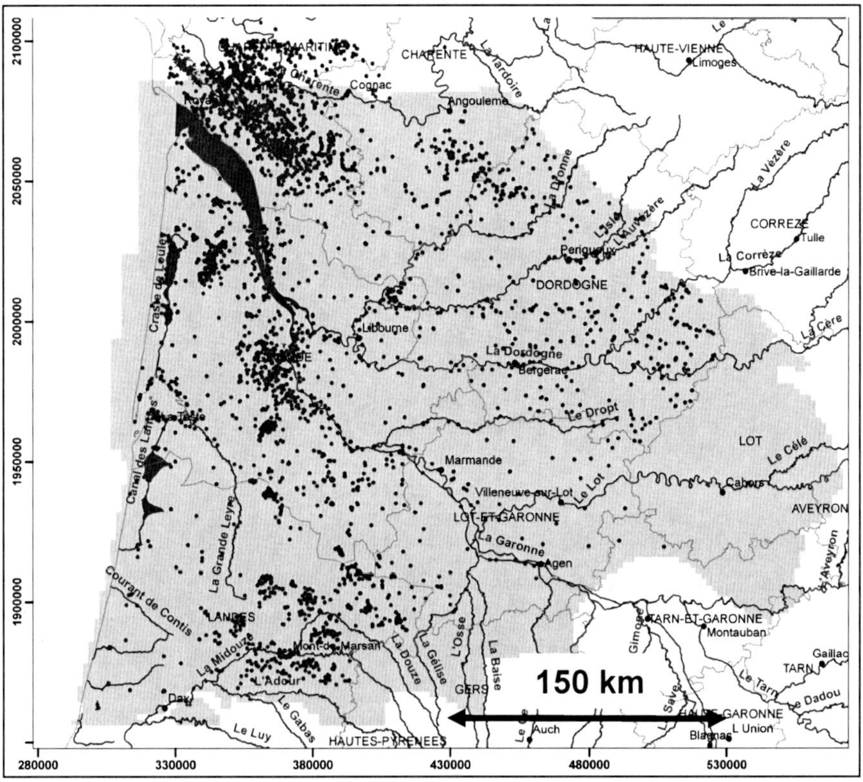

19 Regional Model of Groundwater Management in North Aquitania Aquifer System... 279

Fig. 19.4 Lateral extension of the modeled area and position of pumping wells (*dots*)

aquifer recharge until 2050 in order to identify those likely to best preserve the resources while maintaining minimum groundwater levels in areas identified as "at risk". Different spatial distributions of pumping can lead to the same total volume pumped, but will not produce the same effects on sectors where the maintenance of a groundwater level is deemed necessary. It is therefore essential to properly adjust the model on the current pumping and those projected in the future.

The lateral extension of the model is illustrated in Fig. 19.4, which also shows the location of the 3,760 wells of the aquifer system.

19.2 Extension of the Modeled Area

The model, named MONA for Model of North Aquitania, simulates the groundwater aquifer system in multilayer North Aquitaine Basin. Fifteen major layers were retained in the most recent release of the model: Plio-Quaternary

Table 19.1 List of the 15 geological layers modeled

Name	Layer	Area (km²)
Plio-Quaternary	1	11,350
Helvetian	2	8,900
Aquitanian-Burdigalian	3	12,600
Oligocene	4	16,200
Upper Eocene	5	15,600
Middle Eocene	6	17,300
Lower Eocene	7	17,300
Campano-Maastrichtian	8	17,350
Conacian-Santonian	9	24,900
Turonian	10	26,700
Cenomanian	11	23,000
Tithonian	12	7,300
Kimmeridgian	13	13,500
Bathonian-Callovo-Oxfordian	14	31,400
Bajocian	15	24,200

Langhian-Serravalian (Helvetian), Aquitanian-Burdigalian, Oligocene, Eocene, middle Eocene, Lower Eocene, Campano-Maastrichtian, Coniacian-Santonian, Turonian, Cenomanian, Tithonian, Kimmeridgian, Bathonian-Callovian-Oxfordian and Bajocian (Table 19.1).

19.3 Pumping in the Aquifer System

The database currently includes 3,760 wells in the 15 geological formations (Fig. 19.2). The volume pumped in 2007 in these 3,760 wells was 325 millions of m³ per year, with a maximum of 360 million m³ per year in 2003. Some wells from the database, however, are located outside the extension of the model MONA. Within the model MONA, there are 3,250 wells which pumped about 290 million m³ per year in 2007 with a maximum of 320 million m³ recorded in 2003. Figure 19.5 shows that there has been a very large increase of pumping from 1960 to 2007.

Most wells are located in the Miocene, Oligocene, Eocene and Turonian aquifers. These aquifers are also the most exploited in terms of flow pumped. Note also that a large number of wells simultaneously pump several aquifers.

Formations of the Upper Jurassic (Kimmeridgian and Tithonian) are only marginally exploited in their confined part because they have low productivity.

The share of water for industrial use has declined steadily since 1972 (currently about 3%), while at the same time, pumping for agriculture and domestic water have increased sharply from 1980 to 1990. The evolution of pumping is much more stable since the mid-1990s.

19 Regional Model of Groundwater Management in North Aquitania Aquifer System... 281

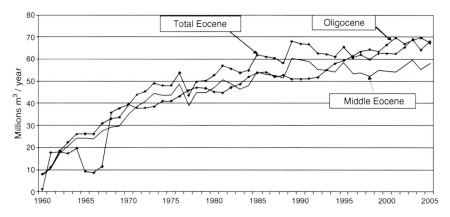

Fig. 19.5 Evolution of withdrawal rate in Eocene and in Oligocene

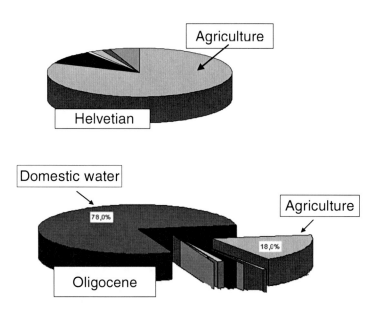

Fig. 19.6 Distribution by use of withdrawal rate in selected formations. Top: Helvetian (Miocene); bottom: Oligocene

The domestic water usage is more than 70% in confined aquifers. This percentage however masks significant heterogeneity since this usage is only 10–20% in the Miocene aquifers mainly used for agriculture, and over 90% in the middle Eocene and Bathonian-Callovian-Oxfordian aquifers (Fig. 19.6). The Eocene Oligocene and aquifers are by far the most commonly used for domestic water.

19.4 The Groundwater Flow Computer Code

MARTHE code, developed at BRGM [9, 10], has been used for the calculation of flows in the multilayered aquifer system. The main characteristics of this computer code for flow simulation are the following:

- Finite Volumes using irregular parallelepipeds
- 3D and Multilayer geometry
- Millions of cells possible
- Integrated hydroclimatic balance: Rainfall, Evapo-Transpiration, Runoff, Infiltration
- Fully coupled aquifer with river networks and drain networks
- Fully coupled nested grids system.

MARTHE has also the following functionalities which were not used in the scope of this project, namely:

- Flow and transport modeling in Saturated and Unsaturated Zone
- Integrated mass transport by advection + dispersion + decay
- Density effects: Salinity, Thermal effects
- Energy transfers
- Multiphase Flow: Water/Gas, Water/Oil, Fresh Water/Salt Water
- Coupling with geochemistry reactions [11]. The reactive calculations result from TOUGHREACT, code developed at LBNL, Berkeley, California [12]
- Automatic calibration of parameters

One important feature is the possibility of true "Multilayer" and/or "Full 3D" geometry. With this geometry, aquifer layers may disappear locally and there may be short circuits between layers (Fig. 19.7).

19.5 Model Description

Figure 19.7 displays the geometry of the 15 layers of the model. The general structure of the model contains 67,000 square cells of size 2 km. In its current version, the model does not explicitly represent the flow in the aquitards but nevertheless simulates vertical infiltration taking account of their permeability and thickness. Five weather stations managed by Météo-France are used to calculate the groundwater recharge used as input by the model. Rainfall and PET (Potential EvapoTranspiration) data are used to calculate the effective rainfall in outcrop areas of every modeled aquifer layer. The effective rainfall is then distributed into runoff and aquifer recharge, using a distribution relation.

Each year, the model simulates the average state of every groundwater formation in response to pumping and recharge accumulated over the year. The model was calibrated in transient state over the period 1972–2007 using 380 observed time series of piezometric levels.

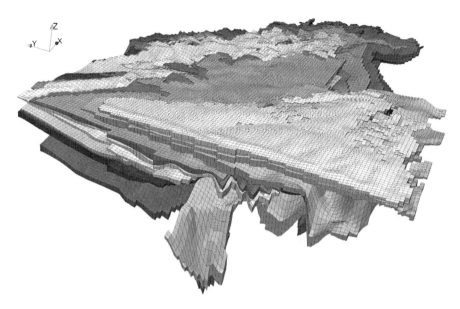

Fig. 19.7 North Aquitania multilayered aquifer system showing local disappearance of layers and short circuits

The data have been updated successively in 1999, 2001 and 2005, which served as validation period for the model and helped to control its robustness.

As a summary, the North Aquitania model has the following characteristics:

- 15 layers: from the Plio-quaternary down to the Bajocian formation
- 67,000 square cells of size 2 km
- Calibration period: 1972 – 2007
- More than 3,250 pumping wells
- 380 time series of piezometric levels for model calibration
- 5 weather stations for calculation of groundwater recharge.

The calibration of the model over the whole period 1972–2007 is satisfactory. In order to illustrate this, the comparison of the observed and simulated water level variations in four wells located respectively in middle Eocene (Cénon Mairie – EOCM), in lower Eocene (Monsegur – EOCI), in Campano-Maastritchian (Léognan – CAMP) and in Coniacian-Santonian (Gontaud-de-Nogaret – COST) is displayed in Fig. 19.8.

19.6 Definition of Scenarios for the Period 2008–2050

After calibration the MONA model was used to test scenarios of pumping and aquifer recharge until year 2050. The aim was to identify the scenarios likely to best preserve the resources, while maintaining minimum groundwater levels in areas identified as "at risk". The scenarios must incorporate simultaneously pumping forecasts and climate predictions.

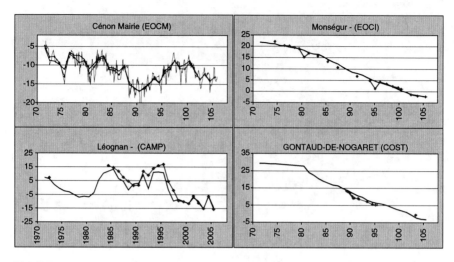

Fig. 19.8 Observed (symbols) and simulated water levels at four wells in middle Eocene (EOCM), lower Eocene (EOCI), Campano- Maastrichtian (CAMP) and in Coniacian-Santonian (COST)

19.6.1 Pumping Scenario

The forecast in future pumping integrate the construction of new wells and the evaluation of pumping rates in 2050. The time evolution of pumped flows was estimated assuming stabilization in the period 2008–2050 of pumping for agricultural and industrial uses, and increased pumping for domestic water corresponding to the expected growth of population from 2008 to 2050 in the area included in the MONA model.

To assess changes in domestic water need, the retained hypothesis was:

- An individual need identical to that of 2007
- A change in population using INSEE "median" scenario, forecast with OMPHALE model based on the 2006 population census.

Figure 19.9 displays the forecast evolution of population by INSEE (Institut National de la Statistique et des Études Économiques) for the period 2006–2030.

Figure 19.10 displays the pumping rates of the 3,250 wells included in the MONA model during the period 1972–2007 and projection until 2050.

19.6.2 Climatic Scenario

The results of simulations of climate change from the IPCC's work (GIEC) have been obtained from CERFACS. These data, derived from Météo-France Arpege model, disaggregated to the cell of 8 km, provide projections of rainfall and PET for

19 Regional Model of Groundwater Management in North Aquitania Aquifer System... 285

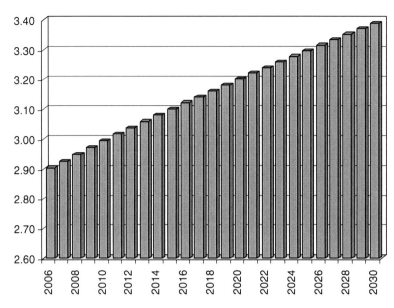

Fig. 19.9 Forecast evolution of population by INSEE for the period 2006–2030. (Population in million inhabitants)

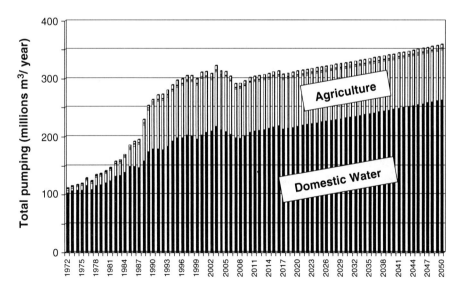

Fig. 19.10 Pumping rates of the 3250 wells included in the MONA model during the period 1972–2007 and projection until 2050

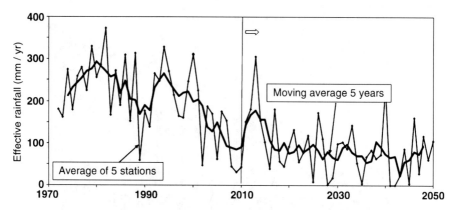

Fig. 19.11 Average effective rainfall computed from observed data in 1972–2007 and from climate change scenario Arpege A1B (Météo-France/IPCC) in 2008–2050

the period 2000–2050. Among the available scenarios, it is Arpege A1B scenario, considered as "moderate", which has been chosen. Figure 19.11 displays the "effective rainfall" computed using the observed data during the period 1972–2007 and using the Arpege A1B data during the period 2008–2050.

19.7 Results and Discussion

This scenario leads to a general decline in groundwater levels that would result in a lasting imbalance in deep groundwater of northern Aquitaine Basin. Near the estuary of the Gironde, the continuous decline of groundwater levels would cause the progressive collapse of the piezometric level crest in the Eocene that could disappear and allow an inflow of brackish water. Maps of water level drawdown from 2007 to 2050 resulting from this scenario have been calculated. These maps are intended only to show the major trends and should not be interpreted in detail at the local level. It should also be noted that the simulation, which neglects the storage in the aquitards, might be somewhat pessimistic especially in the deeper layers. Figure 19.12 displays the water level drawdown in four aquifers. It appears that in every of these four aquifers there are areas where drawdown exceeds 35 m.

Figure 19.13 displays the aquifer water level forecast in the same four wells presented earlier during the calibration phase. During 2007–2050 the aquifer water level in these wells decline in the range 15–20 m.

Although the calibration is quite satisfactory, uncertainties remain because of the following difficulties:

- The role of faults and of vertical hydraulic connections is hypothetic
- There is a poor network for monitoring deep groundwater levels

19 Regional Model of Groundwater Management in North Aquitania Aquifer System... 287

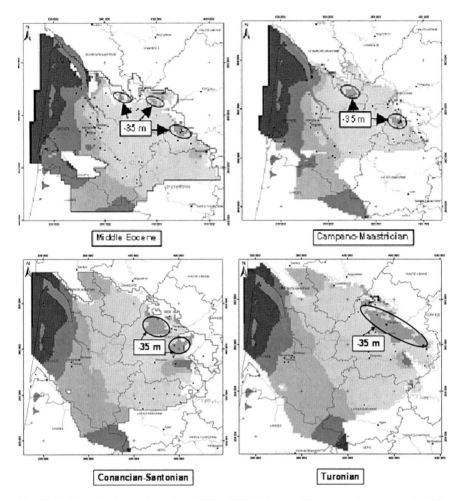

Fig. 19.12 Water level drawdown from 2007 to 2050 in four aquifers using climate scenario A1B

- Some hydraulic parameters are uncertain due to the small number of measurements in deeper formations (Table 19.2).

The following improvements are scheduled:

- A better determination of groundwater recharge by simulation of river flow rates using the calculated runoff
- The use of a finer time step: daily for hydroclimatical balance
- A sensitivity analysis, simulating several Climate Change scenarios.

288

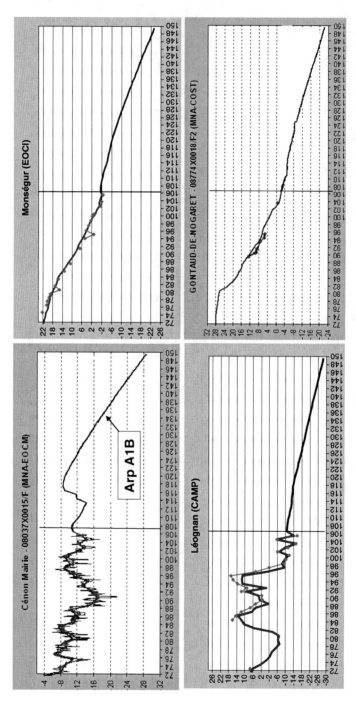

Fig. 19.13 Scenario A1B: Aquifer level forecast at four wells in middle Eocene (EOCM), lower Eocene (EOCI), Campano-Maastrichtian (CAMP) and in Coniacian-Santonian (COST)

Table 19.2 Number of measured hydraulic parameters available in deeper formations

Layer name	Layer number	Number of values Transmissibility	Specific flow
Campanian	8	10	–
Coniacian-Santonian	9	3	9
Turonian	10	2	10
Cenomanian	11	5	3
Tithonian	12	1	3
Kimméridgian	13	3	3
Bathonian-Callovo-Oxfordian	14	12	18
Bajocian	15	1	6

19.8 Conclusion

The regional model of the multilayered aquifer system of North Aquitania, involving 15 layers, was used to predict changes in groundwater levels over the period 2008–2050. A realistic estimate of pumping rates evolution, based on the anticipated increase in population in the region, and incorporating water saving, has been combined with Meteo-France Arpege A1B climate change scenario. Introducing this scenario in the model it appears that the water savings in the Eocene and Oligocene, forecast in 2015–2017, lead to an improvement in local areas where these savings will be achieved. However, the overall increase of pumping, based on the planned population increase, would result in a general decline in groundwater levels in 2050. Near the estuary of the Gironde, the piezometric level crest in the Eocene could disappear and allow an inflow of brackish water. The deep aquifers of North Aquitaine Basin would be affected by a lasting imbalance. Refinements of the model and a sensitivity analysis would reduce uncertainties and to help to improve the water resources management.

Acknowledgments The work described in this article was co-funded by various institutions, among which: BRGM, the Regional Council of Gironde and Dordogne, the Urban Community of Bordeaux, the Water Agency Adour-Garonne, the Region, the FEDER etc.

References

1. Amraoui N, Bichot F, Platel JP, Seguin JJ (1998) Gestion des eaux souterraines en Aquitaine – Année 2 – Évaluation des ressources. Ajout des couches du Turonien-Coniacien-Santonien, du Cénomanien et du Jurassique moyen et supérieur au Modèle Nord-Aquitain. Report BRGM R 40110
2. Amraoui N, Bichot F, Seguin JJ, Sourisseau B (1999) Restructuration du Modèle Nord-Aquitain de gestion des nappes. Réalisation de 6 simulations pour le schéma de Gestion des Eaux du département de la Gironde. Report BRGM R 40224

3. Gomez E, Pédron N, Buscarlet E (2010) Utilisation du Modèle Nord-Aquitain (MONA) pour appuyer la définition des volumes prélevables dans les aquifères profonds du Nord du Bassin aquitain. Report BRGM/RP-57878-FR
4. Pédron N, Seguin JJ, Capdeville JP (2003) Gestion des eaux souterraines en Région Aquitaine - Développements et maintenance du Modèle Nord-Aquitain de gestion des nappes – Module 4 – Année 1. Report BRGM/RP-52602-FR
5. Pédron N, Platel JP, Lopez B (2005) Gestion des eaux souterraines en Région Aquitaine – Développements et maintenance du Modèle Nord-Aquitain de gestion des nappes – Module 4 – Année 2. Report BRGM/RP-53659-FR
6. Pédron N, Platel JP, Bourgine B (2006) Gestion des eaux souterraines en Région Aquitaine – Développements et maintenance du Modèle Nord-Aquitain de gestion des nappes –Module 4 – Année 3. Report BRGM/RP-53659-FR
7. Pédron N, Platel JP, Bourgine B, Loiseau JB (2008) Gestion des eaux souterraines en Région Aquitaine. – Développements et maintenance du Modèle Nord-Aquitain de gestion des nappes – Module 4 – Année 4. Report BRGM/RP-56614-FR
8. Seguin JJ (2002) Gestion des eaux souterraines en Aquitaine - Actualisation du Modèle Nord-Aquitain. Période 1999–2000. Report BRGM/RP-51758-FR
9. Thiéry D (1990) Software MARTHE, Modelling of aquifers with a rectangular grid in transient state for hydrodynamic calculations of heads and flows, Release 4.3, Report BRGM 4S/EAU n° R32210 & R32548
10. Thiéry D (1993) Modélisation des aquifères complexes - Prise en compte de la zone non saturée et de la salinité. Calcul des intervalles de confiance. Revue Hydrogéologie, 1993, n° 4, pp 325–336
11. Thiéry D, Jacquemet N, Picot-Colbeaux G, Kervévan C, André L, Azaroual M (2009) Validation of MARTHE-REACT coupled surface and groundwater reactive transport code for modeling hydro systems. TOUGH Symposium 2009, San Francisco, CA, 14–16 Sept 2009
12. Xu T, Sonnenthal E, Spycher N, Pruess K (2004) TOUGHREACT User's Guide: A simulation program for non-isothermal multiphase reactive geochemical transport in variably saturated geologic media. Report LBNL-55460, Lawrence Berkeley National Laboratory

Chapter 20
Challenges and Strategies for Managing Water Resources in Morocco Comparative Experiences Around the Mediterranean Sea

Mokhtar Bzioui

Abstract Despite some achievements in the mobilization of water, drinking water, and irrigation, Morocco, a country with much of the territory is subject to a semi arid to arid climate, still faces major challenges for the management of water resources, such as scarcity of water, frequent and long term droughts, floods and flooding exacerbated by inappropriate human settlement, impacts of climate change, overexploitation of groundwater resources and pollution. These challenges are exacerbated by rapid changes in water needs, and the difficulty of following this development with adequate financial means. To maintain a proper balance between supply and demand of water, the Moroccan government has recently adopted a strategy consisting of strengthening water supply, while acting on water demand, and scheduling measures to improve governance of water. A comparison is given for the results of the water policy in Morocco with those of some countries around the Mediterranean.

Keywords Mediterranean sea • Morocco • Water management

20.1 The Water Sector in Morocco

20.1.1 Potential of Water

The potential of natural water resources per capita, which expresses the richness or the relative scarcity of the water of a country, is in Morocco already below the limit of 1,000 m³/inhab/year, commonly adopted as the critical point, indicating the appearance of shortages and latent water crisis.

M. Bzioui (✉)
DRPE – ADH, Rabat, Morocco
e-mail: bziouimo@yahoo.fr

But potential of water in watersheds varies from 180 m³/capita/year watersheds located in the south of Morocco, to 1,850 m³/capita/year for watersheds located in the north, with abundant water [8].

The area in the south and east of the Mediterranean Sea is among the regions that are the poorest in water resources. This is, moreover, exacerbated by the problems of shared water resources that affect the eastern countries of the region.

In Morocco the surface flows of water amount to average year in a few millions m³ for driest watersheds in the south of Morocco (30 Mm³), and of billions m³ for the most favoured watersheds which are located north (more than 1,000 Mm³). These flows generally occur in the form of violent and flash floods. These flows are generally recorded during one period estimated on average from 20 to 30 days for the basins of the south and from 2 to 3 months for the basins of North.

The surface water resources are evaluated in average year at nearly 20 billion m³. The guaranteed resources nine years out of ten or four years out of five are largely lower than this average. In dry year, flow of water can decrease to less than 30% of the average.

20.1.2 Mobilization of Water

The priority given to the sector of water since the sixties made it possible to have a patrimony of hydraulic infrastructures, consisting of 130 high dams with 17 Km³ of storage capacity [2, 3].

Dams play an important role in the hydrous and food safety of Morocco and renders invaluable services to the national economy. They contribute to the energy production, with the development of the access to drinking water, protection against the floods, and the stabilization of the agricultural production thanks to the irrigation of 1.4 million hectares, and to the development of the agro-industry.

Ground water constitutes a significant part of the national hydraulic patrimony. Investigations carried out make it possible to estimate ground water potential, in the 80 identified underground sheets, with nearly four billion m³ per annum which can be considered mobilizable in sustainable way. Now, almost the totality of known renewable ground water is entirely exploited.

All in all, the water resources mobilized are evaluated at nearly 14,000 Mm³ per annum, i.e. 70% of the mobilizable water resources.

20.1.3 Irrigation

Irrigation holds a dominant place in water withdrawal: 11,500 Mm³, out of 14,000 Mm³ used, i.e. 82%, are for the agricultural uses [1, 7].

A long time ago, the Moroccan authorities took initiatives to stabilize, at least partly, the agricultural production by the creation of irrigated perimeters. Great efforts were thus developed during the thirty last years for the development of the irrigation in order to satisfy the food needs, the improvement of the living conditions of the rural populations and the contribution to the development of agricultural exports. Precise objectives have been laid down for more than 40 years for the development of irrigated agriculture. At the beginning of the 1960s a policy of dam construction and equipment of irrigated perimeters was launched, with the objective to reach an irrigated area of a million hectares in the year 2000, whereas only 300,000 ha were, at the time, equipped for the irrigation. Now the irrigated area is estimated at 1,400,000 ha.

20.1.4 Drinking Water

During the two last decades, the sector of drinking water in urban zones profited from a great priority for the mobilization, the production, and the extension of the service [4]. The urban population connected with the distribution network of water has increased from 2.8 to 13.5 million inhabitants during the last 30 years. The rate of connection to the networks of water supply has increased in the cities from 55% to 90% in the same period. It should however be noted that the population of the peripheral zones of the cities is served by terminal fountains. Approximately 10% of the urban population are concerned with this type of water supply but efforts are developed to reduce this proportion, in particular by the practice of social connections.

Water supply in rural area did not have a development as significant as that of the cities. The reason can be attributed to the difficulties related to the dispersion of the habitat, the insufficiency of the public investments, and to the weakness of the institutional framework. In 1992, the rural population provided with drinking water by a public system conceived in an adequate way, controlled and managed suitably, was estimated at nearly 14.3%, and only 6% of this population had particular connections. And these are national averages: the rates were much lower in certain provinces. The investigation carried out in the aforesaid master plan also showed the bad sanitary conditions which characterized the intake points of water: 85% of the intake points did not present the acceptable conditions of drink ability.

Otherwise, it was estimated that in 93% of the cases they are the women and the children who were responsible for the water drudgery, and more particularly the girls who, according to this investigation, were responsible for the water drudgery in 70% of the cases. The hydrous diseases, naturally, were supported by the precarious conditions of supply water in the rural area.

A program entitled PAGER was developed to generalize water supply in rural area. This program was conceived to allow the rural populations, estimated at 12 million in 1995, to be supplied with drinking water within 10 year. Now the access to water for rural population is nearly 80%.

Access to drinking water in countries surrounding the Mediterranean Sea is relatively good compared to other regions of the world. It may be noted the delay of Morocco with regard to rural drinking water; delayed caught up in nearly 10 years.

20.1.5 Sanitation

The access to sanitation can be evaluated by the rate of connection to the sanitation network. This rate currently amounts to 70%. If for sanitation networks the performances are relatively good, those of waste water treatment are, on the other hand, very late: only some waste water treatment plants were built during the twenty last years, and the majority of them do not function correctly. This delay is explained by the failure of the local communities, charged institutionally of sanitation, to support the heavy investments which are necessary, and are not structured to suitably ensure the exploitation of the waste water treatment plants.

But the Moroccan government has recently introduced an ambitious program to alleviate this delay. This situation leads to significant pollution of water resources, as only 10% of water is treated, and the rest is discharged directly into inland waters or at sea.

20.1.6 Hydropower

With 1,200 MW installed capacity, which is nearly 32% of the total power, the plants associated with the dams allow an average energy production moreover 2,000 Million kWh per year when the hydrology is favourable. Saving in fuel, which could be carried out if hydrological conditions are favourable, translated into equivalent fuel importation, is estimated at nearly 700,000 tons of fuel. But the average production carried out during the twenty last years is estimated only at 1,000 million kWh, that is to say the equivalent of 50% of the predicted production. It is the result of the important fluctuations in the incomes of water which characterized hydrology for this period. This situation convinced the planners of the energy sector to reconsider their programs the hydroelectric infrastructures of energy production: during the 20 next years no hydroelectric dams are planned. In Morocco the scarcity of water forces to recourse to energy production only on the event of water surpluses compared to the storage capacity, which results in not holding water equipments for energy.

20.2 Water Management Challenges

The development of water resources in Morocco was done in a relatively satisfactory way during the four last decades: water supply for drinking and irrigated agriculture reached a good level of satisfaction of the needs; in addition, thanks to the mobilization of water by the dams, carried out by anticipation on the dates of saturation of the needs, the long periods of droughts, which prevailed during the two last decades, have not affected significantly drinking water supply for the cities. It is also necessary to underline the important part played by the dams for flood protection.

Difficulties persists however; which will take considerable dimensions, with the risk to compromise the sustainability of water resources development if good measures are not taken, at short time, to find appropriate solutions. These challenges are developed hereafter.

20.2.1 Droughts

During the two last decades which were characterized by prolonged droughts, the pluviometric situation was characterized by a generalized deficit having interested the whole of the country [7]. The effect of these years of drought on hydraulic situation of the basins deeply worsened the deficit of the flow noted since 1970, date of the beginning of the overdrawn cycle observed on the national level. The surface water flows, estimated in average year at nearly 20 billion m^3, were reduced to five billion in 1994–95.

The observation of 50 years of droughts revealed a higher frequency and a more important space extension of the droughts during the twenty last years: five episodes of droughts, on the twelve of the century, were listed during these twenty last years. This made the Moroccan authorities aware of the need for henceforth regarding the droughts as a structural phenomenon and not as a conjunctural one.

20.2.2 Floods and Inundations

These last years, serious floods disturbed in a major way the economic activities of certain regions. Considerable damage of the habitat, the basic infrastructures and the agricultural production was noted in the rural zones. Important damage was also recorded in many urban centres crossed by rivers. The causes of the losses caused by the floods are generally due to:

- The non controlled development of the occupation of the grounds (94% of the zones vulnerable to the floods are concerned with this problem)
- The reduction in the capacity of flow of the wadis by the deposit of cumbersome objects
- The predominance of the sectoral vision in the dimensioning of the works (structures, rain roads and motorways, allotments, pluvial networks)
- Predominance of structural measures works dimensioning (bridges, roads, highways, pluvial networks)
- Curative solutions preferred to preventive ones.

A precise diagnosis was made in 2002/2003, which permits to establish a national plan of struggle against floods. This plan draws up an action plan to treat 500 sites vulnerable to inundations and floods. It is in execution but with a very slow rhythm.

20.2.3 Silting of Dams

Hydrous erosion affects the majority of the basins where the dams are. If this phenomenon finds its origin in physical factors such nature of the ground and its slope, the vegetable cover, and the intensity of precipitations, the human activity accentuates it. The clearing, overgrazing, and the inappropriate farming techniques are, indeed, as many factors which worsen the process of erosion.

It is unfortunately very difficult to find solutions to these aggressions of man on the ground because of the problems of a social nature they raise. The grounds concerned with hydrous erosion are, in majority, located in mountainous zone, generally occupied by poor populations who subsist on activities which create the favourable conditions of erosion. The solutions pass by the limitation, if not the prohibition, of these activities. Alternative solutions to these activities are thus to be found which is not always easy because they are not often accepted by the populations concerned.

It is thus a very complex problem whose solutions must combine the technique and participative approaches with the populations concerned, in addition to considerable financial resources which are necessary to arrange important areas to be treated.

Because of this difficulty the authorities take long to consider an effective programme of works in the upstream basins to protect dams against the silting.

The loss of storage volumes of the dams evolves rapidly: 5% of storage capacity is lost annually, that is to say approximately 70 million m^3, or the equivalent of a storage capacity of dam.

20.2.4 Climate Changes

Attempts at approaches to evaluate the impact of the climatic changes on hydrology were made for the zone of North Africa. They predict a reduction of 20% of the of water flows.

If for the water flows it is possible to establish tendencies, it is difficult to do the same for floods, which are more brutal and space localised.

What is almost certain it is that we are located in a tendency of aggravation of the extreme phenomena.

As well as for the reduction in water flows, the recrudescence of floods is verified. Taking into consideration these new hydrous phenomena becomes necessary for the evaluation and the water management; and one can easily imagine the impact of these phenomena on the conflicts of use of water.

20.2.5 Overexploitation of Underground Water Resources

The ground water resources are overexploited in the quasi totality of the known underground water sheets. The agricultural hydro development combined with the impacts of the droughts observed during the twenty last years, generated an increased overexploitation of the sheets. A generalized fall of the piezometric levels resulted in this. The sheet of Saïss, in particular, showed a fall of 60 m in 20 years.

This generalized and continuous fall of the water levels, observed since the years 1970, risk leading to a clear reduction in the water reserves, the drying up of the sheets, and/or a deterioration of the quality of water by marine intrusion. Overexploitation of ground water resources already puts in danger the economic and social development of certain areas (Souss Massa, Saïss, Temara, Haouz and basins of the South Atlasiques) and can lead to a serious ecological situation by accentuating the desertification.

20.2.6 Pollution

The quality of surface and underground waters is threatened by many pollutions whose principal sources are:

- Water discharges used without treatment of a population of more than 26 million inhabitants (for four million inhabitants it is supposed that the waste water is treated or the waste water have no impact). Nearly 675 million m^3 of urban waste water are currently discharged in the rivers or on the ground without treatment.
- The industrial waste water discharged in the rivers are evaluated to 3,3 million equivalents - inhabitant.
- The annual production of domestic and industrial solid waste is evaluated at approximately 4,700,000 tons. A great part of this waste is put in not controlled discharges, often in or on edge of the beds of the rivers and in zones where the water resource are vulnerable and durably affected.
- Fertilizers and plant health products.
- Accidental discharges of polluting products, in particular with traffic accidents. Since 1987, more than 30 major accidents of vehicles were recorded transporting hydrocarbons in the majority of the cases.

The waste water discharges affect in a significant way the water resources quality. Thus, more than 50% of the stations of control pollution present a water of medium to bad quality.

The growth of the nitrate contents, since more than 10 years, in the sheets near the irrigated agricultural perimeters is certainly one of the most alarming problems.

The example of Tadla aquifers illustrates the extent of the fast progression of pollution by nitrates: almost non-existent there about 15 years ago, this type of pollution affects now more than 50% of the water resources of this water sheet.

20.2.7 Water Conflicts

Despite the relatively good management of water resources in Morocco these water resources become rare due to the natural phenomena, but also due to the human pressure, which contributes to the accentuation of conflicts of water use.

The dimension of these conflicts is still on a level where their resolution is possible thanks to a good organization and effective regulation.

But in the near future the water management will be confronted with great challenges; it will be therefore difficult to surmount these challenges with adopting the same methods of management. That is particularly true for the groundwater resources which are overexploited with a worrying rhythm, which will result in the drying up of the majority of the water sheets, and the appearance of conflicts on broad scale.

20.3 What Strategies to Cope with the Challenges?

Despite a difficult context (a rare and undervalued, threatened by climate change, pollution, and increasing demand), Morocco has managed to meet water demand and to support its socio-economic development with real success through effective management of water sector.

Now, Morocco is undergoing a fundamental transition (transition socio-economic, demographic, climate change), towards a future where structural conditions are very different from those that characterized the recent history of the water sector. To cope with this new context, not only important solutions to mobilize water will be needed but also strong solutions for water management. In the absence of voluntary measures, the shortage of renewable resources in Morocco will be at least 1.9 billion cubic meters in 2030, and groundwater will continue to be overexploited over the whole period. The new strategy adopted by Morocco is therefore based on a policy consisting of measures to manage an ambitious policy of water management: water demand management is in the heart of the economic and social strategy of Morocco. Ambitious programs of conversion of agriculture to drip irrigation are undertaken, and drinking water supply is subject to efforts of renovation performance improving. Solutions at large scale for water mobilization requires time; the demand management is therefore the only lever for improving balance in the transition period. However the preparation of the long term is also taken into account.

The action plan of the water strategy in Morocco has notably the following actions [5]:

– One of the more important action of the water strategy in Morocco is to implement programs to protect groundwater consisting of:
  – Limitation of groundwater pumping in the water sheets
  – Strengthening the system of controls and sanctions in cases of over;
  – Measurement and monitoring of groundwater
  – Strengthening the responsibilities of the basin agencies for good management of groundwater
  – Generalization of water contracts in the areas of water sheets
  – Artificial recharge of groundwater.

20 Water Resources in Morocco

Table 20.1 Losses of water

	Sub region			Total	
Sector	North	East	South	km³/year	%
Agriculture	25	24	46	95	87
Drinking	7	4	3	14	13
Total	32	28	49	109	100

Table 20.2 Losses of water

	Drinking	Irrigation	Industry	
		Hypothesis of performance		
	Net 85%	Net 90%		
Sub region	Users 90%	Plot 80%	Recycling	Total
North	4.60	18.20	9.50	32.30
East	1.80	11.30	2.20	15.30
South	1.60	18.40	4.10	24.10
Total	8.00	48.00	16.00	72.00

- The agriculture uses more than 80% of water resources [6]. Therefore the strategy provides programs for conversion to drip irrigation. To encourage farmers to invest in water saving techniques the Government gives grants of up to 80% for the acquisition of appropriate equipment, requiring funding of about 30–35 billion dirhams.
- Programs to build dams and water transfers from areas with surplus of water to areas with deficit are provided in the strategy. More than 60 high dams and 1,000 small dams will be built until 2030. They will require a very substantial funding (35–55 billion dirhams).
- A program of construction of wastewater treatment plants is in progress, the amount of investment needed to achieve it is estimated at around 40 billion dirhams. Several of these plants will be used to reuse wastewater after treatment.
- Actions to improve governance are also planned in this strategy, with measures to improve the legal framework and measures to reform the organizational framework.
- Table 20.1 shows the importance of water losses for various uses in the Mediterranean region. When we know that often the water-saving actions are less costly than the mobilization of new water resources, one can understand the place accorded by the Southern Mediterranean countries on the water conservation programs in their water strategies.

The following table shows estimation, made by PLAN BLEU, of volumes of water that could be saved with realistic hypothesis (Table 20.2).

20.4 Conclusions

Morocco is a country where water resources are scarce in most parts of its territory, and these water resources and more irregular in space and time. Management of water resources must cope with stress as extreme phenomena (droughts and floods) resulting from climate change. It must also face the reduction of water resources in a context already marked by water stress, and even shortage in major part of the territory. So far Morocco has managed to maintain a balance between offer and demand by focusing on offer management. But, due to scarcity of water resources and population growth, Morocco adopted a new strategy, first consolidating the gains in offer management, but on the other hand, increasing recourse to demand management, and by modernizing governance.

Considering the countries around the Mediterranean Sea, it is noted a marked difference between countries rich in water in the north, and the countries poor in water in the South and East. The latter countries face the same problems as those of Morocco. But whatever their richness of water resources, all the countries around the Mediterranean Sea are involved in water saving policies, which are often more interesting than those who consist to mobilize new resources.

References

1. Administration of the Rural Genius (ARG) (1997) The irrigation in Morocco
2. Bzioui M (2000) Policy and strategies of water management in Morocco. Academy of the Kingdom of Morocco, Rabat
3. Bzioui M (2004) Report on the development of the water resources in Morocco. UN Water Africa
4. Direction de la Recherche et de la Planification de l'Eau (DRPE) (1996) Rural drinking water, Morocco
5. Direction de la Recherche et de la Planification de l'Eau (DRPE) (2004) National plan of water, Morocco
6. Ministry of Agriculture and Rural Development (MARD) (2002) Situation of Moroccan agriculture. MARD
7. Ministry of Public Works (MPW) (1997) The situation on the droughts in Morocco
8. State Secretariat of Water and Environment (SSWE) (2008) Water strategy and action plans of water sector in Morocco

Chapter 21
Environmental Management in Bulgarian Agriculture – Modes, Efficiency, Perspectives

Hrabrin Bachev

Abstract This paper presents the evolution of diverse modes of environmental management in Bulgarian agriculture, and assesses their efficiency and likely prospects of development. First, it analyzes the pace of development and the impact(s) on individual behavior of the major modes of environmental governance – that is: (i) institutional environment (distribution and enforcement of property, user, trading etc. rights and rules); (ii) private and collective modes (diverse private initiatives, and contractual and organizational arrangements); (iii) market modes (various decentralized initiatives governed by "free" market price movements and market competition); (iv) public modes (different forms of Government, community, international etc. intervention). Second, it assesses the impact(s) of dominating systems of governance on the state of environment and identifies the major eco-challenges, conflicts and risks such as an increased competition for natural resources, degradation and contamination of farmland, pollution of surface and groundwater, loss of biodiversity, deterioration of (agro)eco-systems services etc. Third, it projects likely evolution of environmental management in the specific "Bulgarian" economic, institutional and natural environment, and estimates its probable effect(s) on environmental security, and suggests recommendations for institutional modernization and public policies improvement.

Keywords Environmental governance • Market • Private • Public and hybrid modes • Bulgarian agriculture

H. Bachev (✉)
Institute of Agricultural Economics, 125 Tzarigradsko Shose Blvd., Blok 1,
Sofia 1113, Bulgaria
e-mail: hbachev@yahoo.com

21.1 Introduction

There has been a fundamental transformation of Bulgarian agriculture since 1989 when the transition from a centrally planned to a market economy started [4]. New private rights on major natural resources (farmland, forestry, water, eco-system services etc.) have been introduced or restored, markets and trade liberalized, new farming structures evolved, and modern public support and regulations introduced. All that has affected enormously the impact(s) of agricultural on and from the state of environment. Nevertheless, with very few exceptions [2, 3] there are no comprehensive studies on environmental management in Bulgarian agriculture during post-communist transition and EU integration.

The goal of this paper is to present the evolution of diverse modes of environmental management[1] in Bulgarian agriculture, and to assess their efficiency and likely prospects of development.

First, an analysis is made on the pace of development and the impact(s) on individual and collective behavior of diverse modes of environmental governance including:

- Institutional environment ("rule of the game") – that is distribution and enforcement of property, user, trading etc. rights and rules
- Private and collective modes ("private order") – various private initiatives, and contractual and organizational arrangements
- Market modes ("marker order") – diverse decentralized initiatives governed by the "free" market price movements and competition
- Public modes ("public order") – different forms of Government, community, international etc. intervention.

Second, an assessment is made on the impact(s) of dominating system of governance on the state of environment as major eco-challenges, conflicts and risks in Bulgarian agriculture are identified.

Third, a projection is made on likely evolution of environmental management in the specific economic, institutional and natural environment of Bulgarian agriculture, and on the probable effect on environmental security.

Finally, recommendations are suggested for institutional modernization and public policies improvement for effective environmental management in Bulgarian agriculture.

The framework of interdisciplinary New Institutional Economics (combining Economics, Organization, Law, Sociology, Behavioral and Political Sciences) is incorporated into the analysis of Bulgarian agriculture to identify diverse modes and mechanisms of governance, and assess their potential to deal with various environmental challenges and risks.

---

[1] *Environmental management* means management of environment preservation and improvement actions of agents. It requires a system of coordination and stimulation of eco-activity and eco-behavior at different levels –individual, group, community, national, transnational.

21.2 Evolution of Eco-Governance During Transition and EU Integration

21.2.1 Institutional Environment

During most of the transition, the rights on major agrarian resources (such as farmland) and the diverse environmental rights (on clean and aesthetic nature, preservation of natural resources, biodiversity etc.) were not defined or were badly defined [2]. Moreover, inefficient public enforcement of laws, and absolute and contracted rights have been common. All that has negative consequences on the development of farming structures and the efficiency of environmental management. For instance, privatization of agricultural land and non-land assets of ancient public farms took almost 10 years to complete. During a good part of that period, the management of critical resources (farmland, water) was in ineffective and "temporary" structures (organizations under privatization, liquidation or reorganization) with no interests in effective and sustainable exploitation. Furthermore, the short-lease of natural resources and material assets was a major form for farm extension.

Out-dated and sectoral system of public policing, regulations and control dominated until recently, which corresponded little to the contemporary needs of environmental management. There was no modern system for monitoring the status of soil, water, and air quality, and credible information on the extent of environmental degradation was not available.

Neither existed the social awareness of the "concept" of sustainable development nor any "need" to include it in public policy and/or private and community agenda. The lack of culture and knowledge of sustainability has also impeded the evolution of voluntary measures, and private and collective actions (institutions) for effective environmental governance.

Before the EU accession, country's laws, standards and institutions were harmonized with Community Acquis. That introduced a modern framework for the environmental governance including new rights (restrictions) on the protection of environment, integrated territory, water and biodiversity management, polluter pay principle as well as corresponding public institutions for controlling, monitoring and assessment (Executive Environmental Agency, Executive Hydro-melioration Agency etc.).

The EU accession introduces and enforces a "new order" – strict regulations and control; tough quality, food safety, environmental etc. standards; financial support for environmental conservation and market instability etc. The huge European markets are opened which enhances competition and lets Bulgarian farms explore their comparative advantages (low costs, high quality, specificity and purity of produce) as well as give strong incentives for investments in modernization of farms and conforming to higher product, technology and environmental standards.

The external demand, monitoring, pressure and likely sanctions by EU lead to better enforcement of laws and standards. For instance, in 2008, EC blocked payments for Special Assistance Program for Agriculture and Rural Development (SAPARD) because of the considerable mismanagement and corruption. Internal collective actions and social demand for a good governance have also got momentum leading to improvement of public management. Recent success of eco-organizations in putting a ban on genetically modified (GM) crops, the timely reaction against violation of eco-standards in protected zones, and revoking unlawful "exchanges" of valuable public agricultural and forestry lands, all are good examples in that respect.

Nevertheless, a good part of the new "rules of the game" are not well-known or clearly understood by various public authorities, private organizations and individuals. Generally, there is not enough readiness for an effective implementation of the new public order because of the lack of information and experience in agents, adequate administrative capacity, and/or practical possibility for enforcement of novel norms (lack of comprehension, deficient court system, widespread corruption). In many instances, the enforcement of eco-standards is difficult since the costs for detection and penalizing of offenders are very high, or there is no direct links between the performance and the environmental impact. For example, although the burning of fields has been banned for many years, this harmful practice is still widespread in the country. Subsequently, a permanent deterioration of soil quality, wasting of accumulated through photosynthesis soil energy, extermination of soil micro flora and habitats, a significant contribution to green-house gas (GHG) emissions, multiplying instances of forests fires, and a diminished visibility, all come out as a result (EEA, Bulgarian Executive Environment Agency).

Modernization of institutions is also associated with new conflicts between private, collective and public interests. However, the results of public choices have not always been for the advantage of effective environmental management. For instance, the strong lobbying efforts of particular agents have led to a 20% reduction in numbers and a 50% reduction in the area of initially identified sites for the pan-European network for preservation of wild flora, fauna and birds NATURA 2000.

21.2.2 Private and Market Modes

During much of the transition, newly evolving market and private structures have not been efficient in dealing with economic and environmental issues. Most farming activities have been carried out in less efficient and unsustainable structures – public farms, part-time and subsistence farms, production cooperatives, huge business farms based on provisional lease-in contracts (Table 21.1). Furthermore, as many as 97% of newly evolved livestock holdings are miniature "unprofessional farms" breading 96% of the goats, 86% of the sheep, 78% of the cattle, and 60% of the pigs in the country (MAF, Bulgarian Ministry of Agriculture and Food). Moreover, farms adjustments have been associated with a significant decrease in the number of unregistered, cooperative and livestock holdings since 1995, without adequate transfer of land, livestock, and eco-system services management to other structures.

Most farms have had little incentives for long-term investment to enhance productivity and eco-performance. Cooperative's big membership makes individual control on management very difficult. That focuses managerial efforts on current

Table 21.1 Evolution of farming structures in Bulgaria

Indicators	Year	Public farms	Unregistered farms	Cooperatives	Agro-firms	Total
Share in total farms (%)	1989	0.13	99.9			1,602,101
	1995		99.7	0.1	0.1	1,777,000
	2000		99.3	0.4	0.3	760,700
	2007		98.6	0.3	1.1	465,084
Share in total farmland (%)	1989	89.9	10.1			100
	1995	7.2	43.1	37.8	11.9	100
	2000	1.7	19.4	60.6	18.4	100
	2007		32.2	24.7	43.1	100
Average size (ha)	1989	2423.1	0.4			3.6
	1995	338.3	1.3	800	300	2.8
	2000	357.7	0.9	709.9	296.7	4.7
	2007		2.2	613.3	364.4	6.8

Source: National Statistical Institute, Ministry of Agriculture and Food

indicators and gives a great possibility for mismanagement. Since most members are small shareholders, older in age, and non-permanent employees, the incentives for long-term investment for renovation of assets and eco-preservation and improvement have been low.

On the other hand, small-scale and subsistent farms possess insignificant internal capacity for investment and potential to explore economy of scale and scope (big fragmentation, inadequate scale). Besides, there are no incentives for non-productive (eco-conservation) spending due to lack of public control on informal sector. Therefore, primitive and labor incentive technologies, and low compliance with modern agronomic, safety and eco-standards are widespread. Dairy farming is particularly vulnerable since only 1.4% of the holdings with 17% of all cows meet EU standards (MAF).

Finally, larger farms operate mainly on leased land and concentrate on high pay-off investment with a short pay-back period (cereals, industrial). They are most sensitive to market demand and institutional regulations since largely benefit or lose from the timely adaptation to new standards and demand. They also have higher capacity to fund and to adapt to new requirements. Nevertheless, survivor tactics rather than a long-term strategy toward sustainability are common among commercial farms (Fig. 21.1).

Smaller size, owner operating and extensive nature of the majority of farms let avoid certain problems of large public enterprises from the past (over-intensification of production, lost natural landscape and biodiversity, chemical contamination, huge livestock and manure concentration, uncontrolled erosion); revived some local (and more sustainable) technologies, varieties and products; and avert emergence of mad cow disease and bird flue epidemic. Private mode has introduced incentives and possibilities for integral environmental management (including revival of eco- and cultural heritage, anti-pollution, aesthetic, comfort etc. measures; investing in eco-system services, origins, labels) profiting from inter-dependent activities such as

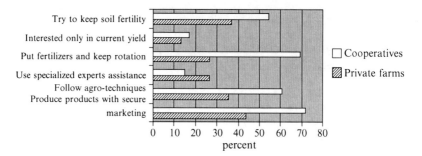

Fig. 21.1 Share of farms implementing different strategies in Bulgaria (percent) (Source: Survey data, 2009)

farming, fishing, agro-tourism, recreation, processing, marketing and trade. There are good examples for private introduction and enforcement of top quality and eco-standards by individual farms (voluntary and trade initiatives), a vertical integrator (dairy and vine processor, food chain, exporter), or direct foreign investor (cereals, oil crops). Private management has been associated with improved environmental stewardship on owned and marketed resources, but less concern to manure and garbage management, over-exploitation of leased and common resources, and contamination of air and water [3].

Since 2001, state irrigation assets have been transferred to newly-evolving Water Users Associations. Around 70 associations have been formally registered servicing 30% of the total equipped for irrigation area. Expected "boom" in efficiency from collective management of irrigation has not materialized because of semi-monopoly situation (terms, pricing) of regional water suppliers, few incentives for water users to innovate facilities and expand irrigation, and uncompleted privatization of state assets. Evolution of farmers and eco-associations has been hampered by the big number of agrarian agents and their diversified interests – different size of ownership and operation, type of farming, preferences, age and horizon.

Market-driven organic farming has emerged and registered a significant growth in recent years but it is restricted to 432 farms, processors and traders, and covers less than 3% of the Utilized Agricultural Area (UAA). There are few livestock farms and apiaries certified for bio-production with highest growth in organic goats and sheep, and a lion share of bees (80%). There are also 242,677 ha approved for gathering wild organic fruits and herbs [7]. Eco-labeling of processed farm products (relying on self-regulation) has also appeared, which is more a part of marketing strategy of certain companies rather than a genuine action for environmental improvement [2].

The organic form has been introduced by business entrepreneurs who managed to organize this new venture, arranging costly certification and marketing of highly specific output. Produced bio fruits, vegetables, essential oil plants, herbs, spices, and honey are predominately for export since only a tiny market for organic products exists in the country. The later is due to the higher prices of products and the limited consumer confidence in the authentic character of products and certification.

21.2.3 Public Modes

During transition public intervention in eco-management was not significant, comprehensive, sustainable, or related to the matter. Eco-policies were fragmented and largely reactive to urgent problems (e.g. floods, storms, drought) with different agencies responsible for individual aspects of natural resources management. In past years a number of national programs have been developed to deal with the specific eco-challenges such as: preservation of biodiversity and environment; limitation of emissions of Sulphur Dioxide, VOC, Ammonia; waste management; development of water sector; combating climate change; developing organic agriculture; management of lands and fights against desertification; agrarian and rural development etc. National monitoring system of environment was also set up and mandatory eco-assessment of public programs introduced. However, bad coordination, gaps, and ineffective enforcement are still typical for the public management.[2]

During the entire transition agrarian long-term credit market was practically blocked while newly evolving farming structures left as one of the least supported in Europe. Aggregate Level of Support to Agriculture was close to zero until 2000, and very low afterward with a small proportion of farms benefiting from the public assistance [1].

Until recently multifunctional role of farming was not recognized, and provision of "environmental service" funded by society. SAPARD measure "Agro-ecology" was not approved by the end 2006, only 201 projects selected, and none funded by the end of 2008. Due to mismanagement SAPARD was suspended by EC in 2008, and a considerable funding lost.

CAP implementation has introduced a considerable support to farming for direct payments, market support, and agrarian and rural development.[3] This amount of resources let more farms get access to public support and fund new essential activities (e.g. commercialization and diversification; introduction of organic farming; maintaining land productivity and biodiversity; agri-environment protection; animal welfare; support for less-favored areas and regions with environmental restrictions; eco-training etc.). Funding for special environmental measures amounts for 27% of the budget of National Plan for Agrarian and Rural Development (NPARD).

There is a mandatory requirement for farms to "keep farmland's good agricultural and environmental status" in order to receive public support. Area-based direct payments also induce farming on abandoned lands and improve environmental situation. However, EU support unevenly benefits different farms as the bulk of public subsidies goes to small number of large farms. Due to the bad design, restricting criteria,[4] complicated and costly procedures, and lack of formal title on

---

[2] e.g. due to organizational and financial reasons Ministry of Water and Environment does not get relevant water information from the institutes of Bulgarian Academy of Sciences.

[3] EU funding is more than five times higher than overall level of support to farming before acceding.

[4] For area-based payments the minimum farm size is 1 ha (for permanent crops 0,5 ha), and for agro-ecological payments 0,5 ha (landless livestock holdings are not-eligible).

Table 21.2 Progression of environmental payments in Bulgarian agriculture

Environmental measures (211, 212, 214)	2007 Farms	2008 Farms	Target (%)	Area (ha)	Target (%)
Less-favored mountainous regions	19,806	20,257	33	237,975	60
Less-favored non-mountainous region	7,273	10,017	–	na	–
Agro-ecology	1,038	1,127	2.6	42,339	26

Source: Ministry of Agriculture and Food

land management, most (small-scale) farms can not participate in public support schemes. For instance, less than 16% of all farms received area based payments and 13% got national top-ups as farms specialized in field crops touch the largest public support [7]. Registered beneficiaries of direct payments with farm's size bigger than 1,000 ha are only 13% but they obtain support for more than 54% of totally subsidized farmland in this group. Similarly, unregistered beneficiaries with farm size smaller than 5 ha are more than 60% but they get payments for merely 9% of supported area in the group.

There has been a considerable progression in implementation of the special environmental measures for less-favorite regions and agro-ecology but it is still bellow the established targets (Table 21.2).

Up to date the level of utilization of funds for agrarian and rural development is merely 10% [7]. Complicated paper work, related high coo-financing and transaction costs, restricting criteria, and huge mismanagement, all are responsible for the slow progress in public support. What is more, all surveys show that many of the specific EU regulations are not well known by the implementing authorities and the majority of farmers. Our recent survey proves that as much as 47% of non-cooperative farms and 43% of cooperatives are still "not aware or only partially aware" with support measures of CAP different from area-based payments. As much as 62% of farms report they will not apply for CAP support due to the "lack of financial resources" (26%), "not compliance with formal requirements" (18%), and "clumsy bureaucratic procedure" (17%).

21.3 Agricultural Impact on Environment

Post-communist development has changed considerably the agricultural pressure and impact on environment.

Market and private governance has led to a sharp decline in all crop (except sunflower) and livestock (but goat) productions[5] while some traditional varieties

---
[5] For potatoes 33%, wheat 50%, corn and burley 60%, tomatoes, Alfalfa hay and table grape 75%, apples 94%, pig meat 82%, cattle meat 77%, sheep and goat meat 72%, poultry meat 51%, cow milk 45%, sheep milk 66%, buffalo milk 59%, wool 85%, eggs 45%, honey 57% (NSI).

and breeds have been recovered. Considerable portion of agricultural lands have been left uncultivated for a long period of time – in some years of transition abandoned land reached one third of the total (MAF). Currently, almost 10% of all agricultural lands are unutilized while fallow land accounts for 9.5% of arable land. Average yields for all major products shrunk to 40–80% of the pre-reform level. The number of cattle has decreased with 61%, pigs with 77%, sheep with 81%, and poultry with 53% (NSI, Bulgarian National Statistical Institute). By 1995 tractors and combines employed in the sector diminished by 54% and augmented to 64% of the 1989 number presently. Now only 5.6% of farms own tractors and 0.7% own harvesters while 40.6% and 30.3% hire or use tractors and harvesters in association (MAF).

All that has relaxed the overall agricultural pressure on environment. However, improper practices also caused erosion and uncontrolled development of some species and suppressing others. Some of the most valuable ecosystems (such as permanent natural and semi-natural grassland) have been severely damaged.[6] Part of the meadows has been left under-grazed or under-mowed, and intrusion of shrubs and trees into the grassland took place. Some of the fertile semi-natural grasslands have been converted to cultivation of crops, vineyards or orchards. This has resulted in an irreversible disappearance of plant species diversity. Meanwhile, certain municipal and state pastures have been degraded by unsustainable use (over-grazing) by private and domestic animals. Reckless collection of valuable wild berries, herbs, flowers, snail, snakes, and fish has led to destruction of some natural habitats. Degrading impacts of agriculture on biodiversity has been significant – all 37 typical animal breeds have been endangered during the last several decades as 6 among them are irreversibly extinct, 12 are almost extinct, 16 are endangered and 3 are potentially endangered [8].

The total amount of chemicals used in agriculture has declined considerably, and now their application per hectare represents merely 22% and 31% of the 1989 level (Fig. 21.2). Currently, N, P and K fertilizers are applied barely for 37.4%, 3.4% and 1.9% of UAA. This trend diminished drastically the pressure on environment and risk of chemical contamination of soils, waters, and farm produce. A good part of farm production has received unintended "organic" character obtaining reputation for products with high quality and safety. Nonetheless, a negative rate of fertilizer compensation of N, P and K intakes dominate being particularly low for phosphorus and potassium. Accordingly, an average of 23595.4 t N, 61033.3 t P_2O_5 and 184392 t K_2O have been irreversibly removed annually from soils since 1990 (MAF). Unbalance of nutrient components has been typical with application of 5.3 times less phosphorus and 6.7 times less potassium with the appropriate rate for the nitrogen used during that period. Moreover, a monoculture or simple rotation has been constantly practiced by large operators concentrating on few crops. All these practices further contributed to deterioration of soil quality and soil organic matter content

Nitrate Vulnerable Zones cover 60% of country's territory and around 7% of UAA. The lack of effective manure storage capacity and sewer systems in majority

---

[6] Approximately 20% of the agricultural lands of Bulgaria are lands of High Nature Value (MAF).

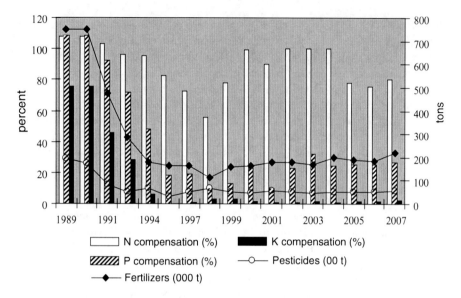

Fig. 21.2 Chemical application and rate of fertilizer compensation in Bulgarian agriculture (Source: National Statistical Institute, Ministry of Agriculture and Food)

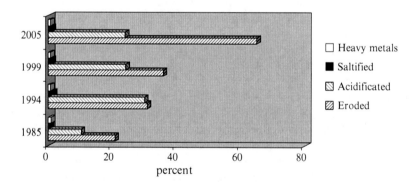

Fig. 21.3 Share of degraded agricultural lands in Bulgaria (Source: Executive Environment Agency)

of farms contributes significantly to the persistence of the problem. Only 0.1% of livestock farms possess safe manure-pile sites, around 81% of them use primitive dunghills, and 116,000 holdings have no facilities at all (MAF). Also decreasing amount of manure has been used for fertilization of merely 0.17% of utilized farmlands in recent years.

There has been a considerable increase in agricultural land affected by acidification (Fig. 21.3) as a result of a long-term application of specific nitrate fertilizers and unbalanced fertilizer application. After 1994 the percentage of acidified soil decreased, but in recent years there has been a reverse tendency along with gradual augmentation of nitrate use. As much as 4.5% of acidified farmlands are with a level harmful for the crops. The fraction of salinized land doubled after 1989 but it is still insignificant part of the total farmland. No effective measures have been taken to normalize soil acidity and salinity throughout the period.

Erosion has been a major factor contributing to land degradation (Fig. 21.3) and its progressing level has been adversely affected by the dominant agro-techniques, deficiency of anti-erosion measures, and uncontrolled deforestation. Around one-third of arable lands are subjected to wind erosion and 70% to water erosion [5]. Since 1990, erosion has affected between 25% and 65% of farmland and total losses varied from 0.2 to 40 t/ha in different years. Soil losses from water erosion depend on cultivation practices and range from 8 t/year for permanent crops to 48 t/year for arable lands. Losses from wind erosion are around 30 t/year and depend on deforestation, uncontrolled pasture, ineffective crop rotation, plowing pastures etc.

Serious environmental challenge has been posed by the inadequate storage and disposal of the expired and prohibited pesticides as 28% of all polluted localities in the country are associated with these dangerous chemicals [5]. Despite progression in management there are still 333 abandoned storehouses in 324 locations for 2,050 t pesticides. Polluted with heavy metals and pesticides soils currently represents bellow 1% of agricultural lands. Re-cultivation of degraded farmlands has been under way, but it accounts for merely 200–250 ha/year [7].

Number of illegal garbage dumps in rural areas has noticeably increased reaching an official figure of 4,000, and farms contribute extensively to waste "production" bringing about air, soil and water pollution [5].

There has been more than 21 folds decline in water used in agriculture comparing to pre-reform level, which contributed to reduction of water stress.[7] In recent years, sector "Agriculture, hunting, forestry and fishery" comprises merely 3.17% of total water use and 0.34% of generated waste waters [9]. Restructuring of farms and agricultural production has been accompanied with a sharp reduction in irrigated farmland and considerable physical distortion of irrigation facilities (Table 21.3). Negative impact of intensive irrigation on overall erosion and salinization diminished significantly after 1990. Nevertheless, primitive irrigation techniques are widespread and augment inefficiency of water use and local soil erosion. Decline in irrigation has also had a direct harmful effect on crop yields and structure of rotation. The level of irrigation depends on humidity of a particular year but has not been effectively used to counterbalance the effect of global warming on farming and degradation of agricultural land.

There has been a considerable amelioration of the quality of ground and underground waters. The nitrate and phosphate content in ground water decreases throughout the transition and currently only 0.7% of samples exceed the Ecological Limit Value (ELV) for nitrate [5]. In drinking water around 5% of analyses show deviation of nitrates up to five times above appropriate level. The later is mostly

---

[7]Bulgaria is among five poorest in water resources European countries and 3d in water availability per capita (EEA). Depending on year's humidity territory accumulates 9–24 billion m^3 water. Since 1990 Water Exploitation Index decline considerably from 55% (2d in Europe) to 33%.

Table 21.3 Evolution and agricultural use of water resources in Bulgaria

Indicators	1988–1992	1993–1997	1998–2002	2003–2007
Total water resources (10^9/m^3/year)	21	21	21	21
Water resources per capita (m^3/inhabitant/year)	2,427	2,562	2,661	2,748
Total water withdrawal (10^9/m^3/year)	14.04	na	8.674	na
Agricultural water withdrawal (10^9/m^3/year)	3.058	0.141	0.144	0.143
Share of agricultural water withdrawal in total (%)	21.78	–	1.66	–
Share of total actual renewable water resources withdrawn by agriculture (%)	14.36	0.66	0.68	0.67
Area equipped for irrigation (1,000 ha)	1,263	789	622	104.6
Share of cultivated area equipped for irrigation (%)	29.17	17.55	17.36	3.18
Area equipped for irrigation actually irrigated (%)	na	5.42	4.96	51.29

Source: FAO, AQUASTAT [6]

restricted to small residential locations but it is also typical for almost 9% of big water collection zones. Improper use of nitrate fertilizers, inappropriate crop and livestock practices, and non-compliance with specific rules for farming in water supply zones, are responsible for that problem.

Monitoring of water for irrigation shows that in 45% of samples, nitrate concentrations exceed the contamination limit value by 2–20 folds [7]. Nitrates are also the most common pollutants in underground water with N levels only slightly exceeding the ecological limit in recent years. Trend for reduction in concentration of pesticides in underground water is reported with occasional cases of triasines over the ELV since 2000.

In recent years utilization of the sludge from purified waste waters was initiated in agriculture and for recultivation of degradated lands comprising accordingly 20% and 7% of the total amount [5].

There has been almost five times reduction of overall GHG emissions from agriculture since 1988 [10]. The N_2O emissions comprise 59% of the total emissions from agriculture and there is a slight enlargement of the share since 2000. Agriculture has been a major ammonia source accounting for two-thirds of the national emission. The majority of NO_2 emissions come from agricultural soils (87%), and manure management and burning of stubble fields (13%). The methane emission from agriculture represents about a quarter of the national. The biggest portion of CH_4 comes from fermentation from domestic livestock (72%) and manure management (24%).

21.4 Prospects of Eco-Management and Policy Recommendations

Deepening the EU integration and CAP implementation will improve the institutional environment for Bulgarian agriculture – the specification and enforcement of various rights and rules, bettering the management of public programs, progression in eco-monitoring and assessment etc.

There will be a significant improvement of sustainability of farming structures as public support will gradually increase (augmentation of area based payments, better utilization of public funds), reach more legitimate beneficiaries, and cover a larger part of farms activity (including eco-management). For instance, according to the plan, the support to unfavorable mountainous regions will cover 60,000 farms and 328,000 ha, agri-ecology measures will involve 40,000 farms with 110,000 ha, area under sustainable use will reach 110,000 ha for maintaining biodiversity and 160,000 ha for improvement of soils quality, contracts for water quality enhancement will expand to 1,000 (MAF).

Experience of EU countries demonstrates that some eco-standards and terms of eco-contracts are very difficult to enforce and dispute. In Bulgaria the compliance rate will be even lower because of the unequal regional capability to introduce and control new rules, ineffective court system, domination of "personal" relations and bribe. Thus, more farms than otherwise would enroll will participate in such schemes (including the biggest polluters and offenders).[8]

Direct costs (lost income) for conforming to requirements of special programs in different farms vary considerably, and they have unequal incentives to participate. Having in mind the voluntary character of most CAP support instruments, the biggest producers of negative impacts (large polluters and non-compliant with eco-standards) will stay outside of these schemes since they have highest eco-enhancement costs. Small contributors will like to join since they do not command great additional costs comparing to supplementary net benefit. Government is less likely to set up high performance standards because of the perceived "insignificant" environmental challenges, a strong internal political pressure from farmers, and possible external problems with EU control (and sanctions) on cross-compliance. Therefore, CAP implementation will probably have a modest positive impact on the environment performance of Bulgarian farms.

There will be evolution and expansion of private and collective modes for environmental management such as voluntary initiatives, codes of professional behavior, eco-contracts and cooperation, quasi or complete integration. Some environmental, infrastructural, and rural development projects requiring large collective actions and coalition of resources will be effectively initiated, coordinated, and carried by the existing cooperatives and business forms. The later will further enhance sustainability of these organizations. Furthermore, some economic or ecological

---

[8]Data for the 1st years of implementation of direct payments demonstrate that the sanctions increase because of violation of eco-requirements and over-statement of land size (MAF).

needs will bring about further changes in farm size, forms and type of eco-governance. For instance, a big interdependency of activities in eco-system services will require concerted actions (cooperation) for achieving a certain effect; asset dependency between livestock manure supplier and nearby organic crop farms will necessitate direct coordination; a high mutual capacity, cite, time of delivery, product specificity etc. dependency between a processor and suppliers will tighten vertical integration; specific needs of drinking water company will justify a private agreement (eco-contract) with farmers etc.

Special governing size or mode is also imposed by the institutional requirements – a minimum scale of activities is set for taking part in public programs like marketing, agri-ecology, organic farming, tradition and cultural heritage; signing a 5 year public eco-contract dictate a long-term lease or purchase of managed land etc. Our recent survey has proved that as much as 41% of non-cooperative farms and 32% of cooperatives are investigating possible membership in professional organizations.

There will be further development of market modes such as organic farming, industry driven eco-initiatives (eco-labeling, standards, professional codes of behavior), protected products and origins, system of fair-trade, production of alternative (wind, manure) energy at farm etc. For instance, significant EU market, lower local costs, and growing national demand create strong incentives for organic and specific productions by the larger enterprises (including joint venture with non-agrarian and foreign capital). According to the plan organic farming will reach 5% of the production and 8% of UAA by 2013 (MAF). Similarly, new incentives for production of bio-fuel and clean energy would induce development of new area of farm activity associated with that new public and market demand.

The process of farms adaptation will be associated with the concentration of natural resource management and the intensification of production. Besides, global climate change would affect severely agricultural development. For instance, by 2030 water availability on more than 50% of the territory will decrease 5–10%, a severe water stress is projected for the South-Eastern parts and a medium water stress in some places of country (EEA). All these will revive or deepen some of the environmental problems unless pro-environmental governance (public order, hybrid mode) is put in place to prevent that from occurring.

Few livestock farms are able to adapt to new EU restrictions, and related reduction of farms and animals and improved manure management will be associated with a drop of environmental burden by formal sector. Besides, newly introduced quota system for cow milk will limit animals increase and direct efforts into less intensive sheep, goat, and buffalo productions.

A few (semi)subsistence farms will undertake market orientation because of the high costs for farm enlargement and adjustment (no entrepreneurial capital and resources available, low investment and training capability of aged farmers, insufficient demand for farm products). The measure "Support to semi-market farms" is having no great effect (insufficient demand, restricting criteria) and its redesign is being considered. For authority it is (technically and politically) impossible to enforce the official standards in the huge informal sector of economy. Thus, massive (semi)subsistence farming with primitive technologies, poor safety, environmental and animal welfare standards will persist in years to come.

Finally, most farm managers have no adequate training and managerial capability, and are old in age with small learning and adaptation potential.[9] The lack of readiness, experiences, and potential for adaptation in public and private sectors alike will require some time lag until "full" implementation of CAP in "Bulgarian" conditions. There will be also significant inequalities in application (and enforcement) of new laws and standards in diverse farms, sub-sectors, and regions of the country.

There is growing interdependency and competition for environmental resources between different industries, social groups, and regions. That will push further overtaking natural resources away from farm management and transfer to urban, transport, industry etc. use. The needs to compete for, share, and sustain natural resources will require a special governance (cooperation, public order, hybrid form) at eco-system, regional, national and transnational scales to reconcile conflicts and coordinate eco-actions.

Having in mind the state, trends and challenges of eco-management in Bulgarian agriculture following policy recommendations can be suggested:

First, environmental policy is to be better integrated in the overall and agrarian and rural development policies while the effective design and enforcement of environmental measures are to get a high priority. Presently most public efforts are put on addressing urgent socio-economic problems while improvement of eco-management is perceived as less important. Accordingly, no measures are taken to mitigate or prevent various environment related risks in agriculture (e.g. likely negative impacts from climate change). Furthermore, there is to be more stability and certainty in eco-policy (a long-term public commitment) in order to induce effective private and collective actions. One of the major reasons for low investments in green energy production has been frequent changes and uncertainty about the long-term policy development in that new area.

Second, proclaimed integral approach of soil, water and biodiversity management is to be completely applied in planning, funding, management, monitoring, controlling and assessment of sustainable use of resources and wastes. Moreover, it is to be extended to integral management of all natural resources in a particular eco-system, territory, and region as all stakeholders (community leaders, farmers, businesses, residents, interest groups, consumers) are to be involved in the decision-making process. Individual elements, aspects and responsibilities of eco-management are usually divided between various agents with poor coordination, conflicting interests, and inconsistency, controversies, gaps and inefficiency of actions. Furthermore, neglected eco-system, eco-system services, life-cycle, environmental accounts, digital water trade, and other modern approaches are to be incorporated into the design of public intervention and program management.

---

[9] Average age of the farm managers is 61 as 70% of them are older than 55 (MAF).

Third, property, user, management, trading, discharge etc. rights on natural resources, eco-system services and wastes are to be better defined, and further privatized, collectivized and regulated as in the case of irrigation water, provision of environmental preservation and eco-system services, supply of renewable energy, (nitrate, GHG) emissions and waste discharge and trade etc. Furthermore, a greater range of diverse public instruments is to be used including appropriate pricing, quotas, public support, taxing, interlinking etc. Prospective incentives (including funding of eco-actions and taxation on overuse of eco-resources) are to be introduced to prevent the over-intensification and support farms adaptation.

Fourth, adequate and internationally comparable environmental data collection, and independent assessment of driving forces, pressure, impacts and responses are to be organizationally and financially secured. There are insufficient data on environment in general, and agricultural linkages with (contribution to, affection by) the state of environment – soil, water and air contamination; waste production and decomposition; total social costs, energy intensity, eco- and water-foot print, benefits from agricultural production; effect on environmental conservation and improvement; renewable energy production; impacts of climate change; existing and likely risks etc. For instance, Bulgaria is second worst in Europe in Total Energy Intensity of GDP and improvement of that indicator is very important. Also mechanisms for comprehensive and timely disclosure of eco-information are to be assured, and effective methods for communication to decision-makers and stakeholders at all levels and public at large introduced.

Fifth, different CAP instruments are to be better adapted to the specific conditions (needs) of Bulgarian agriculture such as: immense small-scale and subsistent farming, domination of small-size livestock holdings, fragmented and dispersed farmlands, low adaptability of farms, domination of tight horizontally and vertically integrated forms in certain sub-sectors, ineffective administration etc. Accordingly, public support to prospective business and non-for profit ventures as well as informal modes complying with size, environmental etc. requirements is to be given; direct payments to (landless) livestock farms is to be institutionalized; access to EU support (area-based payments) of public organizations such as research institutes and universities given; application of the EU criteria for supporting "Semi-market" farms ("enormous" for Bulgaria size of 2,400 Euro) and "Young farmers" is to be reconsidered; funding from unpopular measure "Support to semi-market farms" re-directed to perspective measures like "Young farmers" and "Modernization of farms"; support to restoration of abandoned farmland and organic livestock farming to be introduced (currently only organic forage production is supported); support to eco-innovation, farm adaptation, and mitigation of existing and likely risks provided etc.

Sixth, more hybrid (public-private, public-collective) modes are to be employed given coordination, incentives, and control advantages. (Pure) public organization, funding and enforcement of most environmental, animal welfare, biodiversity etc. standards are very difficult or impossible at all. It is particularly truth for the huge informal sectors of the economy and remote areas of the country. Individual "punishments" often do not work well while overall damages from the incompliance are immense. Policies is to be oriented to market orientation of subsistence farms, support and incentives for diverse (including new, specific, not-traditional) private and collective eco-modes, and eco-programs for informal farms, groups and other ventures. Public support to voluntary environmental initiatives of farmers, rural and community organizations (informing, training, assisting, funding) and assistance in cooperation at grass-root, eco-system, trans-regional and trans-border level will be more efficient in terms of incentive, coordination, enforcement, and disputing costs. Practical involvement of farmers and other stakeholders in priority setting, management, assessment and disputing of public programs and regulations at all levels is to be institutionalized in order to decrease information asymmetry and possibility for opportunism, diminish costs for coordination, implementation and control, and increase the overall efficiency and impact.

Seventh, a special attention is to be given to improvement of agrarian and environmental education and training of students, farmers, administrators, rural entrepreneurs and residents, and consumers. That will require a fundamental modernization of education system and the National Agricultural Advisory Service. The later is to be re-oriented to farmers and rural agents (rather than bureaucracy) needs; reach all agrarian and rural agents though introduction of effective methods of education, advice and information (TV, radio and on line information; demonstration; sharing experiences) suited to the specific needs and capability of different type farmers (prospective, young, semi-market, business) and rural agents; establish a system of continues (life-long) rather than occasional training; include environmental and waste management, and rural development issues; cooperate closely with research institutes, universities, and private and collective organizations; involve farmers and other stakeholders in management, implementation and assessment of its program etc.

Eighth, modernization of eco-management will not be achieved without a significant improvement of the overall institutional environment and public governance – perfection of law and contract enforcement system, combating mismanagement and corruption in public sector, removing restrictions for effective market, private and collective initiatives etc.

Nineth, more support is to be given to multidisciplinary and interdisciplinary research on various aspects and impacts of environmental management, including on factors and forms of eco-governance and their impact on individual and collective eco-behavior and eco-security. Currently, uni-disciplinary approach dominates, and efforts of researchers in Ecology, Technology, Economics, Law, Sociology, Behavioral and Political Sciences are rarely united; most studies are focused on the governance of individual (economic or social or environmental) aspect of sustainability, or on formal modes and mechanisms; they are typically restricted to a certain form (contract, cooperative, industry initiative, public program), or management level (farm, eco-system), or particular location (region); uni-sectoral analyses are broadly used separating the governance of farming from the governance of overall households and rural activities. "Normative" (to some ideal or the model in other countries) rather than comparative institutional approach between feasible alternatives is broadly employed, and significant social costs associated with the governance (the transaction costs) are ignored. Consequently, understanding on institutional, cultural, economic, behavioral, technological, ecological, international etc. factors of agrarian sustainability and security is impeded – spectrum of feasible formal, informal, market, private, public, integral, multilateral etc. modes of governance can not be identified, and the efficiency, potential, complementarities, and prospects of existing and other possible (including imported) modes of management assessed. All these restrict possibilities to assist public policies and the effective design of public intervention, and support individual, collective and business actions for sustainable development.

References

1. Bachev H (2007) National policies related to farming structures and sustainability in Bulgaria. In: Cristoiu A, Ratinger T, Paloma SGy (eds) Sustainability of the farming systems: global issues, modeling approaches and policy implications. EU JRC IPTS, Seville, pp 177–196
2. Bachev H (2008) Management of environmental challenges and sustainability of Bulgarian agriculture. In: Liotta P, Mouat D, Kepner W, Lancaster J (eds) Environmental challenges and human security: recognizing and acting on hazard impacts. Springer, Dordrecht, pp 117–142
3. Bachev H (2009) Governing of agro-ecosystem services. Modes, efficiency, perspectives. VDM Verlag Dr.Muller Aktiengesellscaft & Co. KG, Saarbrucken
4. Bachev H (2010) Governance of agrarian sustainability. Nova Science, New York
5. EEA (Various years) Annual State of the Environment Report. Executive Environment Agency, Sofia
6. FAO (2010) AQUASTAT. Food and Agriculture Organization, Rome
7. MAF (Various years) Agrarian paper. Ministry of Agriculture and Food, Sofia
8. MEW (Various years) Official papers. Ministry of Environment and Water, Sofia
9. NSI (Various years) Statistical book. National Statistical Institute, Sofia
10. Vassilev Hr, Christov C, Hristova V, Neshev B (2007) Greenhouse gas emissions in republic of Bulgaria 1988, 1990–2005, national inventory report 2005. MEW, Sofia

Chapter 22
Autonomous Desalination and Cooperation. The Experience in Morocco Within the ADIRA Project

Vicente J. Subiela and Baltasar Peñate

Abstract Fresh water supply in the world, particularly in developing countries, is becoming a more and more challenging problem and affects many multidisciplinary aspects, such as security, health, development, economics and environment. The increment of population, the climate change and the environmental impacts on the water resources are generating a progressive reduction in the per capita drinking water availability, particularly in developing countries of Africa and Middle East. As a contribution to solve this situation, desalination technologies have been producing fresh water supply for more than five decades; the current capacity of desalinated water technology installed worldwide is over 45 million of daily cubic meters. Nevertheless, this solution has its own disadvantages; one of the main associated problems of these "water factories" is the requirement of energy (heat or electricity) for the desalination processes. This inconvenience can be solved currently for small water demands by the use of renewable energies, as solar or wind energy. The Instituto Tecnológico de Canarias (ITC) has been researching on desalination powered by renewable energies since 1996, installing, operating and testing ten different combinations of renewable energy technologies and desalination processes. Moreover, the ITC has installed five units (in real use, not demonstrative) operating in Africa: one in Tunisia (2006) and four in Morocco (2009). Concerning the environmental security of this kind of installations, there are three main topics to be discussed: (i) gaseous emissions – solar/wind powered desalination is a pollution free system during operation, since there is no need of fuel; (ii) brine disposal – rejected stream of water with high concentration of salts has to be properly treated to avoid a local environmental impact, especially in inland locations; (iii) solid wastes – the

---

V.J. Subiela (✉) • B. Peñate
Water Department, Research and Technological Development Division,
Instituto Tecnológico de Canarias, ITC, Santa Cruz de Tenerife,
Playa de Pozo Izquierdo, s/n 35119 Sta. Lucía (Las Palmas), Spain
e-mail: vsubiela@itccanarias.org

long term consumables of the system, as membrane modules, filters or batteries, have to be disposed in an appropriate area. As an example, this paper deals with the experience of desalting with solar PV energy in Morocco and points out not only the technical aspects of the implementation of this kind of cooperation projects, but also economic, environmental and social aspects.

Keywords Autonomous desalination • Reverse osmosis • Solar photovoltaic energy • Remote areas • Morocco

22.1 Introduction

Water, the most important natural resource for life, is becoming scarcer and scarcer. It is not casual that United Nations has declared this decade as the water period [20]. Moreover, this progressive reduction in drinking water availability is a specific problem that affects many multidisciplinary aspects, such as security, health, development, economics and environment.

The situation is particularly serious in developing countries of North Africa and Middle East wherein the predictions are most critical (see Fig. 22.1).

According to the WHO, child mortality related to lack of water quality or quantity is more than 1.4 million per year [21].

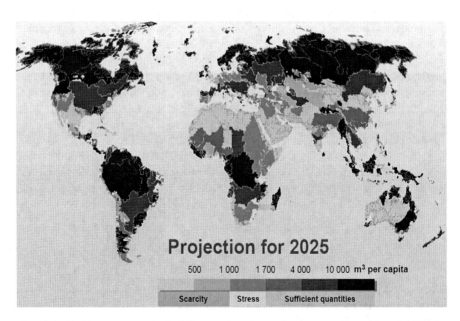

Fig. 22.1 World map that marks in different colors the predicted per capita water availability for the year 2025 [15]

The problem of water supply in dry areas boosted the development of desalination technology, especially in Middle East, wherein most of the desalinated water is produced.[1] Furthermore it has become a very significant industrial activity with a progressive growth; according to the 22nd IDA (International Desalination Association) inventory, the cumulative contracted capacity of desalination plants was 68.5 million of daily cubic meters [9].

Nevertheless, this solution has an important disadvantage: the high energy consumption, as heat for distillation processes, or electricity for membrane processes, like reverse osmosis or electrodialysis. Heat and electricity are mostly generated from fossil fuels, thus the main environmental consequence is the impact in terms of CO_2 emissions; as a rough reference, each cubic meter of desalted water from a sweater RO plant means a generation of 3 kg of CO_2.

Desalination increases energy demand, so it means a higher external dependence and the consequent economic expense in those countries with low energy resources.

Desalination by renewable energies solves both cons: on the one hand it is a pollution-free system; on the other hand, it uses a local energy source as solar radiation or wind energy to power the desalination plant.

This paper presents the idea of autonomous desalination and the experience of four solar driven desalination systems installed in Morocco, within the framework of the ADIRA project (Autonomous Desalination Installations in Rural Areas).[2] There are more details in the website of the project (www.adira.info).

The following text includes not only technical, but also economic, environmental, and social issues.

22.2 The Autonomous Desalination Concept

22.2.1 Generalities

A desalination process is basically a separation process in which a feed stream of salty water is converted into two flows: product or desalinated water, and brine or rejected water. This process is possible thanks to an external energy supply (see Fig. 22.2) in terms of electricity or heat. The amount of energy requirement mainly depends on the type of process and salinity of feed water; the more salinity, the more energy consumption.

If the energy comes from a renewable energy system, then the desalination unit operates off-grid or without external energy supply; in other words, the desalination plant operates in an autonomous mode. There are several possible combinations of desalination processes and renewable energy sources; according to the currently installed combinations, the most used concept is reverse osmosis coupled to solar photovoltaic energy (see Fig. 22.3). The Fig. 22.4 shows a basic diagram of this concept.

---

[1] www.desaldata.com
[2] The ADIRA Project was co-funded by the European Commission, Program Meda-Water.

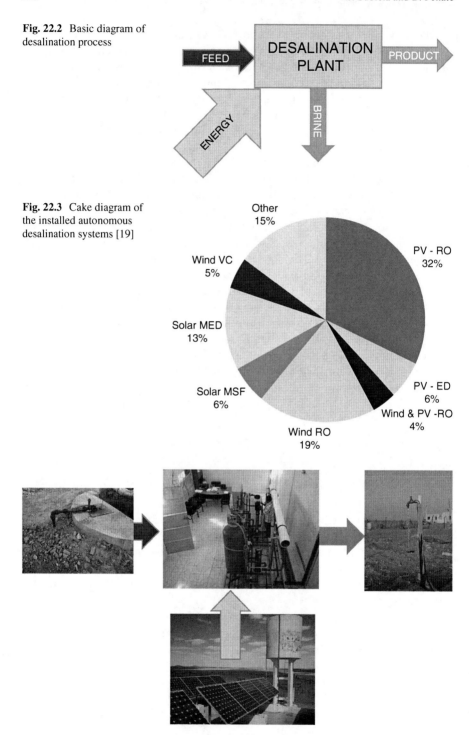

Fig. 22.2 Basic diagram of desalination process

Fig. 22.3 Cake diagram of the installed autonomous desalination systems [19]

Fig. 22.4 Basic diagram and pictures of the main components of a PV-RO unit (feed water well, RO unit, PV system, and product water supply)

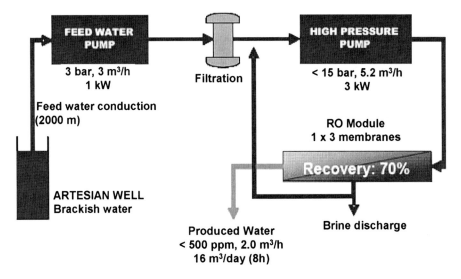

Fig. 22.5 Basic hydraulic diagram of a RO unit [14]

This idea of matching renewable energy resources and desalination is not new, but it is in a progressive development. For further information, it is recommended to consult some of the very interesting published documents on the state of the art of renewable energy powered desalination [4–6, 8, 10, 11, 13]. Moreover, there are several ongoing interesting initiatives focused on this topic [12, 16] and also finished fruitful actions with helpful documentation to be consulted [2].

It is relevant to mention the activities of the ITC, with a long experience in this field (since 1997) and ten different installed and operated autonomous desalination systems [18]; the oldest operating system, since May 2006, is the PVRO unit installed in the inland village of Ksar Ghilène (Tunisia), with a nominal capacity of 2 m^3/h [14].

22.2.2 The PVRO Combination

The combination of photovoltaic energy (PV) and reverse osmosis (RO) is composed of two main parts:

- *Hydraulic part*: a pump in a well boosts the feed water to the RO unit, once there it is firstly filtered and then pressured up and passed through the RO modules or membranes, wherein the reverse osmosis process takes place, and feed flow is separated into two outgoing streams: product water and brine. When the feed flow is brackish water, less saline than seawater, then part of rejected flow can be recirculated (see Fig. 22.5). Produced water is normally chlorinated for its preservation and stored for distribution. Brine must be disposed or treated properly; see details in Sect. 3.6.
- *Electric part*: solar radiation received by the PV panels is converted into DC electricity, then it is stored through a charge controller in a set of batteries and

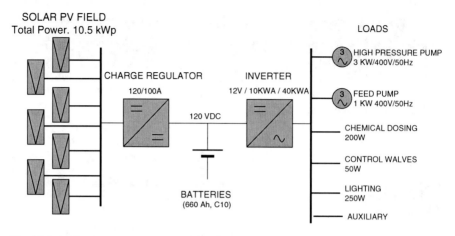

Fig. 22.6 Basic electric diagram of a PVRO system [14]

later transformed into AC electricity by an inverter. This DC/AC converter supplies the electric power to the loads of the system: feed water pump, RO unit, lighting, and other small consumptions (see Fig. 22.6).

22.3 The Experience in Morocco

22.3.1 Generalities

The experience on autonomous desalination in Morocco is part of a wide set of actions developed within the project ADIRA (www.adira.info), under the framework of the EU Program MEDA-Water. The partnership was composed by EU countries (Greece, Germany, Spain) and Mediterranean countries (Morocco, Turkey, Jordan, Egypt).

The main objective of ADIRA project was the analysis of the local reality of certain target countries (Morocco, Jordan, Egypt, Turkey and Cyprus) in order to install and operate autonomous desalination systems in rural areas. The project included complementary elements as the development of decision support tools: a simulation software and a handbook and a collection of useful information in data bases: installed systems, suppliers, experts and so on. Details of the project, recommended references and actions in the target countries can be found in the website [1].

22.3.2 Reality of Morocco. Basic Information

Morocco is a country located in one of the water scarcity areas of the world (see Fig. 22.1). It is particularly interesting for autonomous desalination due to the following reasons:

- There is a high percent of population living in rural areas (80–90%), that is generally much disseminated [17].
- Good solar radiation is generally available.
- Most of the villages in the south half of the country lack of access to electricity and high quality water.
- There is a large amount of salty waters (seawater and underground brackish waters).

One of the first ADIRA activities was a regional analysis to identify the most favorable Moroccan regions for the installation of autonomous desalination systems [17].

As a second step, a similar analysis was developed at municipal level to select the most favorable sites for the installation of the systems. Firstly, a questionnaire was prepared to collect the main data from the potential sites; then a set of criteria, including technical, social, environmental and geographical aspects were used in a process to take the decision of the final locations. A theoretical description of the recommended selection process can be consulted in Banat et al. [3].

At the end, six villages were selected for the installation of the systems:

- 2 in the province of Marrakech, process assumed by the Moroccan ADIRA partner (FM21[3])
- 2 in the province of Essaouira, assumed by ITC
- 2 in the province of Tiznit (currently the sites belong to the recently created province of Sidi Ifni), assumed by ITCThe map in Fig. 22.7 presents the location of the villages for the ITC installations.

22.3.3 *Technical Issues*

After the selection of the sites the technical process starts according to the following steps:

- Technical visits and analysis of local data
- Definition of technical specifications adapted to local conditions
- Tender of call
- Contract
- Engineering
- Manufacturing and collection of equipment
- Transport and delivery of equipment
- Installation
- Commissioning
- Operation.

This process lasted about 2 years. Nevertheless, the project did not finish here; afterwards, suitable follow-up and local actions are required to create the basic local

---

[3]Foundation Marrakech, NGO linked to the Semlalia University focused on health and environmental issues.

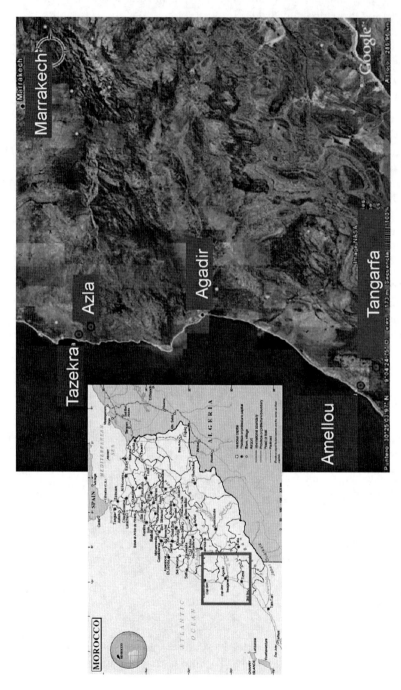

Fig. 22.7 Map with the location of the four sites wherein the PVRO autonomous desalination systems are installed (Tazekra and Azla in the province of Essaouira; Amellou and Tangarfa in the recently created province of Ifni, belonged to Tiznit before this administrative change)

Fig. 22.8 Process of the building construction in Amellou

structure and external support to guarantee the sustainability of the system at long term, by implementing the required maintenance actions.

The enclosed sequence of photos (Fig. 22.8) presents the process for the village of Amellou.

The four autonomous desalination systems installed in Morocco are based on the ITC international patent DESSOL®.[4] The units are composed by the following main elements:

- Monitoring system
- Control system.
- Complementary installations:
  - Feed water pumping
  - Hydraulic installation and piping connections
  - Brine disposal
  - Electric installation.
- Civil works: building and evaporation pond.

The enclosed Fig. 22.9 shows a selection of photos of the main components of the system.

The final view of the four systems is presented in the Fig. 22.10.

A selection of the most remarkable operational parameters is given in the chart presented in Fig. 22.11. The solar radiation (W/m^2), the output power of the inverter (W), and the produced flow (l/h) are shown. The right axis indicates the values for the pressure before the membranes. The RO unit is able to operate in two power operational points, in order to adapt the load to the available power from the solar field. This change happens at 15:30 and affects the analyzed parameters.

---

[4]PCT ES2004/000568.

Fig. 22.9 Photos of the main components of an autonomous desalination system: *1.* PV field on the roof of the building, *2.* RO unit with the horizontal pressure vessels, *3.* Control panel and electric converters, *4.* Set of batteries

Fig. 22.10 General view of the four autonomous desalination systems installed in Morocco: *1.* Case of Amellou, *2.* Case of Tazekra, *3.* Case of Tangarfa, *4.* Case of Azla

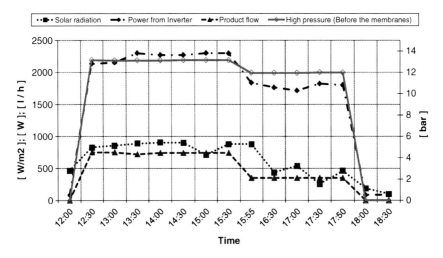

Fig. 22.11 Daily evolution of parameters (2008, PVRO unit operating in Tazekra)

22.3.4 Economic Issues

The complete installation of the four autonomous desalination systems was possible thanks to the economic contribution of the following entities:

- The European Commission, through the Program MEDA-Water
- The Canary Islands Cooperation
- The Governments of the two Moroccan provinces: Essaouira and Tiznit
- ITC.

This shared funding allowed covering all the required capital expenses to carry out the project. The local economic contribution meant a higher involvement of the Moroccan authorities.

The distribution of investment costs is presented in the Fig. 22.12. As a reference parameter, the specific capital cost to install a complete system is estimated in about 4,100 € per nominal daily cubic meter.

Concerning the operation and maintenance costs, a set of elements were considered: the manpower, the consumables, a replacement fund and diesel for feed water pumping. The O&M costs must be covered on a long term basis, to guarantee the sustainability of the project; therefore, it is necessary an external fund. It would be ideal that the final users of water assumed part of the running costs according to their economic possibilities, however, this option has not been considered for the moment and the water is given for free.

The total cost of water, considering capital and running costs, is estimated in 5.45 €/m^3. This value is quite high in comparison with the common costs of desalinated water from conventional seawater RO units, that lies in the range 0.5–1.32 $US/m^3 [7]; actually the difference would be even higher if the autonomous system used

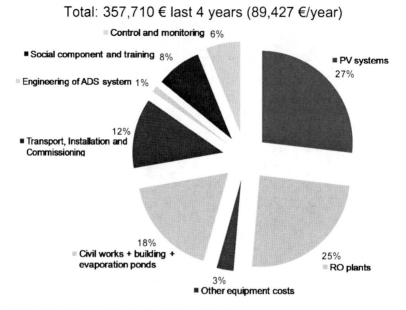

Fig. 22.12 Cake diagram with the distribution of the investment costs

seawater as feed flow instead of brackish water. A seawater PV-RO system requires higher quality materials in the RO unit components and a larger PV area to cover a more intensive power demand, since the more salinity in the feed water, the more operation pressure, and consequently, the more energy is demanded.

So the obvious question is *"Does it make sense to consider autonomous desalination with so high costs?"*. The low costs of conventional desalination are due to two main factors: the scale factor and the connection to a constant and cheap source of electricity. But this situation will probably change in a short term, when oil prices rise and rise. On the other hand, there are remote areas wherein the cost of conventional water and energy supply is higher than the costs of a renewable energy powered desalination option. Thus, there is a current real commercial market for autonomous desalination.

22.3.5 Social Issues

As any international cooperation project focused on local development, the social aspects are one of the key factors for the success of the project. In some situations, social issues are even more important than technical ones, and require special attention.

The social issues are all those elements related to the actors of the project and their specific role and responsibility along the activity. The Fig. 22.13 shows a basic

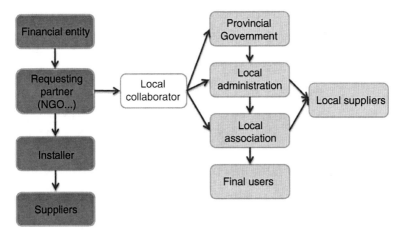

Fig. 22.13 Diagram of the grid of actors

diagram with the grid of the actors involved in this project. The dark boxes represent the external partners and the clear ones represent the local partners.

The intermediate actor or local collaborator (white box) is the most critical one since has the role of contacting directly to the local actors to receive their necessities and visions and also to transmit the points of view of the external partners. The intermediate actor must know the local reality, from the cultural, socioeconomic, and political point of view. This is only possible if this entity has a minimum previous experience in the target country.

In the case of the experience of the ITC within the ADIRA project, different companies assumed this role, with different actions in the main stages of the project:

– Beginning of the project: Contacts with the local authorities for the analysis and selection of the potential sites
– Before the installation of the systems: Negotiation of the content and further signature of the agreements between the local authorities and the ITC to close the commitments of both parts along the whole project.
– During the installation: follow-up of the local process and first phase of the local training
– After the installation: negotiation of the content and further signature of the ownership transfer agreements; a mandatory step established by the EC, by which the final owner was the municipality wherein each system is installed
– Further operation (1 year after the commissioning): follow up of the operation of the systems, identification and correction of failures, elaboration of a field study to identify potential sites to replicate the project.

Concerning the role of the local actors, it was necessary to include them from the initial steps of the project so that to take into account their opinions and suggestions.

More specifically, the involvement of Moroccan actors during the ADIRA project could be summarized as follows:

- Provincial authorities (Essaouira and Tiznit provinces): they had a small but very relevant participation, since all the important decisions concerning the local commitments had to be approved by these entities previously. On the other hand, it was very valuable the involvement of the technical staff of both provincial governments
- Municipal authorities (four entities, one for each site): they assumed the follow-up of the responsibilities of the local suppliers, in collaboration with the local associations
- The local associations: they were the representation of the final users and informed them about the project
- The local suppliers: part of the capital costs were expended at local level, with the associated benefit to the local economy. In this project the civil works, the buildings and the evaporation ponds, were constructed by local companies
- The final users: despite there was no almost direct contact, but through the local associations, they were an essential actor at practical level since they decided with their behavior and real acceptance how good was the water quality and how interesting was the installed system for them.

22.3.6 Environmental Issues

Solar energy is a pollution-free and also a local recourse: avoids CO_2 emissions and does not mean external energy dependence. This is the main point in providing high quality fresh water by the use of a renewable energy resource within autonomous desalination systems.

Nevertheless, there are other environmental aspects related with the systems: the appropriate disposal of the rejected water, or high salt concentration outgoing flow, called brine, and the management of the used elements, which become wastes after their lifetime (filling of filters, old membranes, damaged or used batteries, which cannot be correctly recharged any longer, and other small materials).

Concerning the inland brine disposal, there are the following alternatives:

- Mixing with feed water and use for watering special crops adapted to high salinity soils, as the case of some olive tree varieties
- Disposal in a filtering well as long as there is no underground connection that could pollute the feed water well
- Disposal in an evaporation pond. This option requires an important surface, and it depends strongly on the local solar radiation, humidity and wind speed
- Disposal in another close location, as wetlands, provided that it is guaranteed the capacity of that ecosystem to receive the brine flow.

Regarding the consumable materials, it will be necessary to find a suitable place or landfill for the final disposal, or a suitable management assumed by the suppliers.

22.4 Main Results

After a long and intensive period of work, full of many different actions in order to overcome difficulties, the four systems started to operate normally along 2009.

The main facts can be summarized as follows:

- Four solar autonomous PV-RO units operating and producing high water quality (concentrations in terms of dissolved salts):
  - unit in Amellou: 500 ppm
  - unit in Tangarfa: 330 ppm
  - unit in Azla: 340 ppm
  - unit in Tazekra: 630 ppm
- The solar photovoltaic generation for desalinated water production avoids the emission of 0.7 kg CO_2 per cubic meter of produced water.[5]
- About 1,200 persons are provided with high quality water (estimated per capita consumption rate of 20 l/day)
- Four local people trained in the operation and maintenance of the system
- Ownership of the four systems transferred totally to the municipal authorities
- Identification of two villages to replicate the project.

22.5 Lessons Learned and Recommendations

The ADIRA project and the particular experience of the ITC in Morocco has been a set of multi-disciplinary activities, including technical, social, economic, and environmental aspects. This section points out the extracted conclusions and the most recommendable indications for future similar actions, synthesized in the form of bulleted lists.

22.5.1 Technical Issues

– Complete identification of the potential sites and further careful selection process of the final location, taking into account the points of view from the local actors
– Adaptation of the concept and the design of the system, and also the rest of the elements of the technical solution to the local conditions, and not the opposite, i.e., tailor made solutions. Special attention should be paid to the type, quality and available flow of raw water

---

[5]Considering that the average specific energy requirement of the autonomous desalination systems is 0.76 kWh/m³, and that 1 kWh from a conventional thermal power plant produces 0.9 kg of CO_2

- The concept of a photovoltaic system coupled to a reverse osmosis unit is the most used in autonomous desalination since both are mature technologies and there is a long experience; nevertheless there are other technology concepts that can be considered as alternatives in the initial part of the technical study
- Reconsideration of the initially planned technical solution if it is necessary after the results of the field visits. The most appropriate final solution can be even quite different; desalination by RE sources is an option and not the best one necessarily
- Preparation of a good control strategy and system addressed to minimize the energy consumption, maximize the water production and increase the lifetime of the components
- Selection of high quality and tough materials in all the equipment, despite a higher investment cost
- Inclusion of a specific training phase addressed to the local technicians who will assume the local follow-up actions: coordination, preventive maintenance, corrective maintenance
- In the case of PV-RO systems, it is strongly recommended the use of batteries; if feed flow is seawater, then the RO plant should include an energy recovery system.

22.5.2 Economic Issues

- This kind of projects require an important initial investment, thus it is normally necessary to request the economic support from different financing entities
- It is very important the economic compromise of the local authorities, as part of their participation, and according to their real economic availability
- Contracting of part of the services and supplies to local providers
- Definition, in collaboration with the local partners, of a strategy or process for the economic sustainability of the project at long term in order to guarantee the assumption of the running costs along the whole life of the system. This requires an external funding, but as in the case of the capital expenses, it is very recommendable to reach an agreement for the local economic contribution. Ideally the system should be under a complete self-management.

22.5.3 Social Issues

- Involvement of all the local actors from the starting point of the project, in parallel with the technical process. A cooperation project means an activity in which two parties operate in coordination or "co-operate"

- Identification of an agreed process with the local actors to set the different stages of the system and to share the different responsibilities of the involved actors accordingly
- Transmission of an appropriate vision in the final users: the system should not be considered as just a present, but part as a local useful service that provides a required resource
- Consideration of the cultural, political and administrative aspects of the target site as the local framework wherein the project will be developed
- The agreements with the local actors are recommended but do not mean necessarily its complete fulfillment. In other words, the signed documents must be complemented with close and continuous contacts among the parties.
- Selection of a known and experimented entity to assume the role of the intermediate actor.

22.5.4   Environmental Issues

- Evaluation of the different options for the brine disposal and selection of the best one according to the local characteristics of the location of the system and considering the opinions of the local partners
- Evaluation of the positive environmental impact linked to the installation of a renewable energy driven desalination plant considering at least two facts: integration of a clean energy option and provision of a high quality water
- Evaluation of the impact related to the civil works
- In case of inland locations wherein the well has a limited and variable flow, assessment of the possibilities, related to costs and environmental impact, of re-digging the borehole.

22.5.5   Further Actions

- Autonomous desalination is a set of technologies, which include multidisciplinary issues, thus it is required to consider the contribution of different entities and experts. On the other hand it is necessary to continue the development of multi-partner R&D initiatives
- Promotion of exchange actions and links between companies, between R&D centers and any other private and /or public bodies in order to create networking groups because successful results are always thanks to collective contribution of many participants
- Fresh water is a more and more scarce resource in developing countries, wherein there is an important availability of solar and other renewable energy resources.

Consequently, future actions should be also focused on cooperation projects, since it is a clear link between autonomous desalination and international cooperation
- Identification of the barriers for the development of autonomous desalination and the related strategies and solutions to overcome them. For further details concerning this, it is interesting to consult a roadmap, as part of the findings of a currently ongoing project [16]
- Training is a key question for successful initiatives in developing countries, thus specific initiatives addressed to local engineers and other educated persons should be developed.

Acknowledgments The ADIRA Project was sponsored by the European Community (MEDA-Water Program) under contract number ME8/AIDCO/2001/0515/59610.

The ADIRA project has received also the economic contribution of the Government of the Canary Islands and the ITC.

All the involved Moroccan partners: the NGO FM21, the local technicians, the Government of Essaouira, the Government of Tiznit, the municipal authorities, the local associations, and the final users, have played a key role in the different stages of the project.

All the activities developed by the ITC within the ADIRA project were possible thanks to the implication of a wide set of more than ten people of the Water and Renewable Energies Departments. A special mention is done to the memory of Mr Manuel Leandro Reguillo, Ph D Engineer.

References

1. ADIRA project (2003–2008). www.adira.info. sponsor: EC, MEDA-Water Programme
2. ADU–RES project (2004–2006) (www.adu-res.org) (sponsor: EC, 6th Framework Programme)
3. Banat F, Subiela VJ, Qiblawey H (2007) Site selection for the installation of autonomous desalination systems (ADS). Desalination 203:410–416
4. Bellesiotis V, Delyannis E (2000) The history of renewable energies for water desalination. Desalination 128(2):147–159
5. Delyannis E (2003) Historic background of desalination and renewable energies. Sol Energy 75(5):357–366
6. García-Rodriquez L (2002) Seawater desalination driven by renewable energies: a review. Desalination 143(2):103–113
7. Ghaffour N, Venkat Reddy K (2009) The true cost of water desalination: review and evaluation. In: Proceedings of the IDA world congress, Atlantis, The Palm, Dubai, 7–12 Nov 2009
8. Hanafi A (1994) Desalination using renewable energy sources. Desalination 97(1–3):339–352
9. IDA Desalination Yearbook (2009–2010) Water Desalination report. Media Analytics Ltd. Okford, UK
10. Kalogirou S (2005) Seawater desalination using renewable energy sources. Prog Energy Combust Sci 31:242–281
11. Mathioulakis E, Bellesiotis V, Delyannis E (2007) Desalination by using alternative energy: review and state-of-the-art. Desalination 203(1–2):346–365
12. MEDIRAS project (2008–2011). http://www.mediras.eu. sponsor EC, 7th Framework Programme
13. Papapetrou M, Epp C, Tzen E, (2006) Autonomous desalination units based on RE systems. A review of representative installations. In: NATO Seminar, Solar desalination for the 21dt Century, Tunisia

14. Peñate B, Castellano F, Ramírez P (2007) PV-RO desalination stand-alone system in the village of Ksar Ghilène (Tunisia). In: Proceedings of the IDA world congress-Maspalomas, Gran, Canaria, 21–26 Oct 2007
15. Philippe Rekacewicz, UNEP/GRID-Arendal (2002). http://maps.grida.no/go/graphic/renewable_freshwater_supplies_per_river_basin
16. PRODES project (2008–2010) www.prodes-project.org. sponsor EC, Intelligent Energy for Europe Programme
17. Sánchez AS, Subiela VJ (2007) Analysis of the water, energy, environmental and socioeconomic reality in selected Mediterranean countries (Cyprus, Turkey, Egypt, Jordan and Morocco). Desalination 203:62–74
18. Subiela VJ, De La Fuente J, Piernavieja G, Peñate B (2009) Canary Islands Institute of Technology (ITC) experiences in desalination with renewable energies (1996–2008). Desalin Water Treat 7:263–266
19. Tzen E (2005) Proceedings of the seminar successful desalination RES plants worldwide, Hammamet, Sept 2005. Available at http://www.adu-res.org/pdf/CRES.pdf
20. United Nations (2005). http://www.un.org/waterforlifedecade/
21. WHO (2010) Water sanitation and health. http://www.who.int/water_sanitation_health/mdg1/en/index.html

Chapter 23
Desalination Technologies as a Response to Water Strategy Problems (Case Study in Egypt)

Magdy AbouRayan

Abstract Availability of fresh water is the prime mover of the human life activities. In fact, the development in general depends on fresh water availability. The average per capita renewable water worldwide is 7,700 m^3/year, this average dropped to less than 1,000 m^3 capita/year in MENA countries. As an example, in Egypt it is 780 m^3/capita. The advances in desalination technologies have shown clearly that desalinated water can be used as a substitute to fresh water renewable resources. A breakthrough in reverse osmosis costs has been reached; recently, several studies showed that desalinated water for development of isolated areas is competitive to transported fresh water by pipe line. This fact is clear in developing coastal areas in Egypt and particularly in Sinai, which suffers from severe water shortage. The introduction of solar energy to power desalination equipment has given a new dimension to the expansion of this technology, as several studies have shown in MENA countries. This solution is particularly promising for the supply of fresh water for isolated areas. In the case of Egypt, the renewable water supply is constant at 60 billion m^3/year from three sources: Nile River, rainfall and extraction from ground water reservoir (Nile basin aquifer). The demand is increasing annually at a rate of 1.25%. The forecasting studies show that water balance will suffer from 21 billion m^3/year, in the year 2025. The water resources security is a vital issue for Egypt, and presents a real threat to the economic and social development. Another alarming issue is the water quality deterioration. The pollution of water ways threatens the environment. For Egypt, an IWRM (Integrated Water Resources Management) including desalination options to secure development and land reclamation is a must. The objectives of this chapter are to: (i) present an overview of desalination technologies; (ii) discuss the importance of desalination to respond to water scarcity problems.

Keywords Desalination technologies • Egypt's water resources

M. AbouRayan (✉)
Mansoura University, Mansoura, Egypt
e-mail: mrayan@mans.edu.eg

23.1 Introduction

Desalination is a treatment process that removes salts from water. Saline solutions other than sea water are typically described as brackish water with a salt concentration from 1,000 to 11,000 ppm TDS (Total Dissolved Solids). Normal sea water has a salinity of 35,000 ppm up to 40,000 TDS or more, mostly sodium chloride.

A typical desalination plant consists of a water pretreatment system, the desalination unit, and a post-treatment system. A desalination plant, as depicted in Fig. 23.1, may be considered as a "black box" through which streams of water and energy flow.

Several desalination processes have been developed, but not all of them are reliable and in commercial use. On the basis of their commercial success, desalination technologies have been classified into major and minor desalination processes, as indicated in Table 23.1. The major desalination processes are split into two main categories, thermal (or distillation) and membrane processes (Fig. 23.2).

The thermal processes for sea water desalination are: Multistage-Flash Distillation (MSF), the Multi Effect Distillation (MED or ME), the Vapor Compression (VC) process and the solar distillation [5].

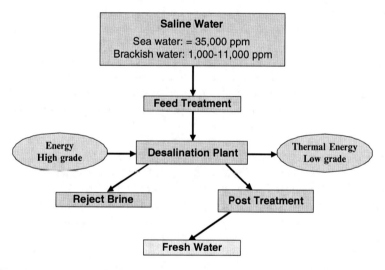

Fig. 23.1 Water and energy flow diagram of a desalination unit [9]

Table 23.1 Commercially available desalination processes

Major processes	Minor processes
• Thermal	
– Multistage-flash distillation	– Freezing
– Multi effect distillation	– Membrane distillation
– Vapor compression distillation	– Solar humidification
• Membrane	
– Reverse osmosis	
– Electrodialysis	

23 Desalination Technologies

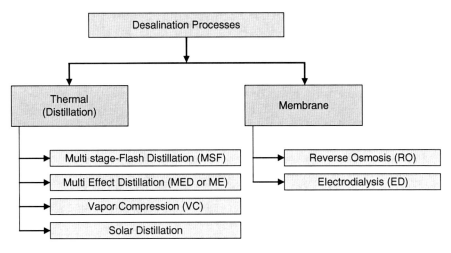

Fig. 23.2 Desalination processes [3]

Table 23.2 WHO standards for potable water [6]

Constitutes	Concentration (limited values)	ppm TDS (max. allowed values)
Total dissolved salts	500	1,500
Cl	200	600
SO_4^{2+}	200	400
Ca^{2+}	75	100
Mg^{2+}	30	150
F	0.7	1.7
NO_3^-	<50	100
Cu^{2+}	0.05	1.5
Fe^{3+}	0.10	1.0
Ph	7–8	6.5–9

Membrane processes consist of Reverse Osmosis (R.O.) and Electro dialysis (ED) processes. Electro dialysis is confined to the desalination of brackish water, while Reverse Osmosis can be used for both, brackish and sea water desalination.

Two critical parameters of the desalination processes are the quality of the produced water and the energy required. In general, desalination is an energy intensive technology. The energy input to the plant may be thermal, (defined in terms of units of water produced per unit of steam or per 2,500 kJ used), mechanical or electrical (expressed in kWh/m^3).

The quality of the produced water depends on the desalination process. In fact, distillation processes produce water around 20 ppm TDS, while membrane processes are usually designed to produce water of 100–500 ppm TDS. Potable water for human consumption should comply with the World Health Organization (WHO) limits, Table 23.2, and should not be totally devoid of salts.

In the following paragraph a short summary of each commercial technology will be presented.

23.2 Different Desalination Technologies: Major Processes

23.2.1 Thermal Processes

23.2.1.1 Multiple-Stage Flash Distillation

Multiple Stage Flash (MSF) distillation is the most widely used desalination process, in terms of capacity.

In this process, as well as in all distillation processes, the sea water is heated, producing water vapor that is in turn condensed to form fresh water. The water is heated to the boiling point to produce the maximum amount of water vapor.

There are two configurations concerning MSF process. The first one, the "Once Through" configuration, consists of two sections:

– Heat rejection section.
– Brine heater.

The second MSF configuration is called "Brine Recirculation" (Fig. 23.3), and consists of the three following sections:

– Heat rejection section
– Heat recovery section
– Brine heater.

An MSF plant can contain from 4 up to 40 stages. Increasing the number of stages reduces the heat transfer surface that is required, reducing the capital cost. This has to be offset against the cost of providing extra stages. As a consequence, complicated optimization calculations have to be undertaken where the main parameters are capital cost versus operating cost.

The Multi Stage Flash distillation process has played a vital role in the provision of water in many areas, particularly in the Middle East. The installed capacity of the process has grown considerably over the last 25 years.

MSF has been developed and adapted to large scale applications, usually greater than 5,000 m^3/day. At present, the largest MSF plant, contracted or in operation, has a water production capacity of 60,000 m^3/day (Eljubil, Saudia Arabia).

The process is widely used in the Gulf countries with 75% of the global total installed capacity. In Europe, the MSF process is mainly used in Italy and in Spain.

23.2.1.2 Multiple Effect Distillation

Multiple Effect Distillation (MED or ME) was the first process used for sea water desalination. It is widely used in the chemical industry where the process was originally developed. The MED process is similar to the MSF process since it also operates in part by flashing. Moreover, in this process the majority of the distillate is produced by boiling.

23 Desalination Technologies

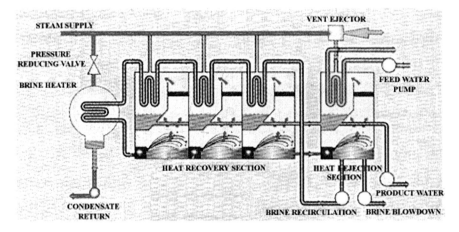

Fig. 23.3 Typical flow diagram of multi-stage flash distillation plant

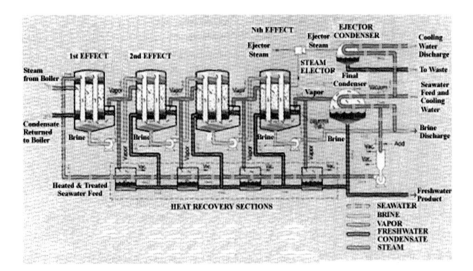

Fig. 23.4 Typical flow diagram of multi-effect distillation plant – horizontal falling film plant

MED, like MSF, takes place in a series of vessels (effects) and uses the principle of reducing the ambient pressure in the various effects. This permits the feed water to undergo multiple boiling without supplying additional heat after the first effect. The principle of its operation is shown in Fig. 23.4. Plants tend to have smaller number of effects than MSF stages. Usually 8–16 effects are used in typical large plants, due to the relation of the number of effects with the performance ratio (which cannot exceed the number of effects of the plant).

As in an MSF plant, special attention is required concerning the operating temperature, to avoid scaling and corrosion of materials. Also, extra care is required concerning the control of the brine level in each effect.

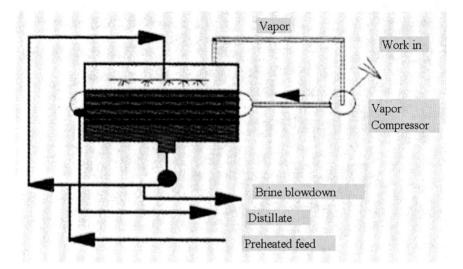

Fig. 23.5 Typical flow diagram of vapour compression plant

23.2.1.3 Vapor Compression

There exist two Vapour Compressor (VC) processes. The first configuration is Mechanical Vapour Compression (MVC), in which a mechanical compressor is used. The second, is the Thermal Vapour Compression (TVC), in which a thermo compressor or ejector is used to increase the vapour's pressure. Both types are widely used.

The fundamental concept of this process is inherently simple, in that after vapour has been produced it is then compressed to increase its pressure and consequently its saturation temperature before it is returned to the evaporator as the heating vapour for the evaporation of more liquid.

The main equipment used in the VC compression process is the evaporator, the compressor, pumps and the heat exchanger.

In this process, the feed water is preheated in a heat exchanger or a series of heat exchangers by the hot discharge of the brine (Fig. 23.5).

The power consumption of the compressor, and therefore the efficiency of the process, is dependent on this pressure difference. Thus, the compressor represents the main energy consumer in the system.

Extra care is required with the control of the brine level in the evaporator and the proper maintenance of the compressor. Some manufacturers use compressors that rotate at very high speeds. Operation at low temperatures minimizes the formation of scaling and corrosion of materials.

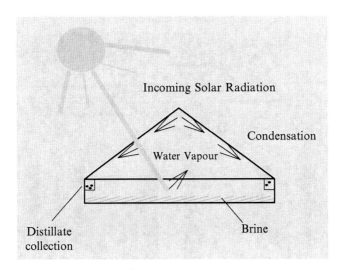

Fig. 23.6 A typical solar distillation plant design

23.2.1.4 Solar Distillation

Solar distillation is a process in which the energy of the sun is directly used to evaporate fresh water from sea or brackish water. The process has been used for many years, usually for small scale applications.

In solar distillation plants the solar radiation is trapped in the solar still by the greenhouse effect. A simple solar still consists of a shallow basin of brine, lined with some black material to get good radiation absorptivity, and covered by a vapour-proof, transparent roof designed to act as a condenser (Fig. 23.6).

Solar distillation is usually suitable for small scale applications. In remote areas, where low cost land is available and solar radiation is high, solar distillation can be viable.

23.2.2 Membrane Processes

23.2.2.1 Reverse Osmosis

Reverse Osmosis (R.O.) is the most widely used process for sea water desalination. R.O. process involves the forced passage of water through a membrane against the natural osmotic pressure to accomplish separation of water and ions. The principle is shown in Fig. 23.7 [1].

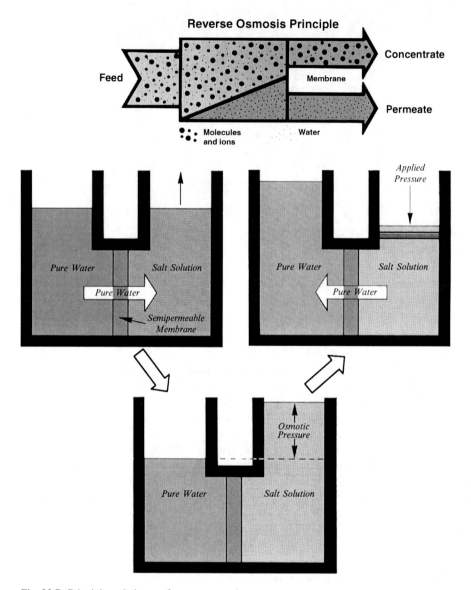

Fig. 23.7 Principle and phases of reverse osmosis

A typical R.O. system consists of four major subsystems (Fig. 23.8):

- Pre-treatment system
- High pressure pump
- Membrane modules
- Post-treatment system.

Feed water pre-treatment is a critical factor in the operation of an R.O. system, due to membranes sensitivity to fouling. Pre-treatment commonly includes feed

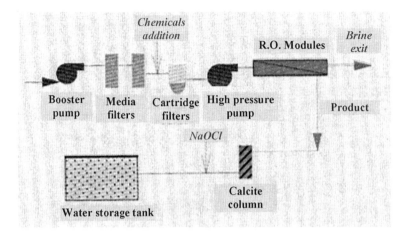

Fig. 23.8 Typical flow diagram of reverse osmosis plant

water sterilization, filtration and addition of chemicals in order to prevent scaling and bio fouling. The post-treatment system consists of sterilization, stabilization and mineral enrichment of the produced water.

Two types of R.O. membranes are used commercially. These are the Spiral Wound (SW) membranes and the Hollow Fiber (HP) membranes. SW and HP membranes are used to desalt both sea water and brackish water. The choice between the two is based on factors such as cost, feed water quality and water production capacity.

Due to the R.O. unit operation at ambient temperature, corrosion and scaling problems are diminished in comparison with distillation processes. However, effective pre-treatment of the feed water is required to minimise fouling, scaling and membrane degradation. In general, the selection of the proper pre-treatment, as well as the proper membrane maintenance is critical for the efficiency and life of the system.

As a general rule, a sea water R.O. unit has a low capital cost and a significant maintenance cost due to the high cost of the membrane replacement. The cost of the energy used to drive the plant is also significant. The major energy requirement for Reverse Osmosis desalination is for pressurizing the feed water; Energy requirements for SWRO have been reduced to 4.0 kWh/m^3, for large units with energy recovery systems. For brackish water desalination the energy requirement is between 1 and 3 kWh/m^3.

A breakthrough in Technology and cost have been realized for large units of 100,000 cubic meters per day (examples in Larnaka, Cyprus and in Ashdod, Israel).

23.2.2.2 Electrodialysis

ED is an electrochemical process and a low cost method for the desalination of brackish water. Due to the dependency of the energy consumption on the feed water salt concentration; the Electrodialysis process is not economically attractive for the desalination of sea water.

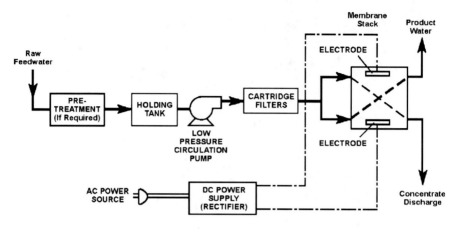

Fig. 23.9 Basic components of an ED plant

In Electrodialysis (ED) process, ions are transported through a membrane by an electrical field applied across the membrane. An ED unit, as seen in Fig. 23.9, consists of the following basic components:

- Pre-treatment system
- Membrane stack
- Low pressure circulation pump
- Power supply for direct current (rectifier)
- Post-treatment.

The principle of electrodialysis operation is shown in Fig. 23.9. When a pair of electrodes is placed in a container of saline water and connected to an outside source of direct current (like a battery), electrical current is carried through the solution, with the ions tending to migrate to the electrode with the opposite charge. Positively charged ions migrate to the cathode and negatively charged ions migrate to the anode. If between the electrodes a pair of membranes (cell), formed by an anion permeable membrane followed by a cation permeable membrane is placed, a region of low salinity water (product water) will be created between the two membranes. Between each pair of membranes, a spacer sheet is placed in order to permit the water flow along the opposite faces of the membranes and to induce a degree of turbulence. One spacer provides a channel that carries feed (and product water) while the next carries brine. By this arrangement, concentrated and diluted solutions are created in the spaces between interlaced membranes.

23.2.3 Technology Comparison

The choice of one process over the other is very site specific and depends on the conditions of the feed water, energy source, demographic distribution, etc.

23 Desalination Technologies

Table 23.3 Characteristics of the major desalination processes [11]

Process	Feed water Type	Energy source	Product water quality (ppm TDS)	Typical max plant Capacities (m³/day)
MSF	Sea water	Steam	~10	5,000–60,000
MED	Sea water	Steam	~10	5,000–20,000
VC	Sea water	Electricity	~10	2,400
SWRO	Sea water	Electricity	~350–500	128,000
BWRO	Brackish	Electricity	~350–500	98,000
ED	Brackish	Electricity	~350–500	45,000

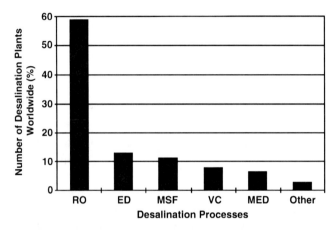

Fig. 23.10 Percentage of the number of desalination plants worldwide [12]

In order to select a desalination process, a number of factors need to be taken into consideration (see Table 23.3). The main points can be discussed by looking at the table data and at the worldwide distribution of desalination processes, as reported in Fig. 23.10.

From the past brief review of desalination technologies it is clear that the technology is mature and commercially wide used. A break through is expected in Reverse Osmosis technology in point view of cost both installation and operation. Another important dimension in desalination is the use of renewable energy to drive desalination units. There are available now in the market commercial desalination units that use renewable energy, suitable and competitive for remote areas [7]. The actual water demand, particularly in southern Mediterranean countries, is critical [2, 8].

Many countries suffer from water scarcity and the annual per capita water renewable resources is below 500 cubic meters. The other countries suffer from water dispute on shared rivers basins. Recourse to desalination is obvious. In the following paragraph the case study of Egypt will considered, in order to show the necessity of desalination in order to balance the water demand in Egypt.

23.3 Egypt Water Supply

The main source of renewable water supply in Egypt is the Nile river. According to the signed treaties between Egypt and other Nile basin countries, Egypts share is 55.billions of cubic meters per year. The publishing figures of water demand and use in Egypt is contradictive. In the following tables published by the ministry of water resources shows that the deficit in water balance will start from the year 2017. Infact physically the deficit started from the year 2009. Some areas in upper Delta suffered from water insufficiency.

The official report shows a deficit of water balance started from year 2017 and foresees ways to overcome it based on:

- The use of modern irrigation methods which will change the pattern of water demand.
- The use of treated sanitary drainage (expected four billion cubic meters).
- Enhancing the re-use of agricultural drainage to the maximum of eight billion cubic meters.
- Increasing the share of desalination part in remote areas, and as a complementary irrigation in areas as northern Mediterranean coast.

The northern Mediterranean coast has a precipitation of 10 cm during the winter time. It is concentrated in a narrow strip. This area of approximately three million Fedanns (1 Hectare ~ 2 Fedanns) is not efficiently used, due to water scarcity. A complementary irrigation by desalination of brackish water will result in an efficient use of this area, taking to a 37% of increment of the cultivated areas in Egypt. The average salinity of underground water is 5,000 ppm. The cost of desalination of brackish water is now down to 0.2 $ per cubic meter (Table 23.4).

Reviewing the development of water demand shows that the demand was increasing during the years 1990s at a rate of 2%. There is no accurate forecast for the average increment in water demand in the coming years after 2010. But the data from the year 2000 to the year 2010 is 3% approximately. This taking into account

Table 23.4 Volume of reuse-source vis-à-vis water supply in Egypt [4]

		Current (billion m^3)	Projected for 2010 (billion m^3)
Reuse source	Municipal wastewater	0.4	1.6
	Industrial effluent	0.4	1.8
	Desalinated	N/A	0.5
	Drainage water	4.5	7.0
	Groundwater (Reuse)	3.8	5.8
	TOTAL of reuse-resource	9.1	16.7
Supply	Nile River	55.5	57.5
	Groundwater	0.7	1.2
	TOTAL of supply	56.2	58.7
	Augmentation %	16	28

Table 23.5 Water Resources and Extraction Year 2009

Type of water resources	Billion cubic meter per year
Nile river	55.5
Precipitation	1.8
Renewable groundwater extraction	4.0
Fossil groundwater extraction	1.0
Agricultural drainage reuse	7.5
Water desalination	0.2
Total water supply	72.9

Table 23.6 Future water resources in year 2017–2030 (Source: Ministry of water resources and irrigation report, [10]) in Billion cubic meters per year

Type of water resources	2017	2030
Nile river	57.5	60
Waste water reuse	2.5	5
Agricultural drainage reuse	8.4	10.5
Ground water renewable	7.4	7.5
Fossil ground water extraction	3.5	5
Precipitation	1.8	2
Cropping pattern improvement	7	10
Desalination	0.2	4
Total expected resources	88.1	104
Egypt population	88	110

an annual demographic increment of 1.9% and average economic growth of 5–6%. The economic growth based mainly on service industry and tourism which require water (Tables 23.5 and 23.6).

As it may be clear from the table, there are uncertainties regarding the Nile water. A forecasting based on Nile water increment is uncertain. It is worth to mention that there projects waiting to be implemented south of Sudan to increase the share of Egypt and Sudan. The political instability in this region prevents the starting of these projects. The only way to substitute this amount of 4.5 BCM is desalination. This will raise the figure of desalinated water to 8.5 BCM, which requires a heavy investment.

23.4 Concluding Remarks

Desalination is the response to water shortage problems. Conventional desalination technology is fairly well developed and some processes are considered quit mature. In particular, a breakthrough has been achieved in RO technology in terms of cost and power consumption.

Desalination using renewable energy does offer the potential of providing a sustainable source of potable water for small communities in rural isolated areas.

In Egypt, the use of desalination to satisfy water demand in remote areas is a must. The development of agriculture in coastal areas may be based on brackish water desalination as a complementary irrigation source to precipitation.

References

1. AbouRayan M, Benchikh O (1995) Technologie de dessalement de l'eau. UNESCO publications, Paris
2. AbouRayan M, Benchikh O (2000) Mediterranean cooperation for water desalination policies in the perspective of a sustainable development. No. IC18-CT97-0142. Prepared by, Mansoura University Mechanical Engineering, Mansoura Department, Mansoura
3. AbouRayan M, Djebedjian B, Khaled I (2003) Database establishment for the evaluation of the effectiveness and performance of desalination equipment in A.R.E. Prepared by, Mansoura University, Mansoura
4. Chitale MA (1997) The watsave scenario. ICID Publication, New Delhi, India
5. Darwish MA (1995) Desalination processes, a technical comparison, vol 1. IDA Congress, Abu Dhabi, pp 149–173, Nov 1995
6. Delyiannis E, Belessiotis V (1995) Methods and systems of desalination, Greece, Sept 1995
7. European Commission (1998) Desalination guide using renewable energies, energy technology information base 1980–2010 – renewables sector: solar photovoltaic power, final draft, ATLAS project. THERMIE programme (technical and economic analysis of the potential for water desalination in the Mediterranean region), APAS-RENA-CT94-0063, ARMINES, France
8. Genthner K, El-Dessouky H (2000) Desalination strategies in the southern Mediterranean countries. In: Conference session on capacity building strategies for desalination in the Middle East and North Africa. Sponsored by, The Middle East Desalination Research Center and UNESCO
9. Hanbury WT, Hodgkiess T, Morris R (1993) Desalination technology. University of Glasgow, Glasgow
10. National Water Resources Plan (NWRP) (2005) Water for the future. NWRP 2017 Project, final report, Ministry of Water Resources and Irrigation, pp 2–12
11. Ribeiro J (1996) Desalination technology – survey and prospects. IPTS, Seville, Aug 1996
12. Rodriguez-Girones PJ, Rodriguez Ruiz M, Veza JM (1996) Experience on desalination with renewable energy sources. Work Package 1, RENA-CT94-0063. Universida de Las Palmas de Gran Canaria, Las Palmas, March 1996

Chapter 24
Advanced Water Treatment System: Technological and Economic Evaluations

Artak Barseghyan

Abstract The supply of potable water from polluted rivers, lakes, unsafe wells, etc. is a problem of high priority. One of the most effective methods to obtain low cost drinking water is desalination. In this chapter, an advanced water treatment system, based on electrodialysis for water desalination and purification, is suggested. Technological and economic evaluations and the benefits of the suggested system are discussed. The Advanced Water Treatment System proposed clears water not only from different salts, but also from some infections, thus decreasing the count of diseases which are caused by the usage of non-clear water.

Keywords Water treatment system • Solar power system • Electrodialysis • Potable water

24.1 Introduction

The supply of clean drinking water is a problem of high priority.Every year there are two million diarrhoeal deaths related to unsafe water, sanitation, and hygiene the vast majority among children under 5. More than one billion people lack access to an improved water source [4]. There is great demand of improvements in drinking water quality. In many areas the only source for water are polluted rivers, lakes, unsafe wells, etc.

One of the effective methods obtaining low cost drinking water is desalination. There are various methods for desalination of water, such as evaporation, reverse

A. Barseghyan (✉)
Engineering Academy of Armenia, Teryan 105, bld. 17, Yerevan 0009, Armenia

AREV Scientific Industrial CJSC, Yerevan, Armenia

ECOATOM LLC, Yerevan, Armenia

Youth Academy, Yerevan, Armenia
e-mail: artakbarseghyan@yahoo.com, website: www.youth.academy.am

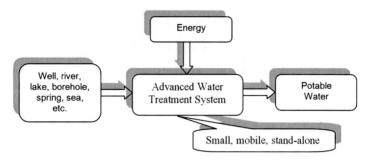

Fig. 24.1 Advanced water treatment system

osmosis, ultra-filtration, ion-exchange, electrodialysis, etc. The market for water desalination has witnessed a significant upturn during the last years and many countries in the world have already created water desalination plants for water supply.

Most of these desalination plants are designed for big cities and provide drinking water more than 50,000,000 l/day. These plants are using complex systems and are not cheap, thus not suitable for use in small rural areas. There is great demand to create water treatment systems (Fig. 24.1), which will be small, mobile, and stand-alone [1]. In many small villages and far locations where drinking water is absent also the electric power is unavailable. So, an alternative energy source such as a solar photovoltaic system, is very useful for creating stand-alone system.

Where is a lack of drinkable water in arid and semi-arid regions often we meet high solar insolation figures for the use of solar energy.

For example in Armenia, which is considered as a semi-arid region, there are very favorable conditions for the exploitation of solar energy, being the average daily insolation as high as 700 W/m^2. This implies the high potential of utilization of photovoltaic energy for water desalination.

24.2 Drinking-Water Worldwide Problem

Diarrhoea is caused mainly from unsafe drinking-water, contaminated food or unclean hands. 88% of cases of diarrhoea worldwide are connected with unsafe water and inadequate sanitation. Diarrhoeal disease is the second leading contributor to the global disease burden.

Unsafe water, inadequate sanitation and insufficient hygiene contribute to 64 million Disability-Adjusted Life Years (DALYs) and ranked fourth in the list of leading health risk factors in the world, behind childhood underweight, unsafe sex and alcohol use.

More than 2.2 million deaths of children per year could be prevented by the reduction of diarrhoeal and malnutrition impacts related to unsafe water, inadequate sanitation or insufficient hygiene (Fig. 24.2).

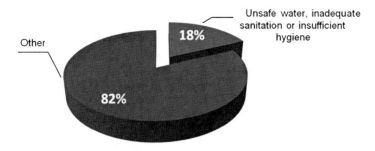

Fig. 24.2 Percentage of deaths of children (0–14 years) (Source: [12])

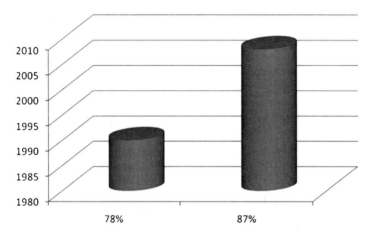

Fig. 24.3 On 2008, 87% of the world's population or approximately 5.9 billion people where using drinking-water from improved sources, compared with 78% in 1990

24.2.1 Drinking-Water Improved Sources

Approximately 1.8 billion people gained access to drinking-water from an improved source starting from 1990 to 2008 (Fig. 24.3). Nearly 900 million people do not use drinking-water from an improved source, this is anyway lower than the 1990th result, which was 1.2 billion.

Global coverage data suggest large disparities between rural and urban areas in terms of the use of improved drinking-water sources. With the rapid urbanization that took place between 1990 and 2008, the urban population not using water from an improved source increased by 40 million [4]. A comparative representation of the availability of improved drinking-water sources for urban and rural areas, in terms of percent coverage for 1990 and 2008 years, is shown in Fig. 24.4.

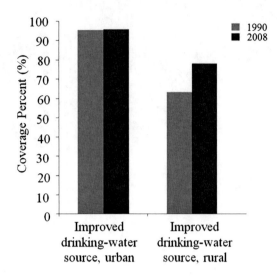

Fig. 24.4 Global coverage levels of improved drinking-water sources for urban and rural environments, comparing 1990 and 2008 years

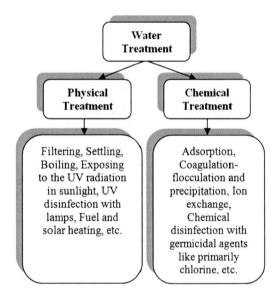

Fig. 24.5 Water treatment physical and chemical methods

24.3 Treatment Technologies

There are various physical and chemical treatment methods for the improvement of water quality. Some of the physical and chemical treatment methods are described in Fig. 24.5.

Some water treatment systems use expensive chemicals that are difficult to obtain, and require complex systems, which are not suitable for small villages or other rural areas. Such expensive and complex technologies may be rational for big cities.

Various treatment methods, such are solar disinfection, boiling, UV disinfection with lamps, and the combined treatments of chemical coagulation-filtration and chlorination are used for bacteria, viruses and in some cases protozoan reduction. However, the ability of some of these methods to remove or inactivate a wide range of known waterborne pathogens has been inadequately investigated and documented.

Various researches have shown that improving the microbiological quality of household water reduces diarrheal and other waterborne diseases. Reductions in household diarrheal diseases of 6–90% have been observed, depending on the technology, the exposed population, and local conditions [10, 11].

24.4 Electrodialysis

Electrodialysis is a method based on ions selective transfer through ion-exchange membranes under the influence of an electric field [7–9, 13]. As a rule, anion and cation exchange membranes are arranged in an interlaced configuration (Fig. 24.6).

The qualitative leap in the development and use of this method occurred immediately after the development of selective membranes (that is, membranes which are selectively permeable to cations or anions). The ion-exchange membrane is a thin polymer sheet with 0.17–0.65 mm thickness, providing the preferred ion transport for cations (cation exchange membrane) or anions (anion exchange membrane).

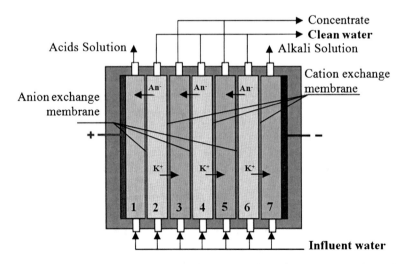

Fig. 24.6 The working diagram of the electrodialyzer with alternating cation and anion exchange membranes

The whole electrodialyzer is created from an electrical insulating polymer material in the form of a rectangular parallelepiped. The working diagram of the electrodialyzer is represented in Fig. 24.6.

From every even section (desalination section) cations migrate toward the negatively charged electrode (cathode) through the cation exchange membrane, which is permeable to them, e.g. from section number 4 cations are passing to the section number 5. Further, their movement is limited by the anion exchange membrane, which is impervious to them.

24.4.1 Method Benefits

Electrodialysis with ion-exchange membranes is a method successfully employed in the past and current practice for producing potable water. A few decades ago, this method was already discussed as a practical tool to process effluent water in order to obtain potable water [2, 8]. Some benefits of the method are listed below:

– Ability to clean up to the Maximum Permissible Concentration (MPC) allowed for the potable water
– Return of the purified water to 60% in a revolving cycle
– Absence of phase transitions in the department of impurities that allows process using small power consumption
– Ability to conduct at room temperature without application or with small additions of chemical reagents
– Simple construction of the equipment.

Additional information can be found in [3, 6, 7, 9, 13].

24.5 Advanced Water Treatment System

A potable water production system (Fig. 24.7) via desalination and purification is suggested [1]. The suggested system (WDPS) is based on electrodialysis to transport salt ions from one solution through ion-exchange membranes to another solution under the influence of an applied difference of electric potential.

The power source for WDPS is the sun. The WDPS size is chosen for having possibility to treat up to 1,000 l of potable water during 1 day. Treated water during daylight hours is stored in tanks, due to the fact that is much cheaper to store water in tanks than accumulate energy in a battery system.

WDPS may also work with batteries, but in this case the price of the equipment increases.

24 Advanced Water Treatment System

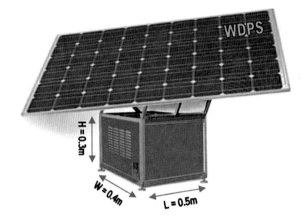

Fig. 24.7 Water Desalination and Purification System (WDPS)

24.5.1 Benefits of the Suggested System

The suggested WDPS has applied economic, social and ecological benefits.

- Practical: WDPS system has a small size, is movable, and, by using solar energy, it can work without mains supply from an electricity network. All the mentioned features make it suitable for a series of application frameworks, such as disaster recovery, rural areas, remote locations, army operations, etc.
- Economic: using the proposed method and technology the price of clear water will be up to ten times cheaper than the current price in some rural areas and far locations in the Republic of Armenia.
- Social: in addition to removing salts, the WDPS is capable to clear water from some infections, thus decreasing the count of diseases which are caused by non clear water usage.
- Environmental: the usage of solar energy in stand-alone locations has incomparable advantages with internal combustion engine, including the atmospheric pollution.

24.5.2 Applications of WDPS

WDPS may be applied for desalination and purification purposes to different contexts of superficial water and groundwater, such as wells, rivers, lakes, boreholes and springs.

It's been conceived for the installation in villages and far locations where electric power and drinking water are absent.

The main advantages of the presented system are:

- low price
- renewable energy supply

Table 24.1 Technical characteristics of the small WDPS model

Weight	120 kg
Energy consumption	150 W
Productivity	1,000 l/day
Raw water source	Well, river, lake, borehole, spring, sea, etc.
Filtration technology	Electrodialysis
Quality of produced water	GOST 2874-82
Drinking water tank dimensions and capacity.	1 m × 1 m × 1 m (1,000 l)
Waste tank	20 l
Equipment dimensions without solar system (Height, length, width)	
Height/length/width	H = 0.3 m/L = 0.5 m/W = 0.4 m
Solar system area	S = 2 m^2 (in presence of electric power supply panels may be removed and the size of the equipment gets much smaller)
Membranes count	N = 50 pcs. (25 pcs. AM, 25 pcs. KM) In case of normal daily usage membranes are guaranteed for 8 years.
Membranes dimensions	0.4 × 0.5 m (50 membranes with a total area of 10 m^2)

- simple schematic structure
- low power consumption, that makes it effective to use with or without photovoltaic power supply
- possibility of free installation and maintenance of a WDPS, getting an income by selling potable water.

24.5.3 Technical Characteristics

The technical characteristics of the WDPS with 1,000 l/day productivity are represented in Table 24.1.

24.6 Market Overview

The global market for water desalination has witnessed a significant upturn during the last years. The world population increases and lacks of fresh water sources, while environment destruction and desertification and some additional factors lead many countries to create water desalination plants for the supply of water.

The global market for water desalination plants/systems increased from $1.7 billion (in 2005) to $2.0 billion in 2008. There are independent forecasts that give $3.6 billion by 2012 [5].

In many countries, e.g. in the Mediterranean region and in the Eastern Asia, the supply of clean drinking water is a problem of high priority. The shortage of drinking water is a major problem in Armenia as well, especially on the Ararat valley and in the Armavir region. Due to our estimation, the market volume of water desalination and purification systems, only for the Ararat valley, is about 500 units. Fortunately the climatic conditions in Armenia, as a semi-arid region, are very favorable for the exploitation of solar energy, where the daily average solar insolation is as high as 700 W/m^2. This implies the high potential of utilization of solar (photovoltaic) energy for water desalination and purification. Furthermore, one of the main pillars of the new national strategy in the field of water is the generalization of the water access, especially in rural areas; for this, the national program of water supply in rural areas was launched some years ago to reach a rural population of 1.5 million inhabitants in 3,000 villages in the year 2015 [1].

24.7 Conclusion

This work presents an advanced water treatment system via desalination and purification, based on electrodialysis. The power source for the suggested system is the sun (standard mains supply can be used alternatively). The presented Water Desalination and Purification System (WDPS) has a small size, is movable, and thanks to the usage of solar energy it can work without electricity network or generators. WDPS may be used in different contexts of superficial and groundwater sources, such as wells, rivers, lakes, boreholes and springs.

All these features make WDPS suitable for a series of application contexts where drinking water and electricity supply network are unavailable, such as disaster recovery, rural areas, remote locations, army operations, etc.

The proposed WDPS clears water not only from different salts, but also from some infections, thus decreasing the count of diseases which are caused by the usage of untreated water.

The main advantages of the presented system are: low price, renewable energy power source, simple schematic structure and low power consumption.

A small WDPS for the production of 1,000 l/day, such as the one shown in Fig. 24.7, draws about 150W from 2 m^2 of photovoltaic panels, and weights 120 kg, making it a very attractive solution for stand-alone applications.

References

1. Barseghyan AR (2009) Testing, evaluation and production planning of advanced stand-alone water desalination systems powered by solar energy. Report on Venture Conference of science and technology entrepreneurship program, Yerevan
2. Bergsma Feike (1962) Dechema monograph, 4.7.7: 449
3. Davis TA (1990) Electrodialysis. In: Porter MC (ed) Handbook of industrial membrane technology. Noyes, New Jersey

4. GLAAS (2010) UN-Water global annual assessment of sanitation and Drinking-Water. WHO Press, Geneva, p 102
5. Helmut Kaiser Consultancy (2008) Water Desalination Worldwide for Sea Water and Brackish Water. http://www.hkc22.com/waterdesalination.html
6. Martoian GA, Barseghyan AR (2010) Novel method based on electrodialysis principle for high activity Liquid Radioactive Waste (LRW) treatment, technological innovations in detection and sensing of Chemical Biological Radiological Nuclear (CBRN) threats and ecological terrorism NATO ASI, Chisinau, Moldova
7. Mulder M (1996) Basic principles of membrane technology. Kluwer, Dordrecht
8. Rauzen FV, Dudnik SS, Gutin EI (1967) Low-level waste waters desalted and purified by ionic membrane electrodialysis. Springer 22(5):491–495
9. Strathmann H (2004) Ion-Exchange membrane separation processes. Elsevier, New York
10. WHO (2006) Household water treatment and safe storage. http://www.who.int/household_water/research/technologies_intro/en/index.html
11. WHO Press (2008) The World health report 2008. WHO, Geneva
12. WHO/UNICEF (2010) Progress on sanitation and drinking-water. http://www.who.int/pmnch/media/membernews/2009/2010_who_unicef_whreport/en/index.html
13. Zagorodni AA (2006) Ion exchange materials: properties and applications. Elsevier, Amsterdam